DYNAMIC FIELDS AND WAVES

Edited by Andrew Norton

Institute of Physics Publishing
Bristol and Philadelphia
in association with

The Physical World Course Team

Course Team Chair	Robert Lambourne
Academic Editors	John Bolton, Alan Durrant, Robert Lambourne, Joy Manners, Andrew Norton
Authors	David Broadhurst, Derek Capper, Dan Dubin, Tony Evans, Ian Halliday, Carole Haswell, Keith Higgins, Keith Hodgkinson, Mark Jones, Sally Jordan, Ray Mackintosh, David Martin, John Perring, Michael de Podesta, Ian Saunders, Richard Skelding, Tony Sudbery, Stan Zochowski
Consultants	Alan Cayless, Melvyn Davies, Graham Farmelo, Stuart Freake, Gloria Medina, Kerry Parker, Alice Peasgood, Graham Read, Russell Stannard, Chris Wigglesworth
Course Managers	Gillian Knight, Michael Watkins
Course Secretaries	Tracey Moore, Tracey Woodcraft
BBC	Deborah Cohen, Tessa Coombs, Steve Evanson, Lisa Hinton, Michael Peet, Jane Roberts
Editors	Gerry Bearman, Rebecca Graham, Ian Nuttall, Peter Twomey
Graphic Designers	Mandy Anton, Steve Best, Sue Dobson, Sarah Hofton, Jennifer Nockles, Pam Owen, Andrew Whitehead
Centre for Educational Software staff	Geoff Austin, Andrew Bertie, Canan Blake, Jane Bromley, Philip Butcher, Chris Denham, Nicky Heath, Will Rawes, Jon Rosewell, Andy Sutton, Fiona Thomson, Rufus Wondre
Course Assessor	Roger Blin-Stoyle
Picture Researcher	Lydia K. Eaton

The editor wishes to thank the following for their contributions to this book: Ch 1: Dan Dubin, Sally Jordan; Ch 2: Sally Jordan; Ch 3; Keith Hodgkinson; Ch 4: Dan Dubin; Ch 5: Carole Haswell. The book made use of material originally prepared for the S271 Course Team by John Walters and Robert Lambourne. For the multimedia packages *Waves*, *Huygens' view of diffraction* and *Virtual ripple tank*, thanks are due to Greg Black, Phil Butcher, Joy Manners, Will Rawes, Jon Rosewell and Andy Sutton.

The Open University, Walton Hall, Milton Keynes MK7 6AA

First published 2000

Written, edited, designed and typeset by the Open University.

Published by Institute of Physics Publishing, wholly owned by The Institute of Physics, London.
IoP Publishing, Dirac House, Temple Back, Bristol BS1 6BE, UK.

US Office: Institute of Physics Publishing, The Public Ledger Building, Suite 1035, 150 South Independence Mall West, Philadelphia, PA 19106, USA.

Printed and bound in the United Kingdom by the Alden Group, Oxford.

ISBN 0 7503 0719 6

Library of Congress Cataloging-in-Publication Data are available.

This text forms part of an Open University course, S207 *The Physical World*. The complete list of texts that make up this course can be found on the back cover. Details of this and other Open University courses can be obtained from the Course Reservations Centre, PO Box 724, The Open University, Milton Keynes MK7 6ZS, United Kingdom: tel. +44 (0) 1908 653231; e-mail ces-gen@open.ac.uk

Alternatively, you may visit the Open University website at http://www.open.ac.uk where you can learn more about the wide range of courses and packs offered at all levels by the Open University.

To purchase other books in the series *The Physical World*, contact IoP Publishing, Dirac House, Temple Back, Bristol BS1 6BE, UK: tel. +44 (0) 117 925 1942, fax +44 (0) 117 930 1186; website http://www.iop.org

1.1

s207book6i1.1

DYNAMIC FIELDS AND WAVES

Introduction

At first sight, the chapter headings in this book may seem to indicate that it will contain a rather disparate set of topics. However, the title of the book, *Dynamic fields and waves*, gives a clue to the underlying unification of these topics.

Chapter 1, 'Fields that vary with time', charts the first step in this unification by describing the process of electromagnetic induction and the various devices that rely on this phenomenon. In it we describe how the nineteenth-century scientists Michael Faraday and Joseph Henry were led to the conclusion that electric and magnetic effects are inextricably linked. In particular their discovery that a changing magnetic field produces an electric field is behind such devices as telephones, electric generators and motors, radio and television transmitters and receivers, and transformers, on which virtually all of modern technology depends. The next bold step in the unification process was that provided later in the nineteenth century by James Clerk Maxwell. He showed that light too is an electromagnetic phenomenon, and can be described in terms of self-propagating electric and magnetic fields. Moreover, visible light constitutes merely one small region of a vast electromagnetic spectrum of radiation, spanning radio waves to gamma-rays.

In Chapter 2, 'Waves and electromagnetic radiation', we explore this idea further. We first use waves on strings and sound waves to illustrate the behaviour of waves in general and to introduce the terminology used to describe them. We then look at electromagnetic radiation in particular, and discuss the phenomena of reflection, refraction, dispersion, diffraction, superposition and interference, which are characteristic of all waves, but are often most easily appreciated in terms of visible light.

In Chapter 3, 'Optics and optical instruments', we extend this discussion to consider the various devices that use lenses and mirrors to exploit the reflection and refraction of light. After introducing the basic building blocks and behaviour of any optical system, we consider in turn how the human eye, microscopes, telescopes and cameras are each used to form images of objects.

Finally, in Chapter 4, 'Special relativity', we return to the unification theme with a discussion of one of the most remarkable theories in the history of science. Using two simple postulates, we show how, in 1905, Albert Einstein essentially overturned Newtonian physics and predicted startling new effects such as time dilation and length contraction for objects travelling at close to the speed of light. The speed of propagation of electromagnetic radiation is central to the special theory of relativity, so it should come as no surprise that Einstein's theory and Maxwell's theory of electromagnetism are intertwined. In fact, unlike Newtonian mechanics, Maxwell's equations already obey Einstein's theory without the need for modification, and electromagnetic induction by motion is shown to arise directly as the result of a transformation between frames of reference.

Open University students should leave the text at this point, and view Video 6 *Waving not drowning*. You should return to the text when you have finished viewing the video.

Chapter 1 Fields that vary with time

1 Current without wires — some examples of induction

Figure 1.1 shows two sealed boxes. One of the boxes has a light bulb holder set into it, with a bulb in place. It is like an ordinary lamp, except that there are no wires connecting it to the domestic electricity supply. The second box has a wire coming out of it, ending in a plug, which is attached to a wall socket. The two boxes are separated by a distance of about a centimetre. When the current is switched on at the socket, the bulb glows, and continues to do so as long as the current continues to flow through the other box. If you were to pass your hand between the two boxes you would feel nothing, and the light from the bulb would not change.

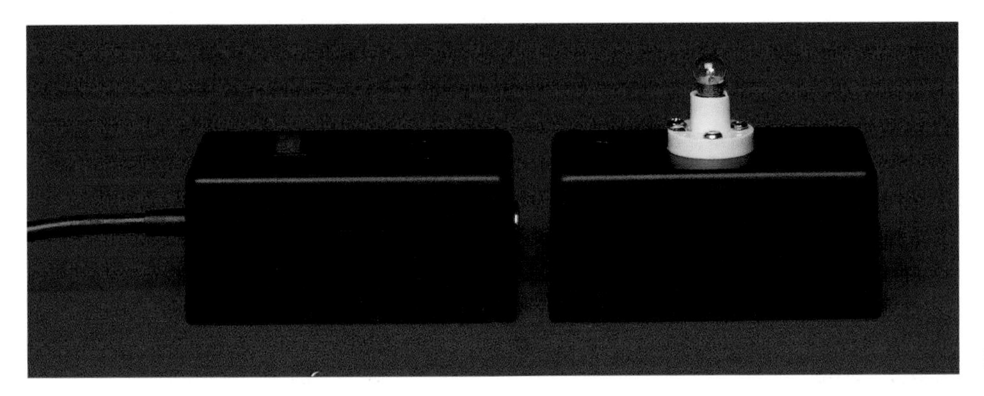

Figure 1.1 A wireless lamp.

How can the bulb light up with no supply of electricity to it? You may think that you have been fooled. Is there a battery inside the box with the bulb? However, as Figure 1.2 shows, all that is inside each box is a coil of insulated copper wire wound around a cylindrical core of iron — a *solenoid*. In *Static fields and potentials* you met solenoids as devices that use electric current to produce a magnetic field. In this chapter you will learn how an effect called **electromagnetic induction** allows a current flowing in one solenoid to 'induce' a current in a second solenoid and thus cause the bulb in the other box to glow.

Now consider something quite different. Suppose you want to melt a small piece of brass or steel, perhaps as part of some jewellery you are making. The melting temperatures of these metals are so high that it is difficult to melt them by direct heating. The standard method of melting uses an *induction furnace*, similar to that

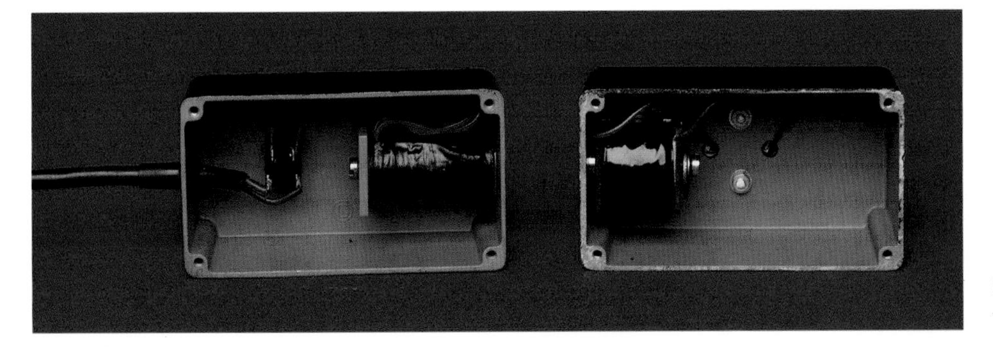

Figure 1.2 Inside view of a wireless lamp.

shown in Figure 1.3a. Superficially, this consists of a box with a hinged door, inside which is a ceramic crucible. A wire leads from the box to an electricity supply. The metal is placed in the crucible, the door is closed, the switch is thrown to complete the circuit, and after a while the metal melts. Taking apart the box to see how it works (Figure 1.3b), you would discover a solenoid encircling the crucible. (The solenoid is contained in a water jacket to keep it from overheating.)

Figure 1.3 (a) An induction furnace; (b) diagrammatic view of the interior of an induction furnace.

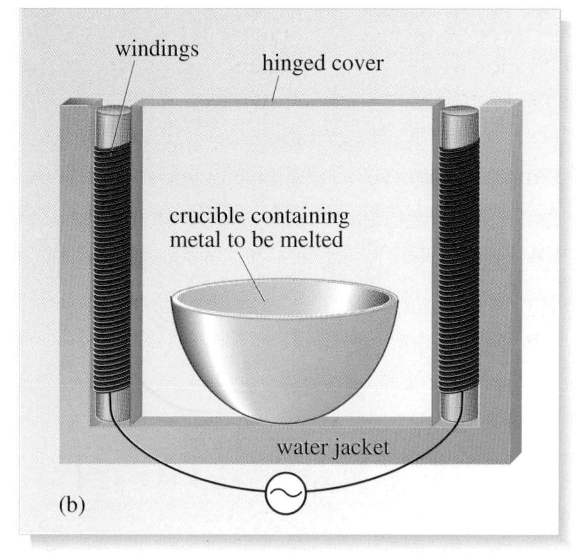

Both the wireless lamp and the induction furnace involve solenoids and therefore magnetic fields. However, the essential common ingredient is that in both cases the current flowing through those solenoids is not constant. Rather, it varies with time and, consequently, so does the magnetic field that each solenoid produces.

Stated simply, *electromagnetic induction* is the effect whereby changing magnetic fields cause, or 'induce', currents. Since fields are involved, these currents can be induced over distances without intervening wires. It is this principle that enables electric razors, food processors, vacuum cleaners, doorbells, washing machines, radios, televisions, telephones, and a host of other devices to work (Figure 1.4). Electromagnetic induction, or just induction for short, is the principal theme of this chapter. In it you will see how currents and magnetic fields which change in time affect each other, and how the devices described above are ingenious arrangements for harnessing these effects for useful purposes.

Figure 1.4 Many everyday devices rely on the principle of electromagnetic induction.

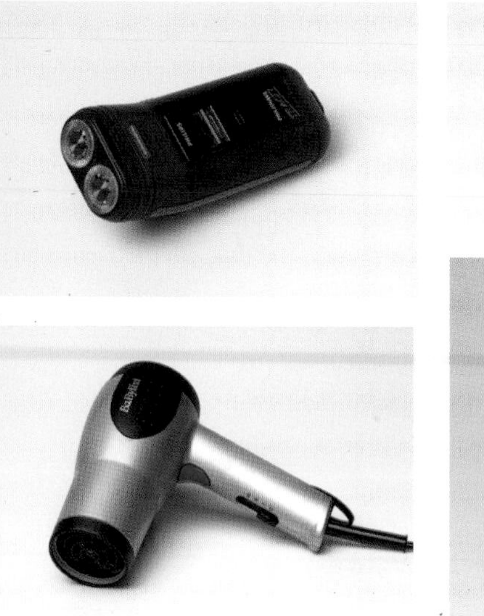

The last part of this chapter will take a new direction. It concerns the work of James Clerk Maxwell, who discovered that electric and magnetic fields can conspire together to create wave-like disturbances that travel at the same speed as that of light. This led to the acceptance that light itself is an electromagnetic wave. This unification of light and optics with electricity and magnetism ranks as one of the major discoveries of science, and is a high-water mark of nineteenth-century physics. It will be further discussed in Chapter 2 of this book.

2 Principles of electromagnetic induction

To set the stage for the discussion, you should recall Oersted's discovery (*Static fields and potentials* Chapter 4), in 1820, that an electric current can affect a magnetic compass. For the first time there was an effect linking electricity and magnetism: electric currents create magnetic fields. Until this time, electricity and magnetism were thought to be entirely separate phenomena.

If an electric current can create a magnetic field, it seems quite reasonable to ask if magnetic fields will have any influence on electric currents. This was the problem that Michael Faraday set himself to solve, though at first with no success.

Michael Faraday (1791–1867)

Michael Faraday (Figure 1.5) was raised in humble circumstances, and received only a rudimentary education. He did, however, learn to read, and when he was later apprenticed to a bookbinder, Faraday took the opportunity to read some of the books that were brought in for binding. By 1813, this interest led him to a job at the Royal Institution, where he remained for the rest of his life. He founded the Royal Institution Christmas Lectures for children in 1826, and gave the lectures himself 19 times.

Faraday worked on the problem of electromagnetic induction from at least the time of Oersted's work in 1820. It is known that he failed to induce currents in 1824, having followed the reasonable, but incorrect notion that magnetic fields would create currents in nearby wires. His breakthrough came in August 1831 when he observed that a *changing* magnetic field was the deciding factor. After nine days of feverish activities he had worked out the basic theory, and in November 1831 he presented a paper on the subject to the Royal Society.

Faraday's achievements were also recognized by the naming of the SI unit of capacitance — the farad — after him.

Figure 1.5 Michael Faraday at work in his laboratory.

When he began his investigations, Faraday did not know the role played by charges in creating electric and magnetic fields. In particular, he did not know the Lorentz force law, which makes it much easier to understand what happens when the fields vary in time. Before you start the next subsection, you should review your knowledge of electromotive force, the nature and production of electric and magnetic fields, and the Lorentz force law, if necessary by referring back to *Static fields and potentials* Chapters 3 and 4. It is particularly important to remember that since electric and magnetic fields are *vector* quantities, they can change because of a change in their *direction*, as well as because of a change in their *magnitude*.

● To be sure that you are ready to proceed, (i) give a definition of EMF and (ii) state what is meant by the Lorentz force.

❍ (i)2The EMF, or electromotive force, is the limiting maximum potential difference between the terminals of a voltage generator (such as a battery or cell) as the current drawn from the generator is reduced towards zero.

(ii) The Lorentz force is given by $\boldsymbol{F}(\boldsymbol{r}) = q[\mathscr{E}(\boldsymbol{r}) + \boldsymbol{v} \times \boldsymbol{B}(\boldsymbol{r})]$. It describes the resultant force experienced by a particle of charge q moving with velocity $\boldsymbol{v}$ at position $\boldsymbol{r}$, under the influence of both an electric field $\mathscr{E}(\boldsymbol{r})$ and a magnetic field $\boldsymbol{B}(\boldsymbol{r})$. ■

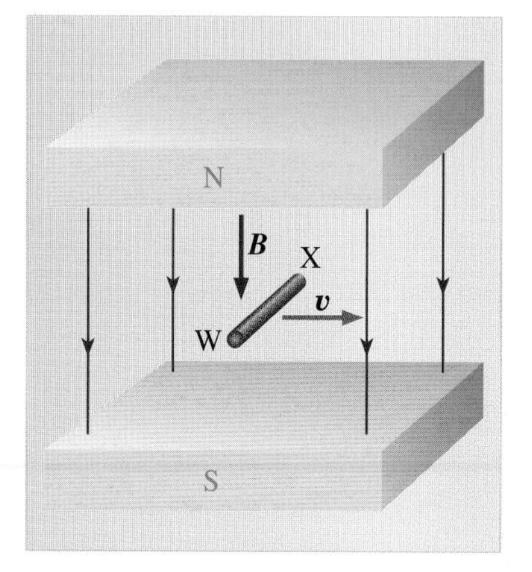

Figure 1.6 A wire element moving in a uniform magnetic field due to a permanent magnet.

2.1 Induction by motion

Conducting wires will be referred to quite a lot in this chapter. As you know, such wires are made from a metal such as copper in which there are a large number of electrons, which are free to move under the influence of an electric or magnetic force. These electrons are known as *conduction electrons*.

The first example to consider is shown in Figure 1.6, where a small piece of wire WX, is seen to be moving with constant velocity $\boldsymbol{v}$ through the magnetic field $\boldsymbol{B}$ of a permanent magnet. It may be assumed that the magnetic field is constant and uniform in the region of the pole faces.

The charges in the wire experience the magnetic part of the Lorentz force, which in a constant, uniform magnetic field may be expressed as:

$$\boldsymbol{F} = q(\boldsymbol{v} \times \boldsymbol{B}). \qquad (SFP \text{ Eqn } 4.2)$$

● What would be the direction of the force on a positive charge in the wire shown in Figure 1.6?

❍ To obtain the direction of a vector cross product, you need to use the right-hand rule. This is shown in Figure 1.7: you can see that a positive charge would experi-ence a force directed towards X. ■

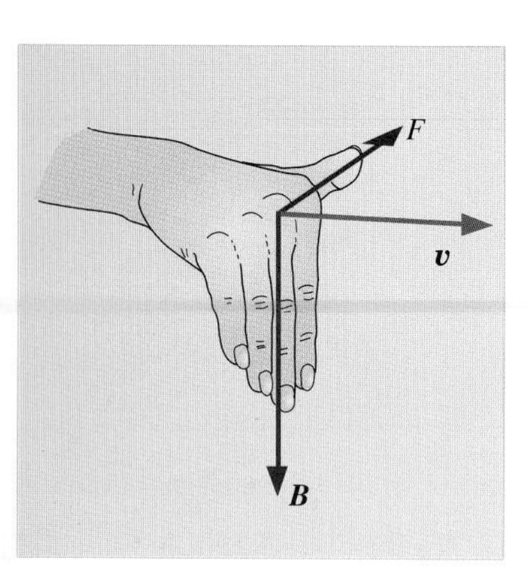

Figure 1.7 Vector directions for Figure 1.6 using the right-hand rule.

Remember that the Lorentz force law always gives the direction of force on a *positive* charge. The force acting on a negative charge will be in the opposite direction, so in this example, a negative charge (such as an electron) would experience a force directed towards W. Note that all the conduction electrons in the wire element feel the same force, and so are going to move in the same direction. This simple observation is important.

Now suppose the wire element is part of a larger circuit, as illustrated in Figure 1.8. Part of the circuit lies in the region of the constant magnetic field, and part of it does not. Because of the Lorentz force on the electrons in the region where the magnetic field is non-zero, they start accelerating towards the conduction electrons ahead of them. From Coulomb's law you know that like charges repel, and so the electrons in front will be 'pushed' in the same direction. The mechanism for the push may be considered as being mediated by an electric field, $\mathscr{E}$, directed from W to X. In terms of this field, each electron experiences a force of magnitude $e|\mathscr{E}|$. This pushing is present throughout the entire loop of wire, even where there is no magnetic field, and creates a *current*, which is registered by the ammeter.

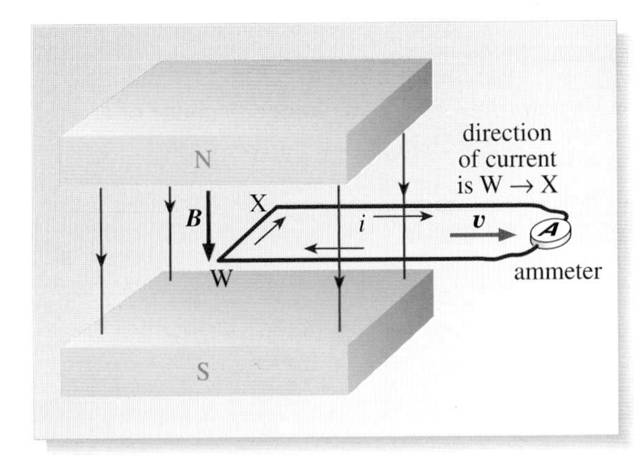

Figure 1.8 Current induced in a circuit by moving it through a magnetic field.

Figure 1.9 Karl Friedrich Gauss (1777–1855) was perhaps the greatest mathematician that ever lived. He made so many important discoveries in mathematics that it would take pages just to list them. His collected work is barely contained in 14 large volumes. A consistent theme in his work is that the theoretical and the practical suggested themselves to him as two parts of a whole. A good example is that as a result of his early employment as a land surveyor, he took the opportunity to create a theory of the differential geometry of curves and surfaces. This subsequently played an important part in Einstein's general theory of relativity.

Moving a wire loop, which is partially in a constant, uniform, magnetic field, induces a current in the wire.

You should remember that current is defined as flowing in the direction a *positive charge* would move, even though (as you know) the current is due to the motion of (negatively charged) conduction electrons.

● In Figure 1.8, which way do the electrons move? Which way does the current flow?

❍ The electrons flow from X to W, or anticlockwise in the loop as seen from above. The current flows from W to X, or clockwise in the loop as seen from above. ■

There are several notable features of this process that you should think about. Perhaps the most important is the fact that a current is only induced when the wire is moving. When the wire is stationary, there is no Lorentz force on the electrons, and so no current.

It has already been mentioned that current flows in the region where the magnetic field is zero, but suppose this part of the wire was several kilometres long. Subject only to a small reduction in the current due to resistance in the wire, the ammeter would register (nearly) the same current as if the wire were, say, only one metre long. Until you think it through, it seems incredible that a localized magnetic field can cause a current to flow in a wire kilometres away from the magnetic field. It is worth emphasizing again the chain of events leading to this result:

- The Lorentz force pushes on the electrons in the region of the magnetic field.
- In turn they push on the electrons in front via an electric field.
- Conduction electrons are easy to push along.
- The push is delivered through an electric field, even where there is no magnetic field.

Two of the pioneers of electromagnetism, the German scientists Karl Friedrich Gauss and Wilhelm Eduard Weber (Figures 1.9 and 1.10) were so amazed at hearing of this effect discovered by Faraday, that they set up an experiment using a very long loop of conducting wire. The wire was placed right across the university town of Göttingen, a

Figure 1.10 Wilhelm Eduard Weber (1804–1891) observed that there could be standardized units for electric and magnetic quantities, which were reducible to combinations of mass, length, time and charge. Now commonplace, this idea was of considerable help in unravelling the origin of electric and magnetic phenomena.

magnet was placed near one end of the wire and a current meter at the other. The wire adjacent to the magnet was moved, and the experiment showed that the current meter reacted during the time the wire was moving, and stopped when the wire was at rest. The fact that such eminent scientists went to the trouble of doing this experiment may seem mildly amusing, but shows that they knew immediately on learning of Faraday's discovery that this was an extremely important result.

In Figure 1.8 the wire was partially in and partially out of the magnetic field region, and so long as the motion continues in this way, current flows. Things change if the loop is entirely within or entirely outside the magnetic field region, and you should now think about what might happen in these cases. To make the situation easier to analyse, consider the situation in Figure 1.11, which shows a square loop of wire whose area is smaller than that of the poles of the magnet.

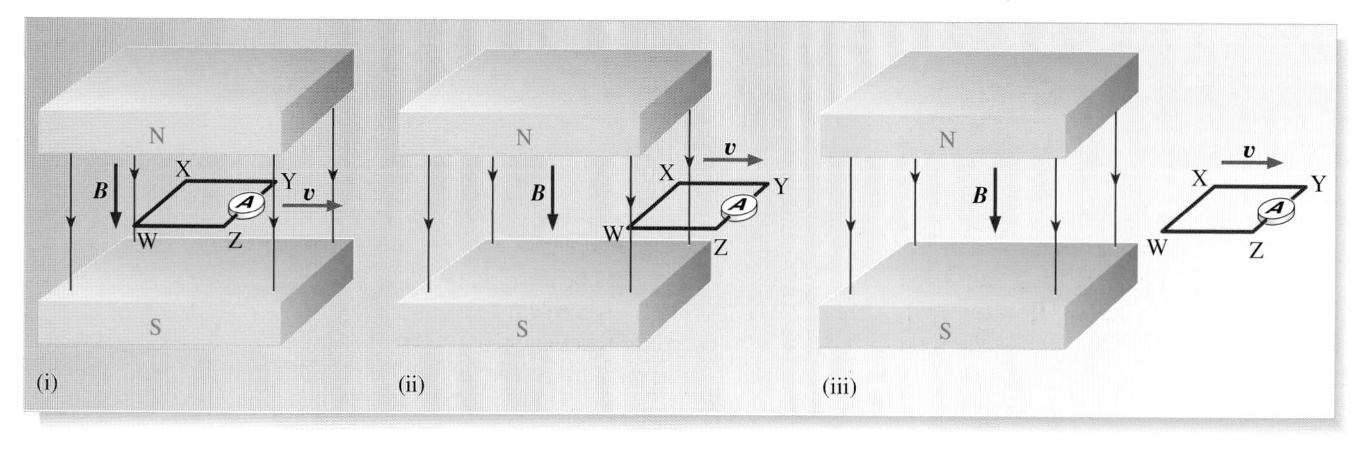

Figure 1.11 (i)–(iii) Three positions of a loop moving in a magnetic field.

Question 1.1 (a) In Figure 1.11(i) a wire loop is moving completely within a magnetic field. What do you think the ammeter would show, and why?

(b) Suppose the loop continues to move. There will be a time when it is partially in and partially out of the magnetic field, as illustrated in Figure 1.11(ii). Later it will be entirely out of the magnetic field as seen in Figure 1.11(iii). What will the ammeter show in each of these situations?

(c) What do you think will happen to the current if the velocity is doubled in magnitude?

(d) What will be the effect on the current if the loop is pulled in the opposite direction with the same speed?

(e) What will be the effect on the current if the magnet is turned upside down?

(f) Finally, what would change if the wire were a circular shape and moving entirely within the magnetic field region? ■

What has been discovered? That motion of a wire in a magnetic field can cause a current to flow in the wire (depending on the geometry of the situation) due to the Lorentz force law, and this causes an electric field to act on the conduction electrons over the entire length of the wire, even in the region outside the magnetic field. Such currents are known as **induced currents** and, since they are the result of the motion of charges through a magnetic field, the process that creates them is known as *induction by motion*.

2.2 Induction by a changing field

Consider the arrangement in Figure 1.8 again. You have studied it when the loop moves through the magnetic field, but what happens if the wire loop is stationary and the permanent magnet moves to the left with a velocity $-\boldsymbol{v}$ as shown in Figure 1.12?

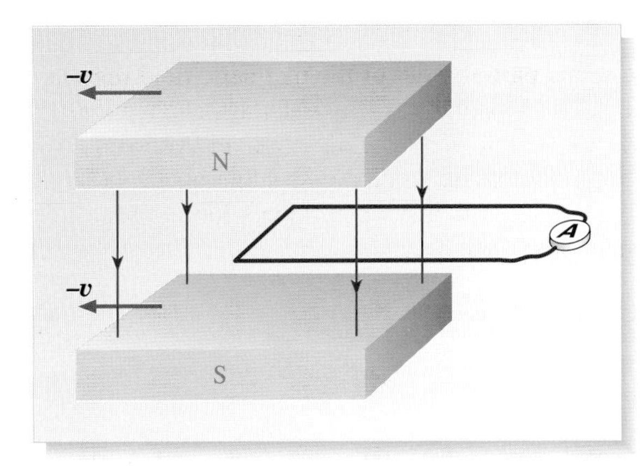

Figure 1.12 A magnetic field moving past a stationary loop of wire.

You may suppose, by direct analogy with the previous section, that a current will flow and there will be a reading in the ammeter. Indeed, if this experiment were carried out, this *is* exactly the result that would be found: there would be an ammeter reading caused by a current flowing. So a current flows when there is a *relative* velocity between the magnet and the loop of wire: it doesn't matter whether it is the magnet or the loop that is actually moving. This follows naturally from the discussion of the Lorentz force earlier, and can be considered as merely a different kind of induction by motion.

● Returning to the situation in Question 1.1 (Figure 1.11(ii)), what would happen if, at the instant when the wire was half out of the magnetic field, the magnet began to move with the same velocity $\boldsymbol{v}$ as the loop?

❍ If the magnet is moved with the same velocity as the loop, the relative velocity between the two would be zero, so no current would flow. ■

Another way of describing what's happening in Figure 1.12 might be to suppose that the changing magnetic field (due to the moving magnet) is creating an electric field, which then causes a current to flow in the wire. In other words:

> A changing magnetic field produces an electric field.

This is quite a different explanation for the phenomenon illustrated in Figure 1.12 than that derived using the Lorentz force law. However, if it is true, it predicts that electromagnetic induction should be observed in a variety of other circumstances, where there is *no* relative motion between the two parts of the apparatus. In fact the statement that a changing magnetic field produces an electric field has been tested again and again, in many different circumstances. It has gained the status of a physical law and it shall be adopted as such here. It is worth emphasizing that, whereas the induction by motion discussed in Section 2.1 was a direct consequence of the Lorentz force law and so should not have been too surprising to you, the induction of a current by a changing magnetic field is a new effect, not predicted by anything you have considered previously.

Moving permanent magnets around can be cumbersome, and changing their field strengths is impossible. To test whether the hypothesis is correct, more complicated magnetic field changes will have to be considered than those obtained by simply moving magnets about. The easiest way this can be done is to produce a magnetic field by running a current through a loop of conducting wire. By changing the electric current the magnetic field can be varied. A second loop can then be brought into near proximity to the first loop and the outcome can be observed. This is the sort of arrangement that was used by Michael Faraday at the Royal Institution in London in his experiments on electromagnetic induction. Independently of this, Joseph Henry (Figure 1.13) discovered that he too could generate currents in the second loop in this way.

Figure 1.13 Joseph Henry.

Joseph Henry (1797–1878)

Joseph Henry was an American physicist who, in a similar way to Faraday, first became interested in science as a result of a book on natural history he came across as a 16-year-old apprentice bookbinder. He set himself the goal of becoming a medical doctor, teaching in rural schools to pay for his tuition at the Albany Academy. His goal changed as a result of his being employed to survey land for a new State road in 1825, and in 1826 he was appointed to teach mathematics and physics at the Academy. His academic career took him to Princeton University (1832) and then to the Smithsonian Institution as Secretary (1846).

His interests ranged widely, and among other things he did work on sunspots, founded the US Meteorological Service, and was instrumental in setting up the system of free international exchange of scientific ideas via journals.

In the area of electromagnetism, he developed insulated wire coils and spooled windings, an electromagnetic telegraph, relays for telegraphs, earthing for circuits, an electric motor and transformers, and investigated self-induction. Historical research shows that, independent of and slightly prior to Faraday's experiment, Henry succeeded in producing induced currents. However, the burden of teaching and administrative work led him to delay publication, and so it is that Faraday's name is attached to the results.

Henry's work was acknowledged by naming the SI unit of inductance after him.

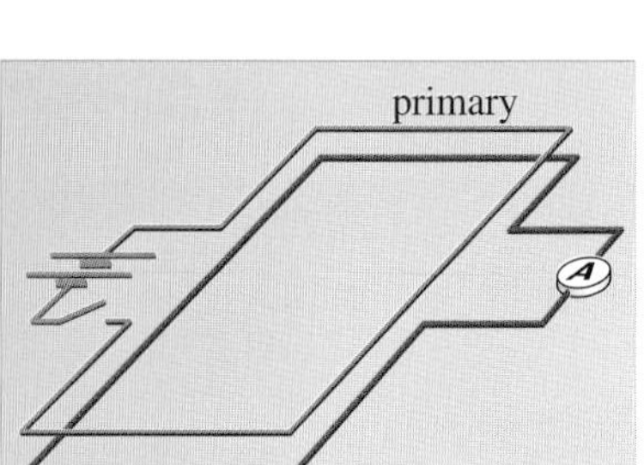

Figure 1.14 Two circuits in close proximity. The primary circuit has a battery and a switch; the secondary circuit has an ammeter to register any induced current. Changing the current in the primary circuit causes a current to flow in the secondary circuit.

It is convenient to call the loop through which the magnetic field is generated the *primary circuit*, and the other loop the *secondary circuit*. The apparatus is illustrated in Figure 1.14. The experiment begins when the switch in the primary circuit is closed, causing a current to flow around it. Careful observation of the ammeter shows that there is an initial surge of current in the secondary circuit, but that it quickly falls to zero and stays there.

How can this observation be understood using the hypothesis that 'a changing magnetic field produces an electric field'? When the switch in the primary circuit is closed, a current quickly builds up in the primary circuit, so there is (briefly) a changing magnetic field produced by the primary circuit. According to the hypothesis, this produces an electric field in the secondary circuit, so causing a current to flow. When the current in the primary circuit has reached a steady value, it is creating a *constant* magnetic field. The hypothesis predicts no induced current in the secondary circuit since there is no *change* of magnetic field, and that is what is observed.

Question 1.2 (a) What will be observed when the switch in the primary circuit in Figure 1.14 is opened? (b) Suppose that the voltage of the battery (or more precisely the EMF) can be varied. After the switch is closed again and a steady current is established in the primary circuit, what will be observed if the voltage is turned up from 2 V to 3 V, and then remains constant? ■

As you have seen, no induced current will be observed in the secondary circuit if the current (and hence the magnetic field) in the primary circuit is constant. This is true whatever the constant value of the voltage supplied to the primary circuit—even if it is so great that the wire's insulation begins to smoke, as was discovered by Faraday!

To summarize, you have seen that currents are produced either by moving conducting wires through magnetic fields, or by changing the magnetic field that threads a conducting wire. The current obtained in either way is called an *induced current.*

2.3 Magnetic flux

Think back to Question 1.1 (Figure 1.11). The current was generated only when a *changing area* of the circuit was crossing the steady field. No current flowed when the whole circuit moved within a uniform field. Somehow the loop area as well as the magnetic field is involved in the process of creating an induced current. Experiment reveals that for a loop of conducting material in a uniform magnetic field perpendicular to the loop, it is the quantity

(area of loop) × (strength of the perpendicular magnetic field)

which is central to the induction of currents. This quantity is a particular example of *magnetic flux*, and is represented by the symbol ϕ. The name 'flux' comes from the fact that a similar quantity is used to study the amount of fluid flowing out of a surface, and this can be used to get an intuitive mental picture of the magnetic flux. Of course a magnetic field is not a stream of fluid, but as a mental picture, an imaginary fluid streaming in the direction of the magnetic field lines is quite useful. The magnetic flux through a given area is a measure of this imaginary 'streaming' through that area.

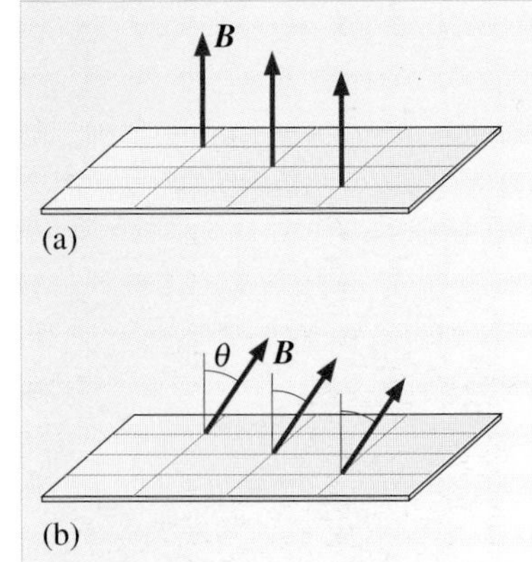

Figure 1.15 (a) A constant, uniform magnetic field $\boldsymbol{B}$ passing through a rectangular surface when $\boldsymbol{B}$ is perpendicular to the surface. (b) A constant, uniform magnetic field $\boldsymbol{B}$ passing through a rectangular surface when $\boldsymbol{B}$ is inclined at an angle θ to the perpendicular.

So far only the case in which the magnetic field is at right-angles to the surface has been considered, as shown in Figure 1.15a. But what if the magnetic field is not perpendicular to the surface? Returning to the picture of a streaming fluid, clearly any field component *along* the surface will not contribute to the streaming effect *through* the surface. So in order to determine the magnetic flux, the component of $\boldsymbol{B}$ which is *perpendicular* to the surface must be determined. Figure 1.15b illustrates how this can be done. In this case the magnetic field is inclined at an angle θ to the perpendicular.

ϕ is the Greek letter *phi*, pronounced 'fie'.

The component of $\boldsymbol{B}$ which is perpendicular to the surface is $|\boldsymbol{B}|\cos\theta$, or simply $B\cos\theta$, where B is used to represent the magnitude of the vector $\boldsymbol{B}$. There is also a component of the magnetic field parallel to the surface, but it does not contribute to the magnetic flux. The general definition of **magnetic flux** is then

$$\phi = AB\cos\theta \qquad (1.1a)$$

where B is the magnitude of the magnetic field, A is the area of the surface through which the magnetic field passes, and θ is the angle between the normal (i.e. perpendicular) to the surface and the direction of the magnetic field $\boldsymbol{B}$. If you recall the definition of the *scalar product* of two vectors, given in Chapter 2 of *Predicting motion*, you will appreciate that an alternative way of writing Equation 1.1a is

$$\phi = \boldsymbol{A}\cdot\boldsymbol{B}. \qquad (1.1b)$$

The vector $\boldsymbol{A}$ has a magnitude equal to the area of the surface, and a direction perpendicular to the surface. So the angle between the vectors $\boldsymbol{A}$ and $\boldsymbol{B}$ is θ as required, as shown in Figure 1.16.

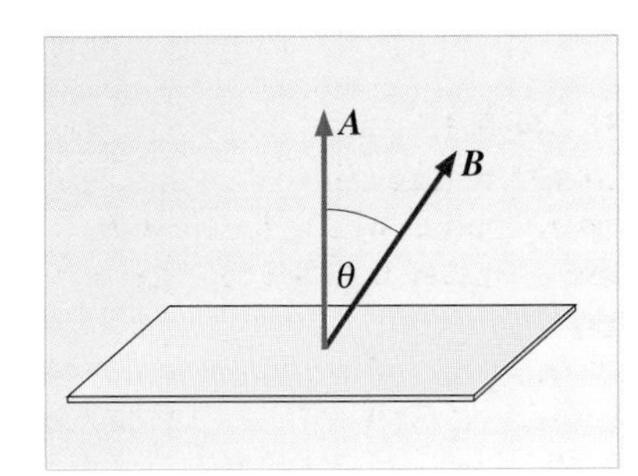

Figure 1.16 The scalar product of the vectors $\boldsymbol{A}$ and $\boldsymbol{B}$ is the magnetic flux ϕ.

What about the SI unit for magnetic flux? Recall that the unit of magnetic field strength is the tesla (T) and that of area is the square metre (m^2). It follows that magnetic flux has the unit of (tesla × square metre), or T m^2 in symbols. This SI unit has its own name, the weber (Wb), after the German physicist Wilhelm Weber, mentioned earlier (Figure 1.10).

$$1 \text{ weber} = 1\,\text{Wb} = 1\,\text{T}\,\text{m}^2.$$

To get a feel for the size of the weber, consider the magnetic flux through a wire loop about the size of your hand, and held at about waist height. The strength of the Earth's magnetic field at the surface of the Earth is about 3×10^{-5} T, and there is no essential difference at waist height. A wire loop the size of your hand would have an area of about 10^{-2} m^2, so the magnetic flux through such a loop would be about 3×10^{-7} Wb. Some other values of magnetic flux are shown in Figure 1.17.

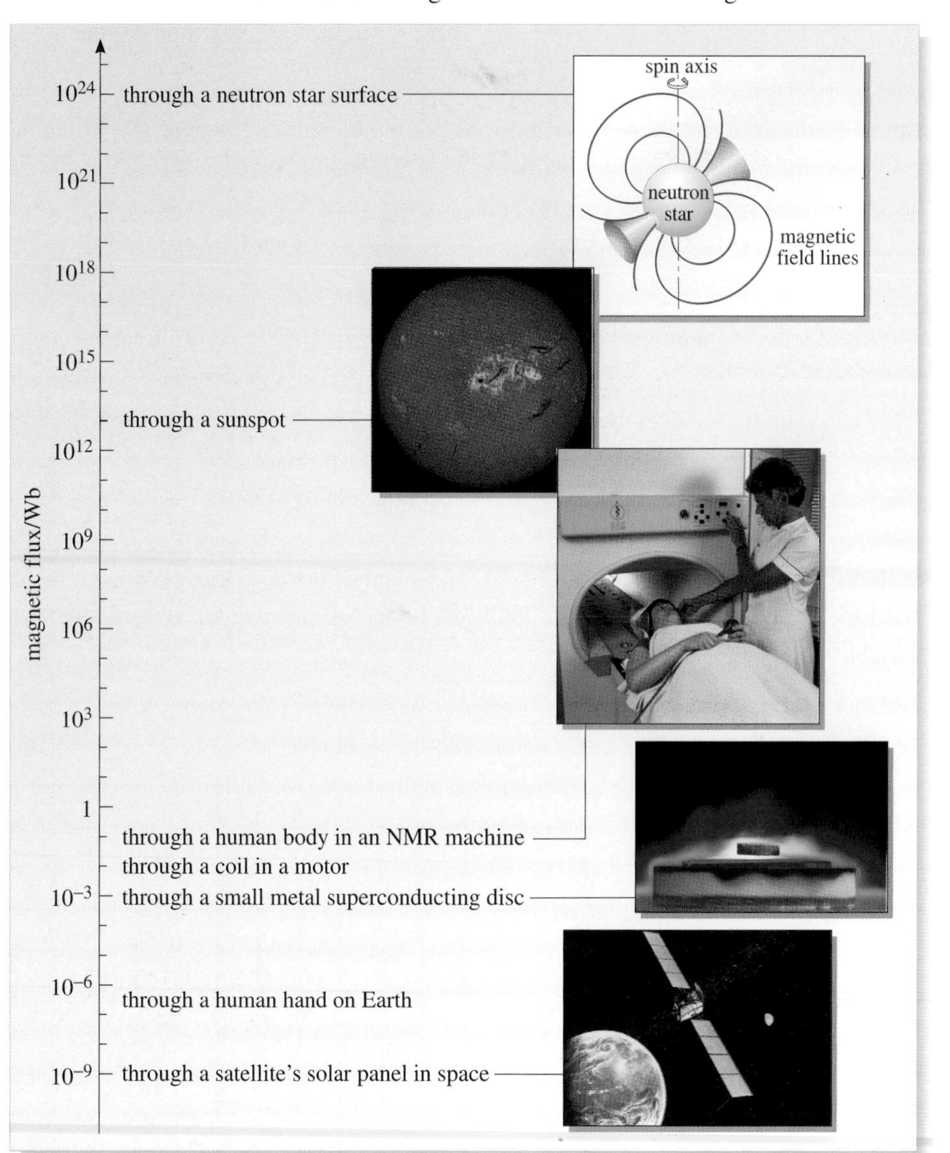

Figure 1.17 Some values of magnetic flux; remember that magnetic flux depends on the magnetic field strength *and* the area.

Question 1.3 Work out the magnitude of the magnetic flux through each of the following circuits: (a) a square circuit with sides of length 0.1 m, with its plane perpendicular to a uniform field of strength 0.4 T; (b) a rectangular circuit of area 10^{-3} m^2, with its normal parallel to a uniform field of strength 10^{-5} T; (c) a circular circuit of area 4.0 m^2, with its normal at an angle of 30° to a uniform field of strength 0.5 T; (d) a circular circuit of area 4.0 m^2, with its plane at an angle of 60° to a uniform field of strength 0.5 T. ■

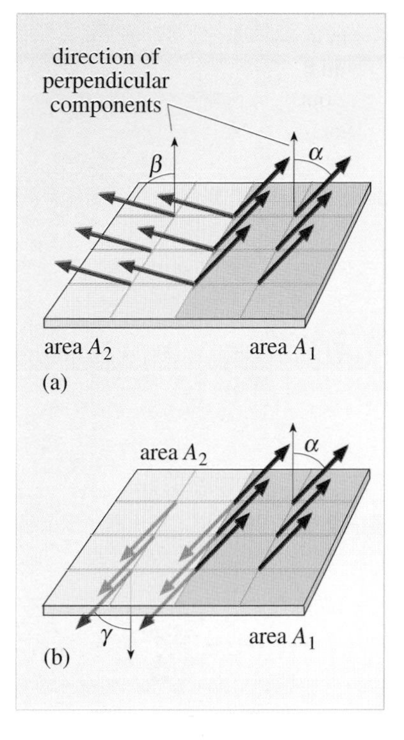

Figure 1.18 A magnetic field in different directions over two parts of a surface: (a) the magnetic field is uniform at inclination β on the left, uniform at inclination α on the right, and perpendicular components are in the *same* direction everywhere; (b) the magnetic field is uniform at inclination γ on the left, uniform at inclination α on the right, and perpendicular components are in *opposite* directions on the two sides.

Although a general definition of magnetic flux has been arrived at, there are two complications that have so far been ignored. First, although magnetic flux is a *scalar* quantity, it still has a *direction* through a surface. So magnetic flux 'streaming in' and 'streaming out' of a surface must be distinguished. If the surface is closed, like a sphere, it is clear what 'in' and 'out' mean, and the magnetic flux 'into' the sphere can be referred to as positive and the magnetic flux 'out of' the sphere as negative. However, if the surface is open, as in the case of a loop for instance, there is no distinguishing direction perpendicular to it. In this case, one of the directions can simply be chosen to correspond to a positive magnetic flux (left to right, say), and the other direction then corresponds to a negative magnetic flux (right to left, say).

The second problem occurs when the magnetic field is not constant across the surface. The correct way to handle this is to break up the surface into small enough pieces so that the magnetic field is essentially constant across each piece. Then the flux can be calculated for each piece, and the total added up. To do this exactly requires integral calculus, and will not be considered in this course. However, it is possible to treat some simple cases without resorting to integration.

A simple example in which the magnetic field does vary is illustrated in Figure 1.18a. Here the magnetic field is constant over part of the surface, inclined at an angle α to the vertical, and then abruptly changes direction to an inclination β over the remainder of the surface. Admittedly, this is a very artificial example, but it illustrates a point. A_1 and A_2 are the areas of the right and left halves of the surface, and there is no problem in calculating the magnetic flux for each piece. What is interesting here is that the normal component of the magnetic field points in the same direction (upwards in the diagram) over the entire surface. The magnetic fluxes must therefore be added to get the total:

$$\phi = A_1 B \cos\alpha + A_2 B \cos\beta.$$

● Figure 1.18b illustrates a variant of the example that has just been considered. The difference is that now the perpendicular component of $\boldsymbol{B}$ on the left points in the opposite direction to the perpendicular component on the right. Find the magnetic flux across the surface. (Assume that the sign convention for magnetic flux gives a positive value for the flux across the right-hand piece.)

❍ The *magnitude* of the flux on the left is calculated essentially as before, but its *direction* is different. So, if the flux on the right is positive, the flux on the left must be negative. Therefore

$$\phi = A_1 B \cos\alpha - A_2 B \cos\gamma. \quad \blacksquare$$

You might be worried that if the areas and the angles were the same, then the total flux would be zero. This would indeed be the case: zero total flux indicates that on average there is the same amount 'streaming in' through the surface in one direction as there is 'streaming out' in the other direction.

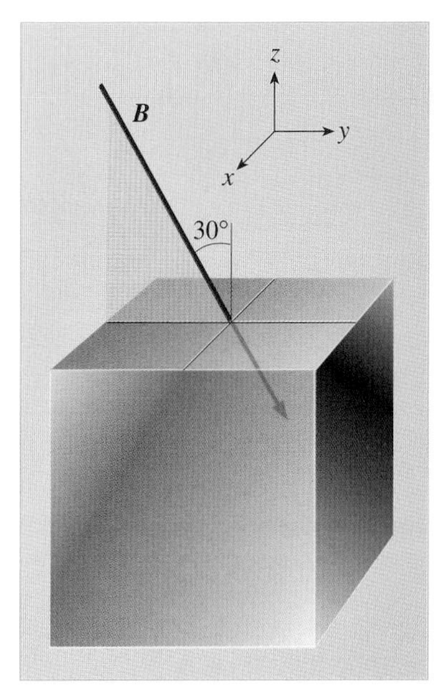

Figure 1.19 A cube sits within a uniform magnetic field. The field $\boldsymbol{B}$ is parallel to the front and back faces of the cube.

Question 1.4 Figure 1.19 shows a cube of side length 0.1 m. There is a constant magnetic field $\boldsymbol{B}$ of magnitude 3.0 T in the direction shown, inclined at 30° from the normal to the upper face of the cube. Find the magnetic flux through each of the six faces of the cube, and then determine the total magnetic flux through the surface of the cube. ■

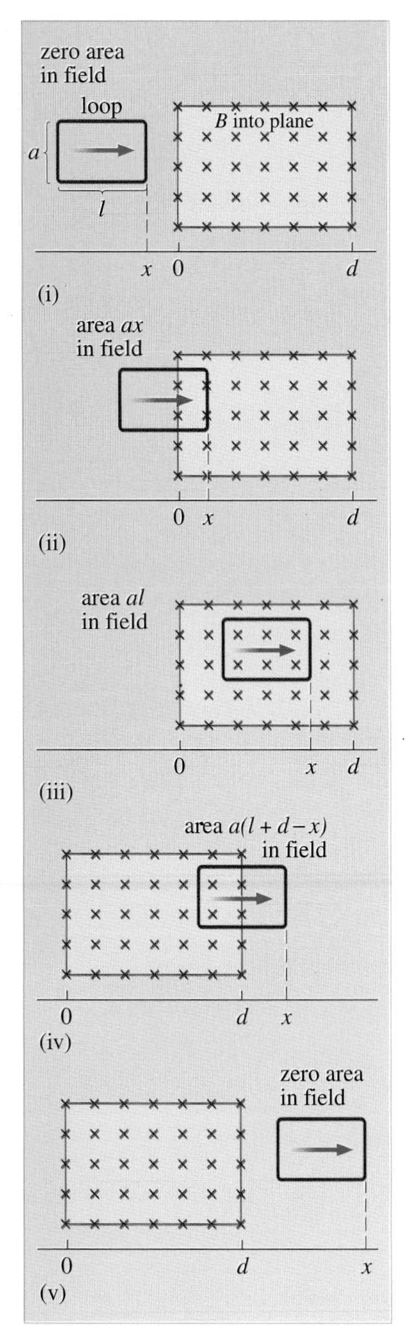

Figure 1.20 A rectangular loop moving to the right at speed v through a magnetic field of magnitude B pointing into the plane of the loop. The loop is shown in five regions: (i) before it enters the magnetic field; (ii) partly within the magnetic field (leading edge in); (iii) entirely within the magnetic field; (iv) again partly within the magnetic field (leading edge out); and (v) after it leaves the magnetic field.

You can probably see that the situation described in Question 1.4 would also apply to any other shape of solid object in a uniform magnetic field. There will be as much magnetic flux streaming into a closed surface as streams out again. In fact:

The total magnetic flux through *any closed surface* is always zero.

This turns out to be true even when the field is not uniform, though that is not apparent from what we have seen so far. In Section 4 this claim will be elevated to one of Maxwell's four equations.

2.4 Changing magnetic flux

The two examples presented below involve situations in which the magnetic flux is changing. The first example considers a situation in which the area of a wire loop within a magnetic field changes with time; the second example looks at a situation in which the angle between a wire loop and the magnetic field changes with time. Later in this chapter these ideas will be developed further in order to quantify the magnitude of the induced current in such wires.

Example 1.1 Magnetic flux through a moving loop

A rectangular loop of conducting wire of length l and width a moves with a speed v in the positive x-direction, through a region of uniform magnetic field of magnitude B and length d, as illustrated in the five frames of Figure 1.20. The crosses in the diagram signify that the magnetic field is directed into the plane of the diagram.

(a) For each situation, (i) to (v), state the possible values for x, the position of the leading edge of the loop with respect to the area of the field.

(b) For each situation (i) to (v), calculate the area of the loop in the magnetic field in terms of a, l, d and x, and hence determine the magnetic flux through the loop for each situation.

(c) Sketch a graph to show the variation of the area within the magnetic field as a function of the position of the leading edge. Sketch another graph to show the variation of magnetic flux as a function of the position of the leading edge.

(d) For each situation (i) to (v), use the fact that $x = vt$ to express the flux through the loop as a function of time, and then differentiate the expressions for magnetic flux to determine the rate of change of magnetic flux with time in each situation.

(e) Sketch a graph to show the variation of the magnetic flux with time. Sketch another graph to show the variation of the rate of change of magnetic flux with time.

Solution

(a) In Figure 1.20(i) the loop is seen before entering the magnetic field. This requires $x < 0$. In Figure 1.20(ii) the loop is partly within the magnetic field. This requires $0 < x < l$. In Figure 1.20(iii) the loop is entirely within the magnetic field. This requires $l < x < d$. In Figure 1.20(iv) the loop is leaving the magnetic field. This requires $d < x < d + l$. In Figure 1.20(v) the loop has fully emerged from the magnetic field. This requires $x > d + l$.

(b) In Figure 1.20(i), when the loop has not yet entered the magnetic field region, there is no area inside the magnetic field. The magnetic flux through the loop is clearly zero, so $\phi = 0$ for $x < 0$.

In Figure 1.20(ii), the area inside the magnetic field region is ax, so the magnetic flux is $\phi = axB$ for $0 < x < l$.

In Figure 1.20(iii), the magnetic flux remains constant at its maximum value as x increases, since the entire loop is within the magnetic field region. This continues to be the case until the leading edge of the loop starts to leave the magnetic field region. The area in the magnetic field is al, so the magnetic flux is $\phi = alB$, for $l < x < d$

Now comes the trickiest part. What area is in the magnetic field in the situation shown in Figure 1.20(iv)? How much of the loop is *outside* the magnetic field? The length of loop outside the magnetic field is $x - d$. So the length inside the magnetic field is the difference between that and the length of the loop — that is, $l - (x - d) = l + d - x$. Thus, the area in the magnetic field is $a(l + d - x)$ and the magnetic flux is $\phi = a(l + d - x)B$, for $d < x < d + l$.

Finally, Figure 1.20(v) is like Figure 1.20(i) so $\phi = 0$, for $x > d + l$.

(c) In Figure 1.21(i) the area of the loop within the magnetic field region has been sketched, showing the five values calculated above. Figure 1.21(ii) shows the magnetic flux as a function of the position of the leading edge, with the appropriate formulae for the magnetic flux.

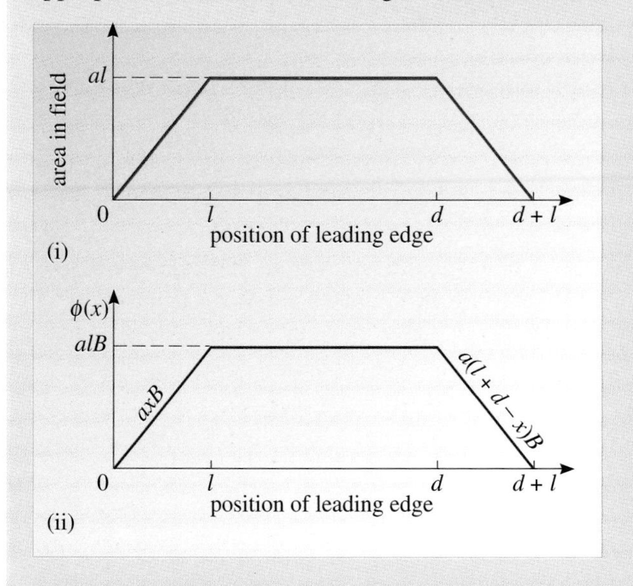

Figure 1.21 (i) The area of the loop inside the magnetic field region as a function of the position of its leading edge; (ii) the magnetic flux as a function of the position of the leading edge of the loop.

(d) If vt is substituted for x everywhere in the formulae for magnetic flux, the result is expressions that can be differentiated, namely:

(i) $\phi(t) = 0$

(ii) $\phi(t) = avBt$

(iii) $\phi(t) = alB$

(iv) $\phi(t) = a(l + d - vt)B$

(v) $\phi(t) = 0$.

Finding the time derivative of the magnetic flux in each situation is now straightforward. It is zero when the magnetic flux is constant, and this takes care of the time when the loop is wholly outside or inside the magnetic field region (situations (i), (iii) and (v)). The magnetic flux is changing only when the loop is entering or leaving the region. The formulae needed have already been written down and the derivatives are not very difficult to do, using the formula

$$\frac{\mathrm{d}(p+qt)}{\mathrm{d}t} = q$$

for any constants p and q.

The rate of change of the magnetic flux is simply

$$\frac{\mathrm{d}\phi(t)}{\mathrm{d}t} = avB \quad \text{in situation (ii),}$$

$$\frac{\mathrm{d}\phi(t)}{\mathrm{d}t} = -avB \quad \text{in situation (iv),}$$

and $$\frac{\mathrm{d}\phi(t)}{\mathrm{d}t} = 0 \quad \text{elsewhere.}$$

(e) The time it takes for the leading edge to get to the point $x = l$ is $t = l/v$. Similarly, the edge gets to the point $x = d$ at time $t = d/v$, and so on. Using this information, the required graph can be plotted showing the variation of magnetic flux with time, as shown in Figure 1.22a.

Using the values for the rate of change of magnetic flux from part (d), a graph of the rate of change of magnetic flux as a function of time can be plotted as shown in Figure 1.22b.

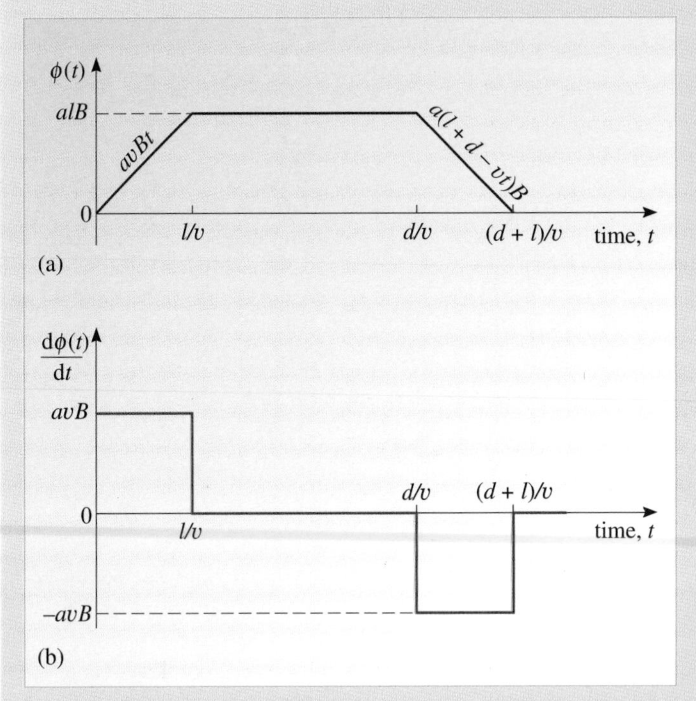

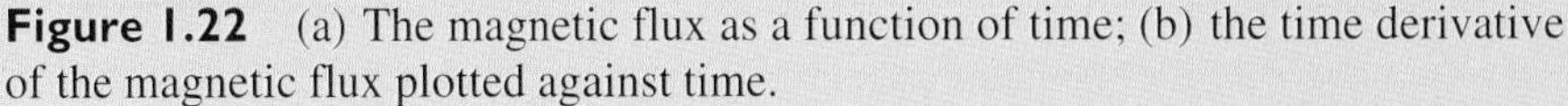

Figure 1.22 (a) The magnetic flux as a function of time; (b) the time derivative of the magnetic flux plotted against time.

In summary, the magnetic flux is zero until the loop starts to enter the magnetic field region. It then increases linearly until it is entirely in the magnetic field region. It stays at this maximum until it starts to leave the magnetic field region, when it decreases linearly until it reaches zero. It then stays zero. Its time derivative is zero except when the magnetic flux is changing, which is in situations (ii) and (iv). In the second situation the magnetic flux is increasing linearly, so its derivative is a positive constant. In the fourth situation it is decreasing linearly, so its derivative is a negative constant. The magnitude of these constants is the same, which is a result of the symmetry of the problem.

In Example 1.2 you are asked to consider what happens when a loop, entirely within the magnetic field of a permanent magnet, *rotates* at a uniform angular speed. You might wish to review some material on uniform rotation before looking at the example. The difference between this example and the others presented earlier is that here the *angle* between the magnetic field and the area is changing in time. This is an important problem because, as discussed later in this chapter, it is the basis of motors and generators. So you should take enough time to be certain that you understand the solution and could reproduce it if asked to do so.

Example 1.2 Magnetic flux through a rotating loop

A rectangular loop of conducting wire of area A is mounted on an axle, which is turned anticlockwise (when viewed from above) at a uniform angular speed ω. The loop is wholly within the uniform field of a permanent magnet as shown in Figure 1.23, and the axle is at right-angles to the direction of the magnetic field. You may assume that the plane of the loop is parallel to the magnetic field when the motion starts.

(a) Find an expression for the magnetic flux through the loop as a function of time, and sketch a graph showing how the magnetic flux varies during one complete period of rotation of the loop.

(b) Determine the rate of change of the magnetic flux as a function of time, and sketch a graph to show how this quantity varies throughout one complete period of rotation.

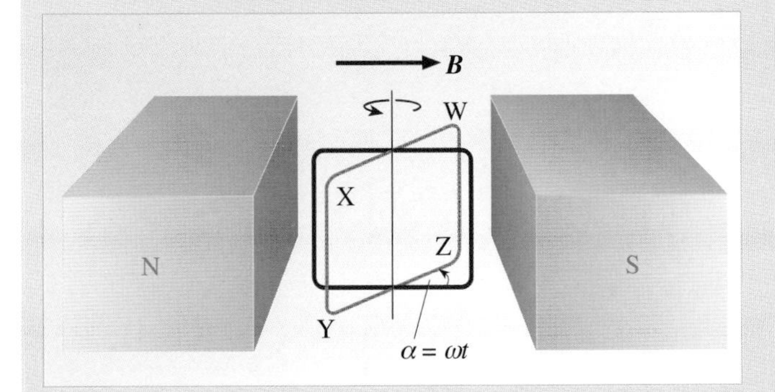

Figure 1.23 A rectangular loop of conducting wire rotating anticlockwise at a constant angular speed ω in the uniform field of a permanent magnet.

Solution

The difficult part of the solution is setting up the trigonometry. Figure 1.24 shows the situation at time t when the loop has rotated through an angle $\alpha = \omega t$. When viewed along the axle from the side of the loop labelled WX, the loop rotates anticlockwise in the magnetic field.

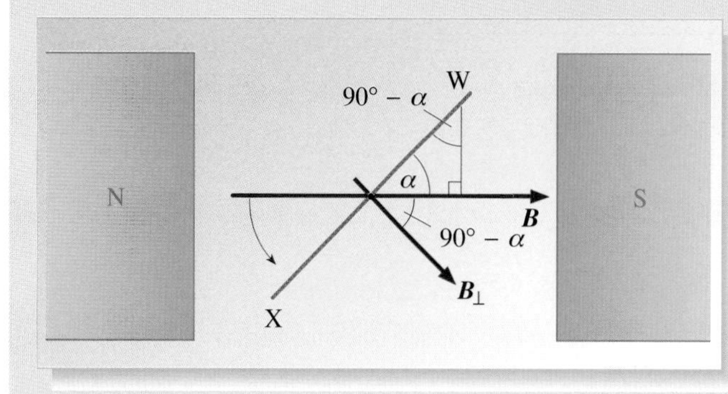

Figure 1.24 The component of the magnetic field perpendicular to the plane of the loop, viewed from above.

(a) Initially the loop was parallel to the magnetic field and there was no magnetic flux through the loop. However, after a time t, the magnetic field will have a component perpendicular to the plane of the loop, whose magnitude is denoted by $B_\perp(t)$. The angle between the magnetic field and the perpendicular to the loop is $\theta = 90° - \alpha$, therefore $B_\perp(t)$ is given by

$$B_\perp(t) = B\cos\theta = B\cos(90° - \alpha).$$

It is a property of sines and cosines that $\cos(90° - \alpha) = \sin\alpha$, so

$$B_\perp(t) = B\sin\alpha = B\sin\omega t.$$

The magnetic field component $B_\perp(t)$ is uniform over the entire loop, and so at time t the magnetic flux through the loop is

$$\phi(t) = AB\sin\omega t,$$

where the notation $\phi(t)$ is a reminder that the magnetic flux is a function of time. A sine function varies smoothly between positive and negative values, so the constant angular speed of the loop gives rise to a periodic variation of the magnetic flux as shown in Figure 1.25a.

(b) To find the rate of change of the magnetic flux with time, you should recall the general rule:

$$\frac{d}{dt}\sin at = a\cos at \quad \text{for any constant } a.$$

Therefore

$$\frac{d\phi(t)}{dt} = \omega AB\cos\omega t.$$

A cosine function also varies smoothly between positive and negative values but is shifted by one-quarter of a cycle with respect to a sine function. The rate of change of the magnetic flux is shown in Figure 1.25b.

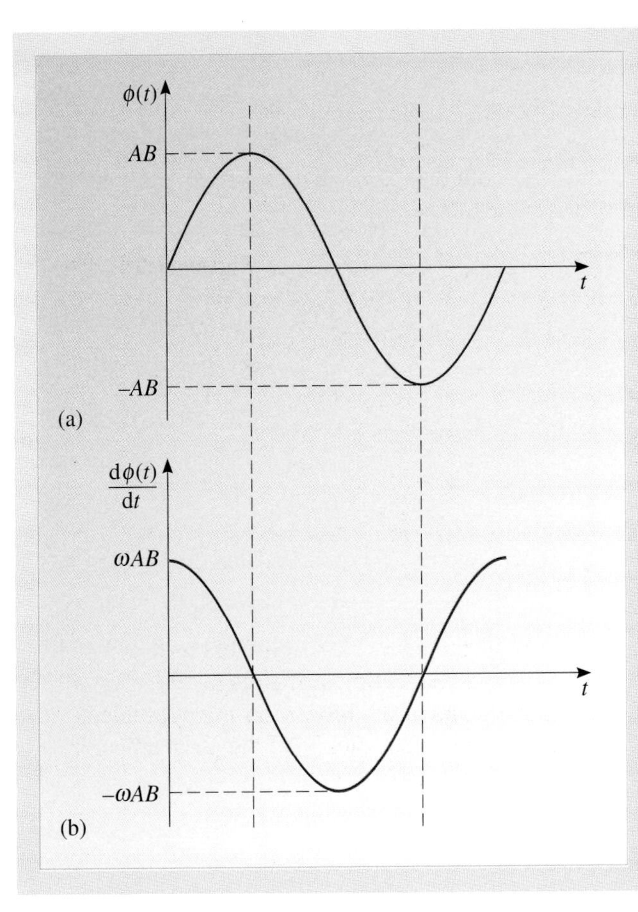

Figure 1.25 (a) The magnetic flux through a rectangular loop of wire as it is rotated in a uniform magnetic field; (b) the rate of change of magnetic flux through the loop of wire as it is rotated in a uniform magnetic field.

2.5 Faraday's law: the magnitude of an induced EMF

You were introduced to electromotive force (EMF) in *Static fields and potentials*. From what you learnt there, you know that if the total EMF around a closed circuit is non-zero then a current will flow in that circuit. This means that an induced current must be due to an **induced EMF**. In fact, this was anticipated earlier in this chapter with the discussion of conduction electrons pushing the electrons ahead of them by means of an electric field, for the EMF in that case is proportional to the electric field along the wire. So if the EMF produced by a changing magnetic field can be determined, then the induced current can be determined too.

The problem is then to determine the EMF induced in a circuit by a changing magnetic field; it was solved by Faraday.

Faraday's law

The magnitude of the induced EMF around the boundary of a surface is equal to the magnitude of the rate of change of the magnetic flux through that surface.

In the specific case of a circuit consisting of a loop of wire, the 'boundary of a surface' is the wire loop and the 'surface' is any area enclosed by the circuit.

Retaining the symbol $\phi(t)$ for the magnetic flux as a function of time and using the symbol $V_{\text{ind}}(t)$ for the induced EMF as a function of time, Faraday's law can be written as the equation:

$$|V_{\text{ind}}(t)| = \left|\frac{\mathrm{d}\phi(t)}{\mathrm{d}t}\right|. \tag{1.2}$$

As noted earlier, magnetic flux has a sign, so the rate of change of magnetic flux and the induced EMF can also be either positive or negative. Nothing has yet been said about the direction: that's the subject of Lenz's law in the next section.

Having determined the magnitude of the induced EMF in a circuit, it is a simple matter to determine the magnitude of the induced current, providing the circuit only contains resistors with a total resistance R. Ohm's law (*SFP* Eqn 3.3) tells us that, in such a case, the magnitude of the induced current is

$$|i_{\text{ind}}(t)| = \frac{|V_{\text{ind}}(t)|}{R}. \tag{1.3}$$

● Show that the units on the right-hand side of Equation 1.2 are consistent with those on the left-hand side.

❍ The units of the right-hand side of Equation 1.2 are those of $\mathrm{d}\phi/\mathrm{d}t$, namely

$$\frac{\text{Wb}}{\text{s}} = \frac{\text{T m}^2}{\text{s}} = \frac{\text{N s m}^{-1}\,\text{C}^{-1}\,\text{m}^2}{\text{s}},$$

since $1\,\text{T} = 1\,\text{N s m}^{-1}\,\text{C}^{-1}$ (*SFP* Chapter 4). So the unit for $\mathrm{d}\phi/\mathrm{d}t$ is $\text{N m C}^{-1} = \text{J C}^{-1} = \text{V}$, as required. ■

● Faraday's law states that a changing magnetic flux induces an EMF, which in turn gives rise to an induced current in a closed circuit. Bearing in mind the definition of magnetic flux, what quantities could be varied in order to induce a current in a wire loop sitting in a magnetic field?

❍ Since $\phi = AB\cos\theta$, changing any one or all of: (i) the area A of the loop in the magnetic field, (ii) the magnitude of the magnetic field B, or (iii) the angle between the loop and the magnetic field θ, would change the magnetic flux and so give rise to an induced current. ■

The most important thing to notice about Faraday's law is that it is the *change* in the magnetic flux that causes the EMF. It is a tempting mistake to mentally equate the magnitude of the magnetic flux with the induced EMF. Even if the magnetic flux is initially small and then decreases a bit, but very rapidly, the induced current surge will be large. Conversely, a high but steady magnetic flux gives no induced current.

Question 1.5 (a) Explain in terms of Faraday's law and induced EMF what happens when the switch in the primary circuit in Figure 1.14 is closed. (b) Contrast what happens when the primary EMF is changed with what happens when it remains constant. (c) Finally, explain in terms of Faraday's law and induced EMF what happens when the switch is opened.

Question 1.6 (a) For the situation described in Example 1.1, suppose that the magnetic field has a strength of $B = 0.5\,\text{T}$, the wire loop has a width $a = 10\,\text{cm}$, a resistance $R = 10\,\Omega$, and the loop moves at a speed $v = 0.5\,\text{m s}^{-1}$. What would be the magnitude of the current induced in the wire loop during the times it is entering and leaving the magnetic field region? (b) For the situation described in Example 1.2, suppose that the magnetic field has a strength of $B = 2.0\,\text{T}$, the loop has an area of $A = 25\,\text{cm}^2$, a resistance of $R = 5.0\,\Omega$, and it rotates at an angular frequency of $\omega = 40\,\text{s}^{-1}$. What would be the maximum current induced in the wire loop? ■

Finally, note that it is of far-reaching consequence that Faraday's law is not simply restricted to describing circuits. Whenever there is a changing magnetic flux through a surface, there is an induced EMF around the *boundary* of that surface. This boundary does not have to be a conducting wire loop. Far from it: it can be anything, even an empty space, as shown in Figure 1.26.

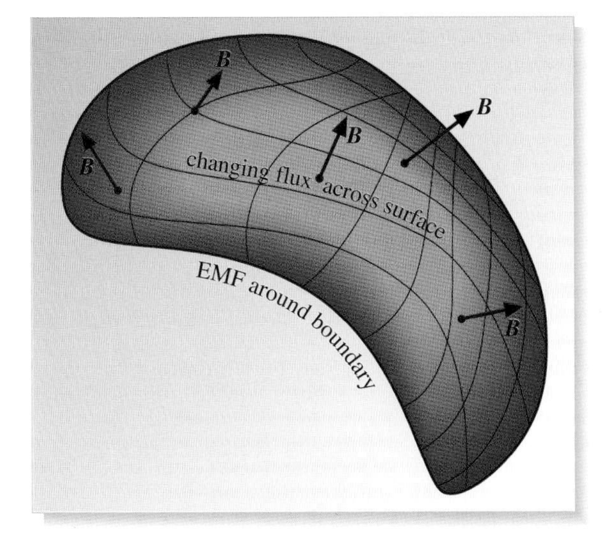

Figure 1.26 Faraday's law is applicable for any surface with a boundary.

2.6 Lenz's law: the direction of an induced current

There is one thing missing from the discussion so far: which way does an induced current flow? The principle that gives the answer is Lenz's law, named after Heinrich Lenz (Figure 1.27), who formulated it in 1833.

Figure 1.27 Heinrich Lenz (1804–1865) was born in the old university town of Tartu, Estonia (then in Russia). He was a professor at the University of St Petersburg, and carried out many experiments to investigate electromagnetic induction, following the work of Faraday.

Lenz's law

When a changing magnetic flux generates an induced EMF, the effect of that EMF is to *oppose* the change that caused it.

Although Lenz's law, like Faraday's law, is phrased in terms of EMFs, remember that the induced EMF will, in general, give rise to an induced current, which in turn will give rise to an induced magnetic field. It is usually the effect of this induced magnetic field that needs to be considered when working out the direction in which the induced current will flow.

As an introduction to Lenz's law, two examples will again be presented. They both relate to situations that have been considered earlier in the chapter, but previously the directions involved could not be determined.

Example 1.3 A moving wire loop in a magnetic field

Consider again the arrangement explored in Question 1.1 and Example 1.1 (Figure 1.20). Use Lenz's law to determine the direction of current flow in the loop in each situation.

Solution

When the loop is outside the magnetic field of the permanent magnet in situation (i), no current flows in it.

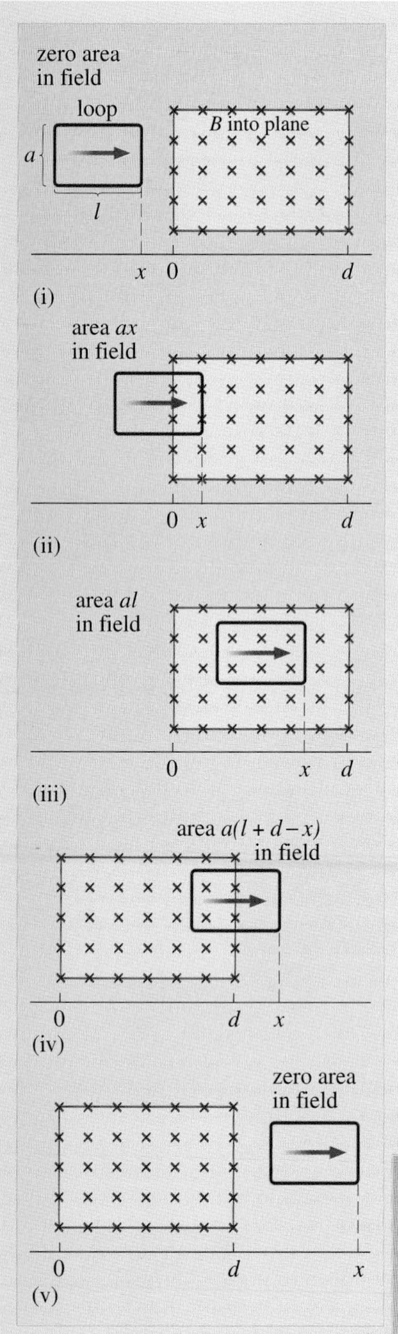

Figure 1.20 A rectangular loop moving to the right at speed v through a magnetic field of magnitude B pointing into the plane of the loop. The loop is shown in five regions: (i) before it enters the magnetic field; (ii) partly within the magnetic field (leading edge in); (iii) entirely within the magnetic field; (iv) again partly within the magnetic field (leading edge out); and (v) after it leaves the magnetic field.

In situation (ii), the loop is partially in the magnetic field region, and the magnetic flux through the loop is increasing. This is the change that is producing the induced EMF. To *oppose* the increase in the magnetic flux through the loop, the current must be in such a direction that it generates a magnetic field that is *opposite* in direction to that of the permanent magnet; Figure 1.20 (repeated here for convenience) shows the direction of the permanent magnetic field being into the plane of the paper. Recalling the **right-hand grip rule**, a magnetic field pointing up out of the plane of the diagram will be created by a current flowing in the *anticlockwise* sense. So this must be the direction of the current produced by the induced EMF.

At a later time the loop is in situation (iii), entirely within the magnetic field region. Now the magnetic flux does not change, and so there is no induced current.

When the loop starts emerging from the magnetic field region, in situation (iv), the magnetic flux through the loop is decreasing. This change will again induce an EMF that will produce a current. To *oppose* the decrease in the magnetic flux, the induced current must produce a magnetic field that is in the *same* direction as that of the permanent magnet. Recalling the right-hand grip rule again, a magnetic field pointing down into the plane of the diagram will be created by a current flowing in the *clockwise* sense.

After this, in situation (v), the induced current falls to zero and remains there.

Note that these answers could have been obtained by the Lorentz force law.

Example 1.4 Two circuits in close proximity

Now consider again the two loops of conducting wire close to each other, illustrated in Figure 1.14 (repeated opposite for convenience), but this time taking account of Lenz's law. Determine the direction of the induced current in the secondary circuit when the switch is closed and when it is opened.

Solution

From the way the battery is connected in Figure 1.14, the primary current flows in the clockwise sense. By the right-hand grip rule, the magnetic field it creates points downward inside the loop, as shown in Figure 1.28. This is true whether the primary current is increasing or decreasing.

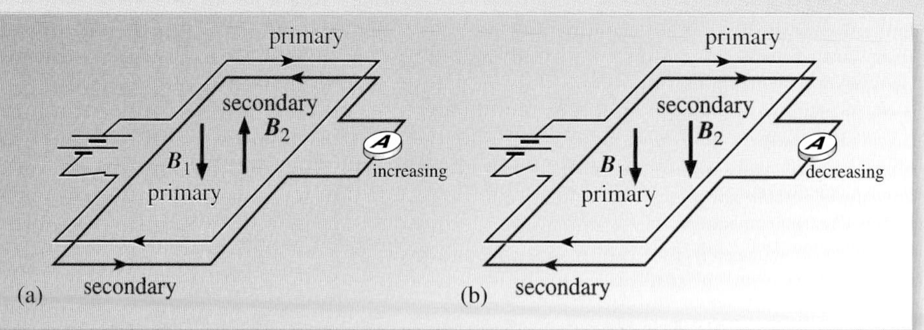

Figure 1.28 Two loops of conducting wire showing the direction of the magnetic field due to the current in the primary circuit. (a) When the primary current is *increasing*, the direction of the magnetic field due to the induced current in the secondary circuit is *upwards*. (b) When the primary current is *decreasing*, the direction of the magnetic field due to the induced current in the secondary circuit is *downwards*.

When the switch is first closed, the primary current builds up, and so the primary field is increasing. This creates an increasing downward magnetic flux through the secondary circuit that will induce an EMF around the secondary circuit and give rise to an induced current flowing in it. According to Lenz's law, the direction of this induced current must be such that it opposes the increasing downward magnetic flux that caused it. Consequently, the induced current in the secondary circuit must flow in the opposite sense to that in the primary circuit; that is, it must flow *anticlockwise*, as shown in Figure 1.28a.

When the switch is opened, the primary current falls to zero, so the primary field decreases. This decreases the downward magnetic flux through the secondary circuit. The secondary field must oppose this decrease, so it must reinforce the primary field. Hence the secondary current must flow in the *clockwise* sense when the switch is opened, as shown in Figure 1.28b.

It cannot be emphasized too strongly that the direction of the magnetic field due to the induced current is always determined by its *opposition to change*. This means that the secondary current is in the opposite sense to the primary current when the primary current is increasing, and in the same sense as the primary current when the primary current is decreasing.

Note that, unlike Example 1.3, here we cannot use the Lorentz force law and must use Lenz's law.

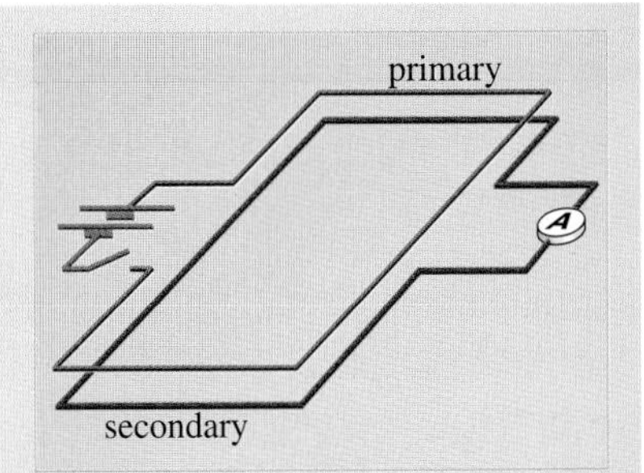

Figure 1.14 Two circuits in close proximity. The primary circuit has a battery and a switch; the secondary circuit has an ammeter to register any induced current. Changing the current in the primary circuit causes a current to flow in the secondary circuit.

Question 1.7 For the set-up described in Question 1.1 and Example 1.1, what would be the direction of current if the loop started outside the magnetic field region on the *right* and moved *leftwards* at a constant velocity?

Question 1.8 Figure 1.29 shows a bar magnet moving towards a coil along its axis. With the poles of the magnet as shown, determine whether the induced current in the circuit flows clockwise or anticlockwise. ■

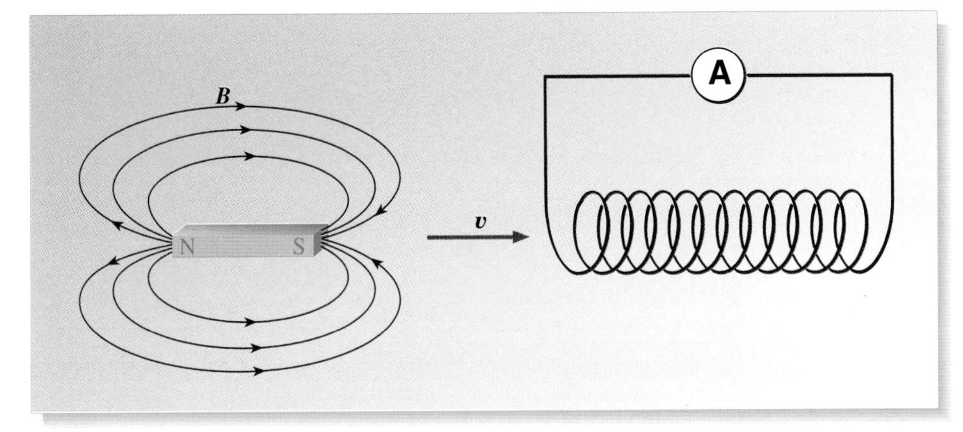

Figure 1.29 A magnet moving towards a coil along its axis.

Now that you have looked at two examples and had a chance to do some questions using Lenz's law for yourself, it is time to clear up some common misconceptions about it, and consider why it is physically reasonable. As emphasized above, the direction of the induced field is always determined by *opposition to change*. When dealing with primary and secondary circuits, it is very tempting to simply say that the induced current flows in the opposite direction to the current in the primary circuit, but you will have gathered from Example 1.4 that this is not always true: it depends whether the current in the primary is increasing or decreasing.

The fact that the induced current always opposes change is a direct consequence of the conservation of energy. If this were not the case, then it would be possible to generate runaway motions. Consider the following example. You know that moving a permanent (bar) magnet near a solenoid causes an induced current to flow in the solenoid; and a current generates a magnetic field in the space around it. Depending on the direction of the induced current, the magnetic field it creates will either enhance or diminish the magnetic field in the region between the magnet and the solenoid. Suppose it were possible for enhancement to take place. This would increase the change in magnetic flux through the solenoid, causing an increase in the induced current, which would enhance the change in magnetic flux still further, causing an increase in the current, and so on. You can see that this is absurd. If it happened, an astronaut shaking a small bar magnet on the far side of the Moon could then cause currents to build up without limit in all solenoids on the Earth. So the current induced in the solenoid must flow in the direction that causes its magnetic field to *oppose* that of the magnet: shaking the magnet creates an induced current in the solenoid, which in turn creates a magnetic field, but this magnetic field opposes that of the magnet and so the magnet gets more difficult to shake.

2.7 Eddy currents

In the various examples considered so far, the emphasis has been on the currents induced in conducting wires. But this is a special case, since Faraday's law says that an induced EMF will be caused in any region in which there is a changing magnetic flux, not only those containing wires.

Consider a piece of 'soft' iron (a ferromagnetic material) to be used as a solenoid core. Soft iron has a crystal structure that enhances the magnitude of the magnetic field threading it, and can also be used to concentrate a field. Now if the iron core is subject to a changing magnetic field, or equivalently, moves through a field, currents will be set up in the iron, since there are conduction electrons available to carry the current. These induced currents are known as **eddy currents** (or *Foucault currents*). Now, iron is not a perfect conductor, so these currents convert electrical energy to thermal energy. In addition, the magnetic fields created by these currents will distort the magnetic field of the solenoid. So eddy currents can be a considerable nuisance, and it may be necessary to reduce them as much as possible.

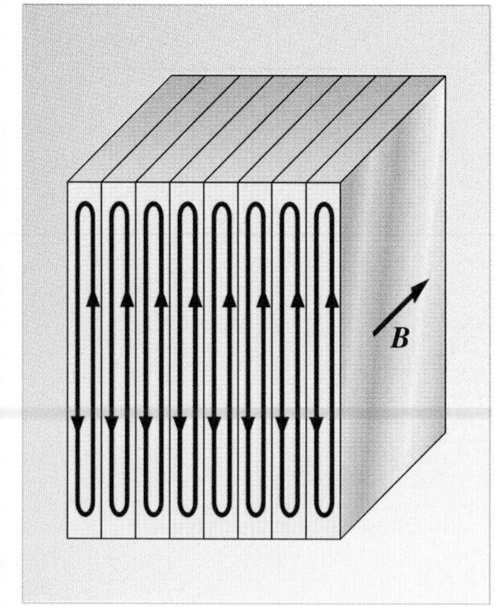

Figure 1.30 Laminated iron block.

The most common way of reducing eddy currents is to use a laminated iron block, as shown in Figure 1.30, rather than a solid block. The block is sliced into thin sheets as shown and the sheets are electrically insulated from each other by a thin coat of varnish or shellac, or even by an oxide coating on the surfaces.

The orientation of the lamination relative to the magnetic field is crucial to the method used to reduce eddy currents. Note that in Figure 1.30 the block is positioned so that the magnetic field is in the plane of each thin sheet. The induced EMF is perpendicular to the magnetic field causing it, so the induced current must flow just in the short thickness of the plate, as indicated in the figure. Such currents will have considerably less effect (both in terms of energy loss and field distortion) than if they circulated in the plane of the sheet.

Eddy currents are important in heavy electrical engineering but they are also significant in the many household devices that require considerably lower voltages than those supplied by the mains. The necessary reduction in voltage is achieved by a transformer, a device whose operation you will study in the next section, but which is especially subject to eddy currents. You may have noticed that the little transformers contained in the power supplies of small electronic devices (such as the one shown in Figure 1.31) get warm even when the device they are attached to is switched off: this is costing you money due to eddy current losses.

Figure 1.31 The transformer in a power supply, connected to the mains, becomes warm even when the device it is attached to is switched off. This is due to eddy currents and costs money in wasted electricity.

It is even more irritating to know that it is theoretically possible to reduce the eddy current losses rather more than is common in household transformers. It turns out that the energy loss in these transformers is proportional (among other things) to the square of the thickness of the laminations, so you can see that thin sheets are at a premium. Unfortunately, cutting and handling such sheets is expensive, and for this reason thicker laminations are generally used.

However, there is one situation where eddy currents are an essential feature. This is in the *induction furnace*, first discussed in the introduction to this chapter as a way of producing sufficiently high temperatures to melt metals. The operation of the furnace was illustrated in Figure 1.3. In this case (and many other cases) the varying magnetic field is provided by means of an *alternating* current. Alternating currents are discussed in more detail in the next section, but for now all that you need to know is that such a current varies with time, creating a varying magnetic field and hence eddy currents in the metal in the central crucible. The resulting heating melts the metal. In terms of the law of conservation of energy, the source of the current supplies energy, and the coil passes it on to the metal. The more energy that is supplied per unit time, the faster the metal will melt. Moreover, there is a minimum temperature below which the metal will not melt, since melting means a break-up of the crystal structure of the metal, and that requires sufficient energy. The energy generated in the coil therefore has to be high. Unfortunately, this has an unwanted side-effect. The eddy currents in the metal in the crucible may be wanted, but those in the core of the solenoid are not. Even with a carefully laminated core, the steady input of energy will heat up the wires to a dangerous level. To reduce the temperature, real furnaces cool the coil in a water jacket, as shown in Figure 1.3b.

2.8 Self-inductance

This section will consider solenoids extensively, so you might like to start by revising what you learnt about the magnetic field in a solenoid in SFP Chapter 4.

When a current in a coil changes, it creates a changing magnetic field throughout the space in and around the coil. This changing magnetic field will in turn give rise to a changing magnetic flux through the coil. By Faraday's law, this changing magnetic flux will induce an EMF in any region in which the changing magnetic flux is non-zero. This will certainly include the original coil itself, so there will be an induced EMF in the coil. By Lenz's law, the direction of this induced EMF will be such that it opposes the EMF due to the original varying current that caused it. Because a self-induced EMF is involved, this process is called **self-induction**.

Evidently the original (time-varying) current must be affected by this self-induced current. But exactly how, and what is the resultant current in the circuit?

Recall from Equation 4.6 in *Static fields and potentials* that the magnetic field strength within a very long cylindrical solenoid, carrying a steady current i, is given by

$$B = \frac{\mu_0 N i}{l},$$

where μ_0 is the permeability of free space, l the length of the solenoid and N is the number of turns. Now, if the current is not steady but instead is a function of time, this expression should be written

$$B(t) = \frac{\mu_0 N i(t)}{l}. \tag{1.4}$$

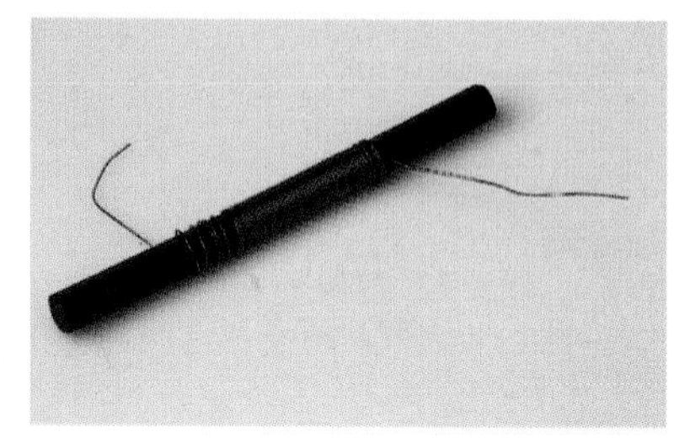

Figure 1.32 A solenoid wound on a soft iron core.

The notation provides a reminder that the magnetic field is a function of time if the current is a function of time. The permeability is that of free space if the core of the solenoid is empty. However, a solenoid typically has a soft iron core (Figure 1.32), as noted previously, so μ_0 must be replaced with the more general μ, the permeability of the core:

$$B(t) = \frac{\mu N i(t)}{l}. \tag{1.5}$$

From this formula, you can see that the use of an iron core increases the magnetic field by a factor μ/μ_0, the ratio of the magnetic permeabilities of iron to free space, and this can be of the order of a thousand.

Now, to find the magnetic flux inside the solenoid, consider initially a one-turn coil. The magnetic flux in the core region would be $\phi(t) = AB(t)$, where A is the cross-sectional area of the core. The total magnetic flux through the core region of a solenoid with N turns is given by

$$\Phi(t) = NAB(t) \tag{1.6}$$

where the capital Greek letter phi, Φ, is now used for total magnetic flux. Substituting for B from Equation 1.5 gives

$$\Phi(t) = NA \times \frac{\mu N i(t)}{l} = \frac{\mu A N^2}{l} i(t) \tag{1.7}$$

(Note that, strictly, all of these equations are only true inside the solenoid and well away from its ends. In real cases, numerical corrections must be applied to take account of the geometric construction.)

Now, if the current in the solenoid is made to vary in time, Faraday's law says that there is an induced EMF in the coil, of magnitude

$$\left| V_{\text{ind}}(t) \right| = \left| \frac{\mathrm{d}\Phi(t)}{\mathrm{d}t} \right| = \frac{\mu A N^2}{l} \left| \frac{\mathrm{d}i(t)}{\mathrm{d}t} \right| \tag{1.8}$$

and Lenz's law says that this opposes the change in the current.

Equation 1.8 can be written more simply as

$$|V_{\text{ind}}(t)| = L \times \left|\frac{\mathrm{d}i(t)}{\mathrm{d}t}\right| \tag{1.9}$$

where the constant

$$L = \frac{\mu A N^2}{l} \tag{1.10}$$

is known as the **coefficient of self-inductance** of the solenoid, and represents all the details of the construction, including the number of turns. It is the constant of proportionality between the magnitude of the self-induced EMF in a coil and the magnitude of the rate of change of current in the coil.

The specific expression for the coefficient of self-inductance in Equation 1.10 is only true for the special case of an infinitely long cylindrical solenoid. What about less regularly shaped coils? It turns out that *any* coil, whatever its shape, whatever core it might have, responds to a changing current through it by a formula similar to Equation 1.9, but the formula for the coefficient of self-inductance, L, will not be the same as in Equation 1.10.

Solenoids are sometimes known as *inductors*, particularly when complex electrical circuits are under consideration, as in radios or televisions. As a shorthand, the terms **self-inductance**, **inductance**, or *inductance of the coil*, are often used for L. The SI unit of inductance is the **henry**, after the American physicist, Joseph Henry who in 1832 observed the effects of self-induction in electrical circuits. One henry is defined to be the inductance of a solenoid in which an EMF of one volt is induced by a current that changes at the rate of one ampere per second. In symbols:

$$1\,\text{henry} = 1\,\text{H} = \frac{1\,\text{V}}{1\,\text{A}\,\text{s}^{-1}} = 1\,\text{V}\,\text{A}^{-1}\,\text{s}.$$

Question 1.9 The solenoid in a small portable radio receiver consists of a soft iron core with 200 turns of wire wrapped around it. The length of the solenoid is 50 mm and it has a cross-sectional area of 40 mm^2. Assuming that the permeability of iron is $1.0 \times 10^{-3}\,\text{T}\,\text{m}\,\text{A}^{-1}$ (where $1\,\text{T} = 1\,\text{N}\,\text{s}\,\text{m}^{-1}\,\text{C}^{-1}$), calculate the inductance of the coil. ■

As Question 1.9 demonstrated, inductances of the order of a few millihenries are typical for small solenoids in household devices.

The current passing through a solenoid results in a magnetic field. A magnetic field stores energy, in the sense that it takes energy to create it, and energy is given back when the magnetic field dies away. It turns out that the magnetic energy stored by a solenoid of inductance L when a current i flows through it is given by

$$\text{magnetic energy} = \tfrac{1}{2}Li^2. \tag{1.11}$$

This energy is drawn from the source (the battery) in order to overcome the opposition to the build up of the current.

- What is the expression for the electrical energy stored by a capacitor? (*Static fields and potentials*, Chapter 2)
- ❍ The electrical energy stored by a capacitor is $\frac{1}{2}CV^2$. There is clearly a similarity between these two expressions for energy. ■

2.9 Transient currents

The next thing to consider is the effect of self-inductance on the current in a solenoid. In problems of this sort it is conventional to idealize everything and assume the solenoid does nothing but supply the induced EMF; that is, it is a pure inductor, and has no resistance. It will also be assumed that the battery is 'ideal' in that it has no internal resistance. Electrical engineers have devised a scheme for representing circuits of ideal elements, and Figure 1.33 shows a circuit consisting of four elements: an inductor, a resistor, a battery and a switch.

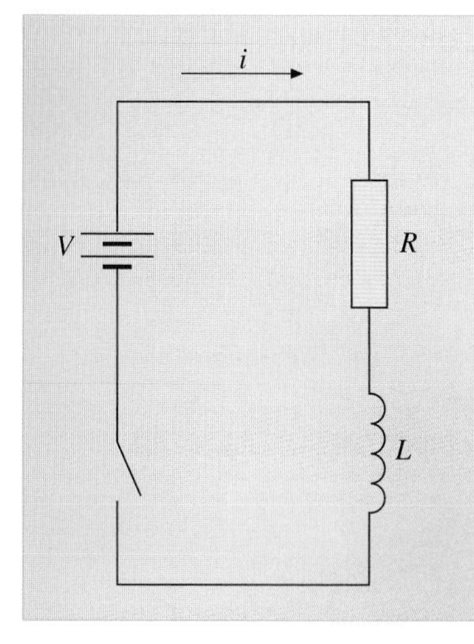

Figure 1.33 Inductive–resistive circuit driven by a battery.

The growth of current in a solenoid

When the switch is closed in the circuit shown in Figure 1.33, current begins to build up in the circuit, but the build up is opposed by the inductance (since the induced current will oppose the change that is causing it). The total EMF in the circuit at any time will be the difference between the battery contribution V_{bat}, which does not vary with time, and the induced EMF, which does. As long as $i(t)$ is increasing, so that $\mathrm{d}i(t)/\mathrm{d}t$ is positive, the induced EMF will have a magnitude $L\,\mathrm{d}i(t)/\mathrm{d}t$, so the total EMF will be

$$\text{EMF} = V_{\text{bat}} - L\frac{\mathrm{d}i(t)}{\mathrm{d}t}. \tag{1.12}$$

The total EMF in the circuit at any instant, according to Ohm's law, is equal to the current multiplied by the resistance. So, while the current is increasing

$$V_{\text{bat}} - L\frac{\mathrm{d}i(t)}{\mathrm{d}t} = i(t)R.$$

Rearranging, and dividing through by R gives

$$\frac{L}{R}\frac{\mathrm{d}i(t)}{\mathrm{d}t} + i(t) = \frac{V_{\text{bat}}}{R}. \tag{1.13}$$

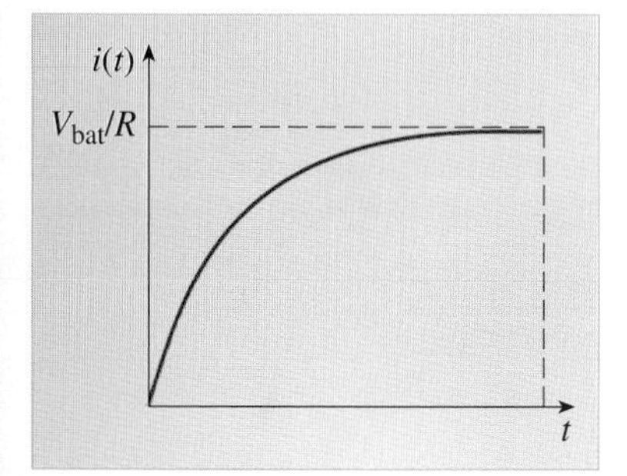

Figure 1.34 The current in the circuit shown in Figure 1.33, as a function of time.

Equation 1.13 states that the sum of the current $i(t)$ and the rate of change of the current $\mathrm{d}i(t)/\mathrm{d}t$ multiplied by a constant L/R is equal to another constant V_{bat}/R. This implies that, when the current is relatively small, its rate of change with time must be relatively large, and vice versa. Equation 1.13 is a *differential equation* that describes the behaviour of $i(t)$ throughout the period when the current is increasing. Its solution introduces an arbitrary constant that can be evaluated using the fact that $i(t) = 0$ when $t = 0$. A graph depicting the solution of Equation 1.13 is shown in Figure 1.34. It shows that when t is small, the current is small, but rising rapidly. At later times the current is larger, but it 'levels off' towards a constant value of V_{bat}/R.

- Referring to Figure 1.34, when is $\mathrm{d}i(t)/\mathrm{d}t$ greatest?
- ❍ The rate of change of the current is greatest initially, since this is where the slope of the graph is greatest.
- Referring to Figure 1.34 again, when is $\mathrm{d}i(t)/\mathrm{d}t$ smallest?
- ❍ The rate of change of the current is smallest at later times, since this is where the graph has the shallowest slope. ■

As you might have realized, the solution to Equation 1.13, displayed in Figure 1.34, is an *exponential function* and mathematically it can be written as

$$i(t) = \left(\frac{V_{bat}}{R}\right)\left(1 - e^{-Rt/L}\right). \qquad (1.14)$$

Figure 1.34 shows that the current builds up towards a value of $i_{max} = V_{bat}/R$. However, in principle, it never quite reaches this maximum value, although in practice it gets immeasurably close to it. Electrical engineers say that i_{max} is the *steady-state current* in this circuit, which is an easily understood terminology. The fact that the current never quite reaches the maximum value means that it is always changing with time (however slowly), so sustaining the induced EMF.

The decay of current in a solenoid

Now consider what happens when the switch in Figure 1.34 is opened. The induced EMF is now the *only* source of EMF in the circuit, so

$$-L\frac{di(t)}{dt} = i(t)R$$

or

$$\frac{L}{R}\frac{di(t)}{dt} + i(t) = 0. \qquad (1.15)$$

Equation 1.15 states that the current $i(t)$ is always equal to the rate of change of the current $di(t)/dt$ multiplied by a constant $-L/R$. However, note that in this situation $di(t)/dt$ is negative. In other words, when the current is relatively small, its rate of change with time must also be small, and when the current is relatively large, its rate of change with time must also be large. As before, the current must satisfy this differential equation at every instant of time, now starting with the known value that $i(t) = V_{bat}/R$ at $t = 0$. A graph depicting the solution of Equation 1.15 is shown in Figure 1.35. This time the solution is an *exponential decay*. It indicates that at early times, the current is large and falling rapidly. At later times the current is smaller and 'levels off' towards a constant value of zero.

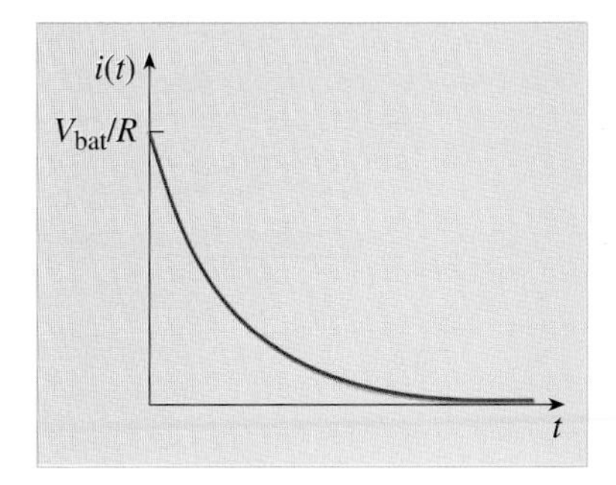

Figure 1.35 The current in the circuit shown in Figure 1.33, as a function of time, after the battery has been disconnected.

- Referring to Figure 1.35, when is $di(t)/dt$ greatest?

❍ The rate of change of the current is greatest initially, since this is where the slope of the graph is steepest.

- Referring to Figure 1.35 again, when is $di(t)/dt$ smallest?

❍ The rate of change of the current is smallest at later times, since this is where the graph has the shallowest slope. ■

Mathematically, the solution to Equation 1.15, displayed in Figure 1.35, can be written as the exponential function

$$i(t) = \left(\frac{V_{bat}}{R}\right)e^{-Rt/L}. \qquad (1.16)$$

- What is the time constant for the exponential decay described by Equation 1.16?

❍ Recalling the discussion of time constants from *Predicting motion* Chapter 2, the time constant for this exponential decay is L/R. The *smaller* this value, the *faster* the current will decay. ■

Notice in Figure 1.35 that, in principle, the decaying current never actually reaches zero, although in practice it gets immeasurably close to it. This again means that the current is *always* varying in time, sustaining the self-inductance.

Question 1.10 Suppose that in the circuit shown in Figure 1.33, the battery has an EMF of 5.5 V, the resistor has a resistance of 22 Ω, and the solenoid (inductor) has an inductance of 32 mH. (a) After the battery is connected, how long will it take for the current in the circuit to reach 50% of the steady-state value? (b) After a few seconds the current is very close to the steady-state value. How much energy is stored in the solenoid? (c) The battery is now disconnected. How long will it take for the current to fall to 50% of the steady-state value? (d) How long will it take to fall to 25% of the steady-state value? ■

The inductance of the solenoid in Question 1.10 is typical of those in a household device. So clearly the time for the current to build up, or decay, in such a circuit is relatively short — of the order of milliseconds. For this reason, such currents are often referred to as **transient currents**. The growing and decaying currents in solenoids are analogous to the transient currents that occur in capacitors which you read about in *Static fields and potentials* Chapter 3. Both processes can be described by similar exponential functions.

3 Applications of electromagnetic induction

This section will now apply what you have learnt about induction to a number of devices, most of which should be familiar to you. You will see that, far from being an abstract physical phenomenon, electromagnetic induction is central to many of the devices that most of us use every day.

3.1 Telephones, microphones and hearing aids

Can you imagine being without a telephone? Life would certainly be rather different! The telephone was invented in 1873 by Alexander Graham Bell (Figure 1.36).

Figure 1.36 Alexander Graham Bell.

Alexander Graham Bell (1847–1922)

To most people, Bell is remembered as the inventor of the telephone and of numerous other devices, including a method enabling a telegraph line to carry multiple messages simultaneously. But he was first and foremost an educator of deaf children, as his father before him had been; his mother had lost much of her hearing in childhood. He met his wife in Boston when she was 15 and he was engaged to tutor her. At that time he had moved from Scotland, where he was born, to a chair in vocal physiology and elocution at Boston University.

Bell's assistant at the time of the invention of the telephone was Thomas A. Watson. The story of the invention of the telephone is that during an experiment on a possible talking machine, Bell spoke to Watson: '*Come here, Mr Watson, I need you*'. Unknown to Bell, Thomas Watson was in the next room, and yet heard the request and realized that it came from the machine they were investigating.

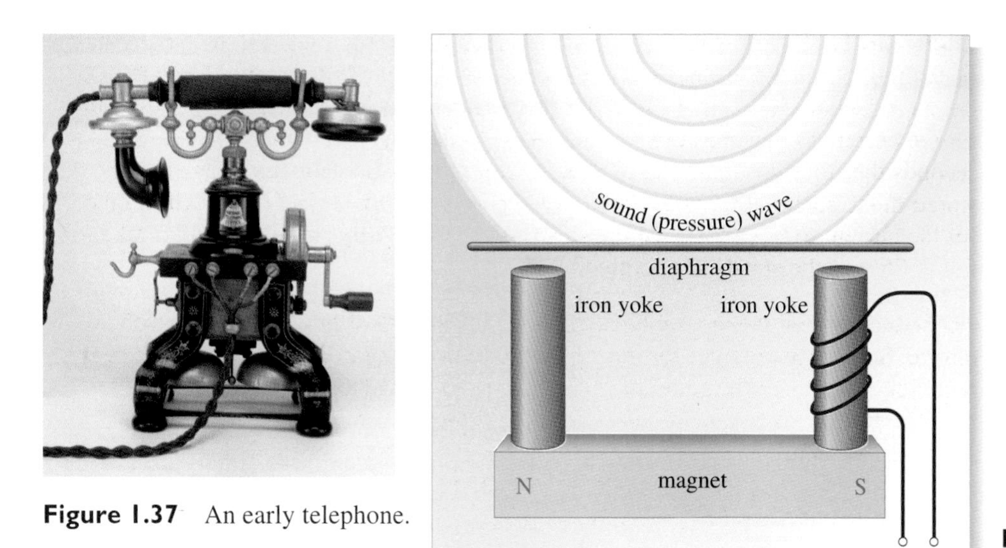

Figure 1.37 An early telephone.

Figure 1.38 The operation of a simple telephone.

Figure 1.37 shows an early telephone and Figure 1.38 gives a simplified representation of the working parts. The device consists of a permanent bar magnet to which are attached two iron cylinders, one of which has a coil wrapped around it, so it acts as a solenoid. The cores are known as *yokes* in telephony, and are magnetized because of their attachment to the bar magnet. A small distance away is a thin iron circular steel plate, known as a diaphragm. The diaphragm is clamped in place at its rim, but is thin enough so that it can bend towards or away from the pole pieces if needs be. The diaphragm is in the magnetic field of the yokes, and so is magnetized too.

Sound travelling through air causes the *pressure* of the air to vary (sound is discussed in considerably more detail in Chapter 2). In the telephone illustrated in Figure 1.38, the resulting pressure changes cause the diaphragm to vibrate. When the diaphragm moves towards the yokes, its field increases the magnetic field enclosing the yokes, and hence the magnetic field threading the coil; but when the diaphragm moves away from the yokes, the magnetic field threading the coils is diminished. Thus, the sound impinging on the diaphragm creates a varying magnetic flux threading the solenoid. This creates a varying induced current in the solenoid, which exactly mirrors the pressure variations of the original sound.

If a second, identical, device is electrically connected to the first, the output current of the first device may be used as the input current of the second. This varying current now causes the second diaphragm to move to and fro exactly as the original diaphragm did, creating new pressure variations in the air — that is, sound. The listener using the second device hears the same sounds as the speaker uttered into the first device.

The currents involved in early telephones were quite small and the diaphragm motions were also small. Before amplifiers were available, this was something of a problem although, with care, small currents can be carried long distances in wires without appreciable distortion. More important is the fact that as a discriminating pick-up device, the ear is marvellously sensitive. Experiments show that the ear is perfectly capable of understanding sound variations of the small magnitudes involved.

Modern telephones and mobile phones are rather more complex machines than the simple device just described. Also note that this consideration of the simple telephone does not touch on the details of a complex *telephone system*, which involves dealing with many different current outputs simultaneously. Since it is impossible to have individual wires from each telephone to every other telephone, multiple signals must coexist on a single wire. The signals must not become distorted due to electromagnetic induction from adjacent signals, and so on. The theory of telephone systems involves both the engineering of the switching and transmission systems, and the mathematics of information theory and coding, which we shall not be considering in this book.

Other devices can be built along the same principles as the simple telephone. A microphone, for example, could operate in the same way. You speak into the diaphragm and this results in an output induced current, whose variations contain the same information as was in the sound that caused the diaphragm to vibrate.

Once sound has been transformed into a current, it is possible to amplify this current. This is how a basic hearing aid works. The device behind the ear contains a tiny diaphragm which picks up sound. This is turned into current, which is amplified, and subsequently turned back into sound by a second diaphragm. The amplified sound is channelled into a tiny tube which has been placed into the ear, resulting in the sending of an amplified and focused sound into the ear canal. Unfortunately, such devices are of limited help to people who are hard of hearing or are suffering from tinnitus. Unless a very expensive device is used, there is no selectivity: all the sound is amplified, including the unwanted background noise. Even in advanced devices, which amplify selectively by frequency, any unwanted noise at the selected frequencies will be amplified. In addition, some types of hearing impairment cannot be improved by simple amplification of the sound: the difficulty lies in getting the useful information to the brain.

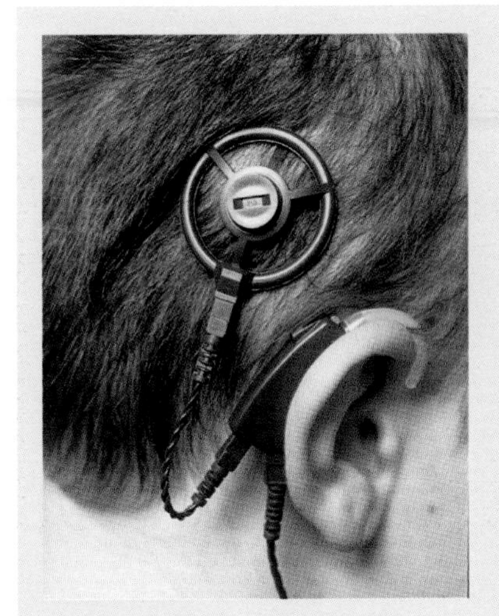

Figure 1.39 A cochlear implant.

Box 1.1 Cochlear implants

Part of the problem with a standard hearing aid is overcome by the cochlear implant (Figure 1.39). In this device an external system transforms the sound input into a varying current, as in a simple hearing aid. The current then travels down a wire inside the ear to an implanted processor, which imparts this signal to the auditory nerve. Unfortunately, no small artificial device is able to function nearly as well as the cochlea (the spiral part of the inner ear, which translates mechanical vibrations into nerve impulses) with its thousands of tiny hair cells. In addition, a seriously hearing-impaired person will likely have only about 30 000 out of an unimpaired 50 000 nerve fibres intact. The result is that cochlear implants deliver a rather fuzzy sound, but with practice the brain learns to discriminate fairly well. For those with sufficient nerve function remaining, and whose brain is active enough to develop further discrimination, this is a minor miracle, which is sure to be improved. Children receive particular benefit from this procedure.

3.2 Generators and motors

It is obvious that the telephone plays an important role in many people's lives, but even more important are the devices used to generate electricity. These too depend on the principles of electromagnetic induction. This section tells the story of how electricity is generated. It is only concerned with principles, and does not consider any of the engineering details of a power station, or linking stations to grids or linking grids to your home. All that is needed is Faraday's law, Lenz's law, and considerable ingenuity! Fortunately, the ingenuity was supplied by the pioneers of the electrical industry.

The story starts with the uniformly rotating rectangular loop considered in Example 1.2, which forms the basis of a simple electrical generator. In that example you saw that the magnetic flux is given by

$$\phi(t) = AB\sin \omega t \tag{1.17}$$

and the rate of change with time is

$$\frac{d\phi(t)}{dt} = \omega AB\cos \omega t. \tag{1.18}$$

Thus, from Faraday's law there is an induced EMF of magnitude

$$|V_{ind}(t)| = \left|\frac{d\phi(t)}{dt}\right| = |\omega AB\cos \omega t| \tag{1.19}$$

and an induced current of magnitude $|i_{ind}(t)|$ (= $|V_{ind}(t)|/R$), is given by

$$|i_{ind}(t)| = \left|\frac{\omega AB}{R}\cos \omega t\right|. \tag{1.20}$$

The induced current is, of course, nothing more than a flow of charged particles (electrons) in the wire loop. So the charges flowing in the arms of the loop will each feel a magnetic Lorentz force ($\boldsymbol{F}_{mag} = q\boldsymbol{v} \times \boldsymbol{B}$). Lenz's law implies that the torque on the loop due to these Lorentz forces must *oppose* the rotation of the loop. Consequently, the direction of the current in the loop at any instant is such that it opposes the change that caused the induced EMF.

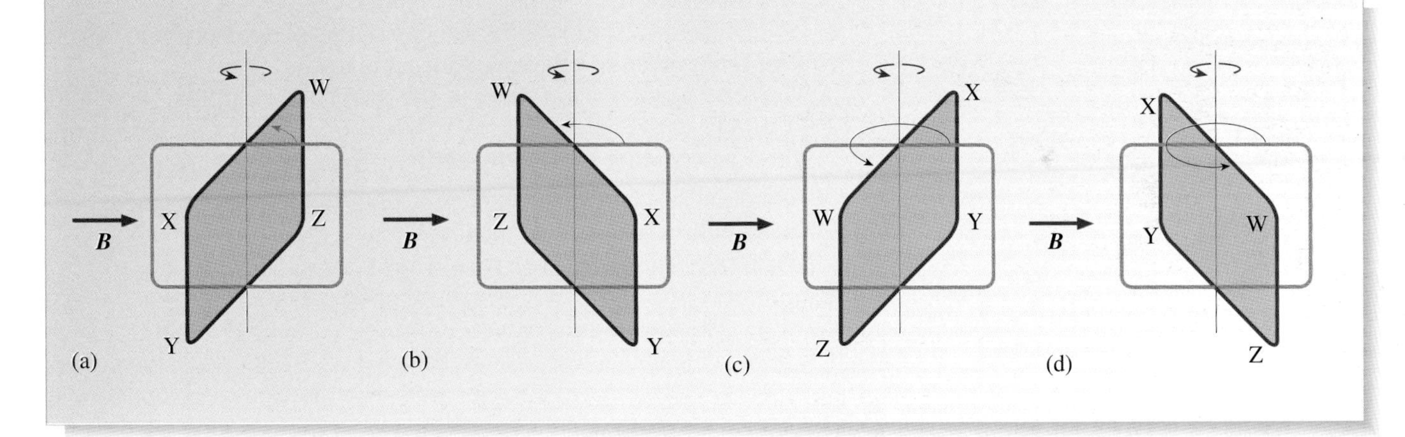

Figure 1.40 A simple generator with the coil in four positions (a–d), one-quarter of a cycle apart. By considering the Lorentz force on positive charges in the arms ZW and XY, the direction of current flow in the loop at each instant can be determined.

Question 1.11 Figure 1.40 shows the rotating rectangular loop of wire in a simple generator at four positions (a–d), one-quarter of a rotation cycle apart. The loop is rotating *anticlockwise* when viewed along the rotation axis from above (where WX is the top edge of the loop). Consider the Lorentz forces on positive charges in the arms ZW and XY of the loop, and so work out the direction of current flow in the loop in each case. ■

For half the cycle, the current flows one way round the loop, and for half the cycle the current flows the other way. It is convenient to introduce the constant $i_{max} = \omega AB/R$, and then write the current as

$$i_{ind}(t) = i_{max}\cos \omega t \tag{1.21}$$

The graph of this function is shown in Figure 1.41, where you can see that i_{max} is the maximum current in the loop. Note also that if the loop had started in a different

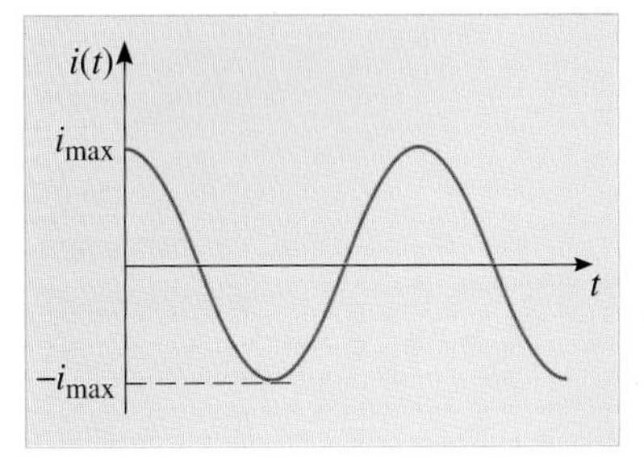

Figure 1.41 A sinusoidally varying alternating current induced by a uniformly rotating rectangular loop in a magnetic field. Note that the induced current is a maximum when the loop is parallel to the magnetic field.

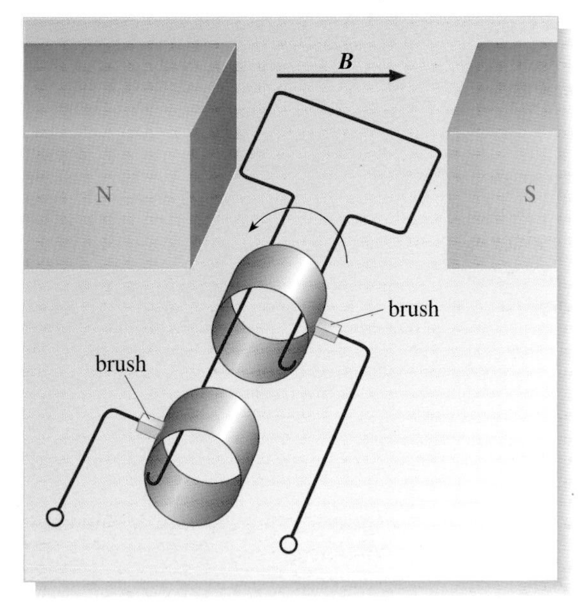

Figure 1.42 The arrangement for drawing a current from a simple AC generator.

position, or if the angles were defined differently, then the final graph might have been shifted horizontally along the time axis, and Equation 1.21 might have been written as a sine function rather than a cosine function. Hopefully you can see that this is rather arbitrary; the important thing is that the variation in current is *sinusoidal*. The current is alternately positive and negative, rising and falling sinusoidally, alternating in sign every half revolution. This current pattern is known as an 'alternating current' (AC).

Provided the external agency supplying the rotation continues at constant angular velocity, the loop rotates periodically, with period $T = 2\pi/\omega$, and so the current continues its periodic pattern. If there were some way to 'tap' the current, the loop could be used as a method of converting mechanical energy into electrical energy. There is such a method, and it is simple and ingenious. The loop is opened and wires led from it to the tapping arrangement. The ends of these wires are fixed to two metal rings, as shown in Figure 1.42. These rings are electrically isolated from each other, and rotate with the loop. A conducting stationary 'brush' rests against each ring, and carries the current away. The construction depicted in Figure 1.42 is therefore a simple **alternating current generator**.

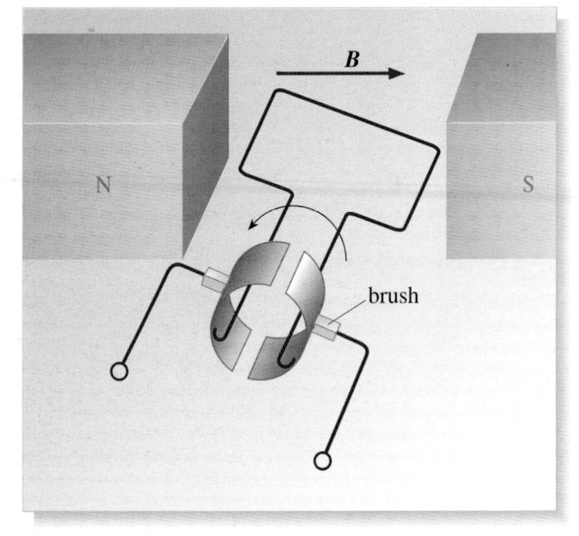

Figure 1.43 The arrangement for drawing a current from a simple DC generator.

Suppose that, instead of an outside agency supplying the motive power, an alternating current were fed in. This would cause the charges to move in the magnetic field, and so experience a magnetic Lorentz force. The same considerations as before imply that the current is creating a torque on the loop, causing it to rotate at a constant angular velocity. Thus, the shaft would turn and the rotation could be used to do useful work. When used in this way, the device is an **alternating current motor**.

Returning to the generator, there might be a situation in which a current is required that does not alternate, but is always in the same direction. This same loop arrangement can be used to obtain direct current (DC) by varying the tap-off arrangements as shown in Figure 1.43. The two rings have been replaced by a single ring, which has been cut in half and separated (the gaps are small). Such an arrangement is known as a *split-ring commutator*. Each wire is connected to one-half of the ring, and brushes rest against the arrangement as before. The current changes sign every half revolution, and the gaps in the ring are half a revolution apart too.

The orientation is such that at the point at which the current in the loop changes sign, each half-ring moves from one brush to the other, thus effectively reversing the connection. Since the output current never changes sign, as shown in Figure 1.44, this arrangement is called a **direct current generator**. The current is said to be *rectified*.

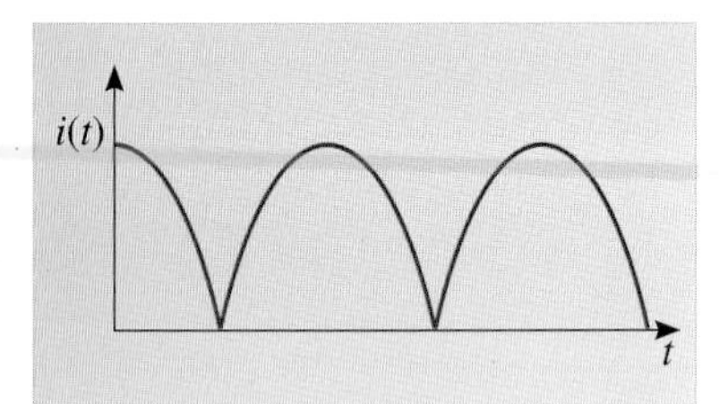

Figure 1.44 The rectified (DC) current output from the split-ring arrangement shown in Figure 1.43.

Note that although the current is rectified, it still varies in magnitude. Direct current is usually wanted at a steady value. The device shown in Figure 1.43 is good for understanding the principles involved, but is clearly not a very practical source of steady direct current. Real AC and DC generators and motors do not look anywhere as simple as the devices just considered. To describe real devices would take us into the realms of the mechanical and electrical engineer's arts, but however complex the arrangements, if you dig deeply enough, they all hinge on Faraday's and Lenz's laws, and all motors on the Lorentz force law.

Question 1.12 (a) How could you turn the DC generator in Figure 1.43 into a **direct current motor**? (b) At two points in its cycle, the torque on the single coil in this motor is zero. Explain where these points are, and suggest how a real motor is able to keep turning without stalling. ■

Another complication is that not only is an electric motor constructed like a generator, but it also *is* a generator. After all, it consists of a coil of wire rotating in a magnetic field. So there will be a changing magnetic flux through the coil, which induces an EMF in the coil to oppose the change. This is known as the *back EMF*, and acts in the opposite direction to the applied EMF from the battery or power supply.

Real power stations (such as those shown in Figure 1.45) generate electricity by using a primary source such as coal, oil, nuclear power or wind to provide the turning torque. The financial, environmental and political arguments regarding which primary sources should and should not be chosen may well override the scientific and technological considerations in any particular case.

Figure 1.45 (a) Generators at a power station; (b) a general view of a power station.

3.3 AC circuits, resonance and radios

Now is the time to return to the study of inductance, which was begun in Section 2. Previously you were asked to consider what would happen if a solenoid was attached to a battery — that is, a source of direct current. You saw that the current approached its final value exponentially, but never quite attained a steady-state value. A similar study can be made with the battery replaced by a sinusoidally alternating source of EMF, that is to say an AC generator. Aside from a resistor and inductor in the circuit, the effect of a capacitor will have to be considered too.

The reason for studying such circuits is because they are the basis of the way that radio, television, radar and other electromagnetic signals are generated and received. How does this come about? As you will see, such circuits cause conducting electrons to *oscillate* back and forth periodically. If part of the circuit is in the shape of an antenna, the oscillating electrons cause energy to radiate outward in the form of electromagnetic radiation. This is what the transmission of a signal is. The frequency of the signal is the same as the frequency of the EMF source, and the strength of the signal is determined by the energy put into the circuit. This signal can be detected elsewhere by a similar circuit, including an antenna. The electrons in the antenna will move sinusoidally in response to the incoming radiation, and the rest of the electrical circuit can be used to decode the input signal. In a sense, the whole process is reminiscent of the telephone, but here electromagnetic signals are used in place of sound.

In order to understand how these circuits operate, it is instructive to first look at circuits containing, respectively, *only* an inductor and *only* a capacitor.

Inductive circuits

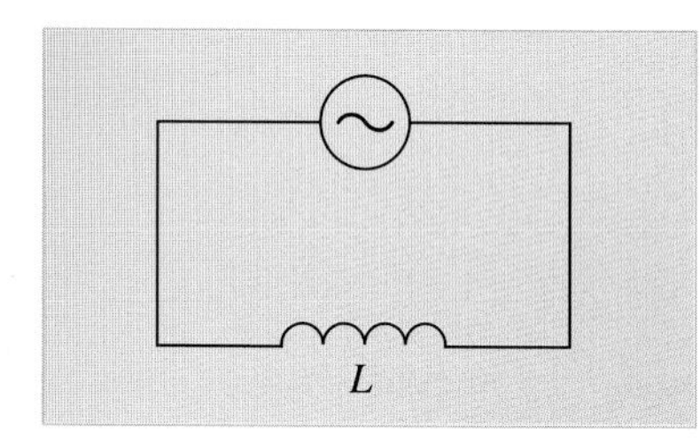

Figure 1.46 An inductive circuit containing a sinusoidal source of EMF (represented by a circle with a sine wave in it) and an inductor.

The first circuit to analyse is shown in Figure 1.46. The AC generator shown here supplies an alternating current of the form

$$i(t) = i_{\max} \cos \omega t \cdot \qquad \text{(Eqn 1.21)}$$

As before, the changing magnetic flux in the inductor generates an induced EMF of magnitude

$$|V_{\text{ind}}(t)| = L\left|\frac{\text{d}i(t)}{\text{d}t}\right|. \qquad \text{(Eqn 1.9)}$$

Remember the rule that

$$\frac{\text{d}}{\text{d}t}\cos ax = -a \sin ax,$$

where a is a constant. Differentiating Equation 1.21 gives $\text{d}i(t)/\text{d}t = -i_{\max}\omega \sin \omega t$, and so in this case Equation 1.9 becomes $V_{\text{ind}}(t) = -i_{\max}\omega L \sin \omega t$, or

$$V_{\text{ind}}(t) = -V_{\max} \sin \omega t \qquad (1.22)$$

where $$V_{\max} = i_{\max}\omega L. \qquad (1.23)$$

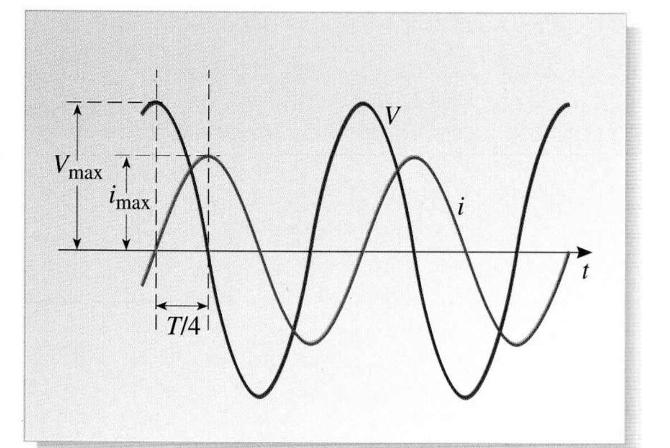

Figure 1.47 Alternating (sinusoidal) voltage and current in an inductor, both with period *T*.

Figure 1.47 shows graphs of $V_{\text{ind}}(t)$ and $i(t)$ against time as described by Equations 1.22 and 1.21, respectively. As you can see, both are sinusoidal with the same period $T = 2\pi/\omega$, but one curve is shifted along the time axis with respect to the other. In particular, for this inductive circuit, the voltage *leads* the current by one-quarter of a cycle. In other words, if the voltage passes through a maximum and then starts to decrease at some particular time, then the current will show qualitatively similar behaviour exactly one-quarter of a period *later*.

Capacitive circuits

The next circuit to analyse is shown in Figure 1.48. The AC generator shown here once again supplies an alternating current of the form

$$i(t) = i_{\max} \cos \omega t. \qquad \text{(Eqn 1.21)}$$

From *Static fields and potentials* Chapter 3, you should recall that the voltage across the capacitor is given by

$$V_{\text{cap}}(t) = \frac{q(t)}{C}. \qquad (\textit{SFP}\ \text{Eqn 3.8})$$

The equation above is in terms of the charge as a function of time $q(t)$, whereas Equation 1.21 is in terms of the current as a function of time $i(t)$. In order to combine these two equations, you should recall that the charge on the capacitor at any instant, $q(t)$, is related to the current by the equation

$$i(t) = \frac{\mathrm{d}q(t)}{\mathrm{d}t}. \qquad (SFP \text{ Eqn } 3.1)$$

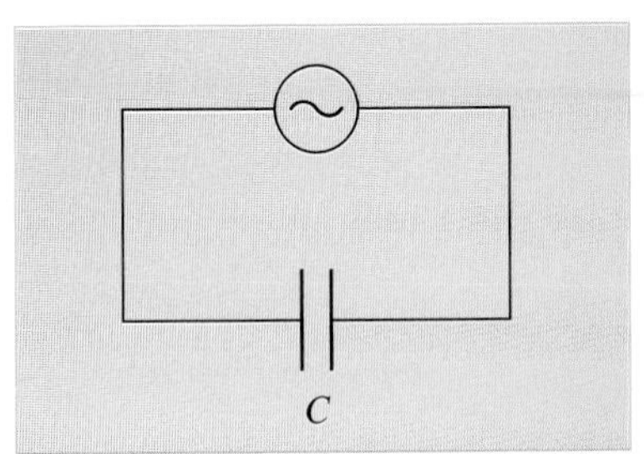

Figure 1.48 A capacitive circuit containing a sinusoidal source of EMF and a capacitor.

So what is needed is an equation for $q(t)$ that, on differentiating, yields Equation 1.21.

The starting point is the general rule that

$$\frac{\mathrm{d}}{\mathrm{d}t}\sin at = a\cos at$$

where a is a constant. So if

$$q(t) = \frac{i_{\mathrm{max}}\sin \omega t}{\omega} \qquad (1.24)$$

then $$i(t) = \frac{\mathrm{d}q(t)}{\mathrm{d}t} = \frac{\mathrm{d}}{\mathrm{d}t}\left(\frac{i_{\mathrm{max}}\sin \omega t}{\omega}\right) = i_{\mathrm{max}}\cos \omega t\cdot$$

which is identical to Equation 1.21. Clearly, therefore, Equation 1.24 is what is needed to represent the charge as a function of time. Now, substituting Equation 1.24 into Equation 3.8 of *Static fields and potentials*, the voltage across the capacitor is

$$V_{\mathrm{cap}}(t) = \frac{i_{\mathrm{max}}\sin \omega t}{\omega C}$$

or $$V_{\mathrm{cap}}(t) = V_{\mathrm{max}}\sin \omega t \qquad (1.25)$$

where $$V_{\mathrm{max}} = \frac{i_{\mathrm{max}}}{\omega C}. \qquad (1.26)$$

Figure 1.49 shows graphs of $V_{\mathrm{cap}}(t)$ and $i(t)$ against time as described by Equations 1.25 and 1.21, respectively. As you can see, both are sinusoidal, with the same period $T = 2\pi/\omega$, but just like the inductive circuit, the two curves are shifted along the time axis with respect to each other. For this capacitive circuit, the voltage *lags* the current by one-quarter of a cycle. Note that this is the opposite result from the inductive circuit of Figure 1.47.

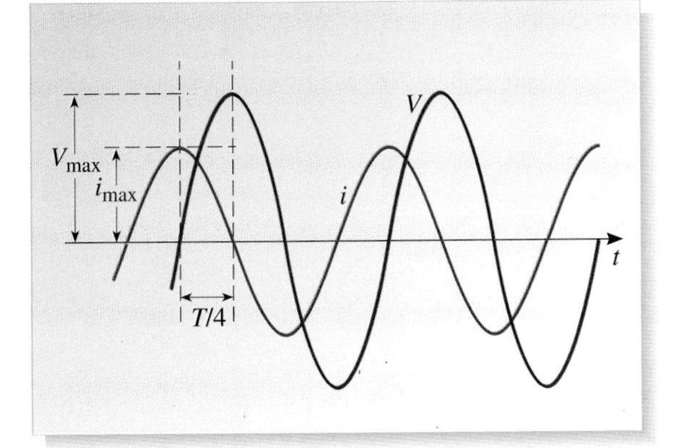

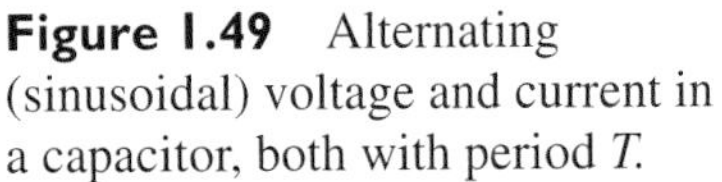

Figure 1.49 Alternating (sinusoidal) voltage and current in a capacitor, both with period T.

Resonant circuits

Now, any real circuit will consist of elements that combine both capacitive and inductive effects, as well as resistive effects. So in a real AC circuit, the relationship between the voltage and the current will depend on a number of factors. In particular, whether the voltage leads or lags the current, and by how much, will depend on the details of all the inductors and capacitors in the circuit. In general, the voltage and current in such a circuit are said to be *out of phase* with each other.

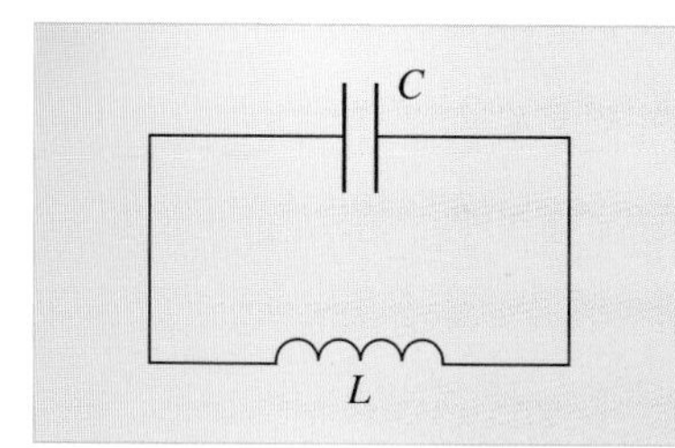

Figure 1.50 An LC circuit containing an inductor and a capacitor.

A relatively simple circuit to analyse is shown in Figure 1.50, which consists of a capacitor connected to an inductor, but no external source of EMF, and no resistance. Such a circuit is often referred to as an ***LC* circuit**, for obvious reasons. Imagine that some separated charge is initially stored on the plates of the capacitor. As the capacitor discharges, a changing current flows in the circuit. This in turn sets up a changing magnetic field, and so a changing magnetic flux, in the inductor. An induced EMF is generated in the inductor, tending to oppose the change, and this in turn causes a current flow to oppose that from the discharging capacitor. The induced current will in turn re-charge the capacitor, and the whole situation will be repeated. There seems to be the possibility of getting charge to flow around this circuit first one way and then the other — an oscillating circuit.

Of course, this is an idealized situation involving no loss of energy, but it is a useful starting point on the road to understanding how a radio receiver works. The behaviour can be analysed mathematically using equations that you have already met in this chapter, and by comparing the situation with that of mechanical oscillations that you met in *Describing motion* Chapter 3 and *Predicting motion* Chapter 2.

To begin, note that there is no external source of EMF in this circuit, so

$$V_{\text{ind}}(t) + V_{\text{cap}}(t) = 0.$$

The voltage across the coil is $V_{\text{ind}}(t) = L\,\mathrm{d}i(t)/\mathrm{d}t$ (Equation 1.9), and that across the capacitor is $V_{\text{cap}}(t) = q(t)/C$ (*SFP* Equation 3.8), so

$$L\frac{\mathrm{d}i(t)}{\mathrm{d}t} = -\frac{q(t)}{C}$$

or

$$\frac{\mathrm{d}i(t)}{\mathrm{d}t} = -\frac{q(t)}{LC}. \qquad (1.27)$$

But current is nothing more than the rate of change of charge, $i(t) = \mathrm{d}q(t)/\mathrm{d}t$ (*SFP* Equation 3.1), so Equation 1.27 becomes

$$\frac{\mathrm{d}}{\mathrm{d}t}\left(\frac{\mathrm{d}q(t)}{\mathrm{d}t}\right) = -\frac{q(t)}{LC}$$

or

$$\frac{\mathrm{d}^2q(t)}{\mathrm{d}t^2} = -\left(\frac{1}{LC}\right)q(t). \qquad (1.28)$$

Now, Equation 1.28 is extremely similar to an equation you met in *Describing motion* Chapter 3 in relation to mechanical simple harmonic motion. There you saw that the following equation described the motion of an oscillator such as a spring

$$\frac{\mathrm{d}^2x(t)}{\mathrm{d}t^2} = -\omega^2x(t). \qquad (DM\text{, Eqn 3.55})$$

From the similarity of these two equations, it seems reasonable to deduce that the electrical system can perform *simple harmonic oscillations*, but in this case it is the charge q that is varying sinusoidally with time rather than the displacement x.

- By comparing Equation 1.28 and *DM* Equation 3.55, write down expressions for the angular frequency and (normal) frequency of the charge oscillations.

- Clearly ω^2 in *DM* Equation 3.55 corresponds to $1/LC$ in Equation 1.28. So the angular frequency for charge oscillations is

$$\omega = \frac{1}{\sqrt{LC}} \tag{1.29}$$

and the frequency of oscillation of the charge is then $f = \omega/2\pi$ or

$$f = \frac{1}{2\pi\sqrt{LC}}. \tag{1.30}$$ ■

Following a similar argument to that used for mechanical simple harmonic motion, a solution to Equation 1.28 is simply

$$q(t) = q_{max} \cos \omega t \tag{1.31}$$

where q_{max} is the maximum value of electric charge stored on the capacitor and ω is the angular frequency of charge oscillation.

- Verify that Equation 1.31 is a solution of Equation 1.28.

- Differentiating Equation 1.31 with respect to time gives

$$\frac{dq(t)}{dt} = -\omega q_{max} \sin \omega t$$

and differentiating for a second time gives

$$\frac{d^2q(t)}{dt^2} = -\omega^2 q_{max} \cos \omega t$$

Substituting for ω^2 from Equation 1.29 and for $q_{max} \cos \omega t$ from Equation 1.31, this becomes

$$\frac{d^2q(t)}{dt^2} = -\left(\frac{1}{LC}\right)q(t),$$

which is Equation 1.28. So Equation 1.31 is indeed a solution of Equation 1.28. ■

Question 1.13 (a) In a mechanical oscillator, such as a pendulum, energy is continually changing from kinetic energy to gravitational potential energy and back again. What do you suppose are the corresponding changes of energy in an *LC* circuit? (b) A mechanical oscillator will lose energy due to damping forces such as air resistance. What is the analogous damping mechanism in an electrical circuit and what is the 'lost' energy transformed into? ■

In the case of a damped electrical oscillation, the differential equation describing the motion of electrical charge is

$$\frac{d^2q(t)}{dt^2} = -\left(\frac{R}{L}\right)\frac{dq(t)}{dt} - \left(\frac{1}{LC}\right)q(t) \tag{1.32}$$

which is directly analogous to the equation for damped mechanical oscillations

$$\frac{d^2x(t)}{dt^2} = -\left(\frac{b}{m}\right)\frac{dx(t)}{dt} - \left(\frac{k}{m}\right)x(t). \qquad (\textit{PM}\text{, Eqn 2.54})$$

Resistance (R) in the case of electrical oscillations is directly analogous to the damping constant (b) in mechanical oscillations. The solution to Equation 1.32 is a combination of a sinusoidal (oscillatory) term and an exponential (decay) term, again in direct analogy with the mechanical case.

● What are the analogues of mass (m) and the force constant (k) when considering electrical oscillations?

○ The analogue of mass is self-inductance (L) and the analogue of the force constant is the reciprocal of the capacitance ($1/C$). ■

The simple LC circuit is the basis of all radios and tuned circuits. Charges will oscillate to and fro in such a circuit at a natural frequency given by Equation 1.30. The significance of the natural frequency lies in the fact that the circuit is *most responsive* to outside signals at that particular frequency — just as mechanical oscillators are most responsive to external driving at their natural frequency. This is another example of the phenomenon of *resonance* that was discussed in *PM* Chapter 2.

When an LC circuit is *driven* at its natural frequency by an external oscillating source, such as a radio wave (see Chapter 2), the effect of the shift in voltage and current due to the inductor is exactly compensated by the shift in voltage and current due to the capacitor. At this frequency, the maximum voltages across the inductor and the capacitor are equal, and so Equations 1.23 and 1.26 can be equated to give

$$i_{\text{max}}\omega L = \frac{i_{\text{max}}}{\omega C}.$$

Figure 1.51 A variable capacitor used to tune a radio receiver.

The maximum current cancels, and the equation can be rearranged to give $\omega^2 = 1/LC$, which is the same as the result obtained above for the natural angular frequency (Equation 1.29). When driven at this frequency, the maximum amount of energy will be transferred from the driving signal to the circuit. This is the condition in which a radio receiver will respond most strongly to an external signal.

To achieve this, you must *tune* your radio receiver so that its natural frequency is equal to the frequency of the transmitter. In practice, radio receivers generally contain a fixed-value inductor and a variable capacitor. By adjusting the capacitance of the capacitor, the natural frequency of the circuit can be altered, and so the circuit can be 'tuned in' to a station broadcasting at a particular frequency. Tuning capacitors have the form of two sets of parallel plates; as shown in Figure 1.51, one set is fixed, and the other can be turned into and out of the gaps in the fixed set. The changing overlap of their surface areas represents the variable capacitance.

● How could an *inductor* be tuned?

○ The answer is in the dependence of the inductance on the *magnetic core* of the inductor. The value of L can be changed by pushing in or pulling out a rod of suitable magnetic material along the central axis of the solenoid. ■

Question 1.14 A radio is to be tuned to BBC Radio 4 long wave, which has a signal frequency of 200 kHz. If the (fixed) inductor in the radio circuit has an inductance of $L = 32$ mH, to what value must the capacitance of the capacitor be adjusted in order to cause the circuit to resonate? ■

3.4 Transformers

The purpose of a **transformer** is to transform EMFs from one value to another. The principle behind such devices can be understood by considering what are effectively two solenoids in close proximity to each other, around a single iron core, as illustrated in Figure 1.52. Assuming no heating losses in the core, this is sometimes known as an *ideal transformer*. Either winding may be connected to an AC source, in which case it is called the *primary* coil; the other winding is then the *secondary* coil, and is connected to whatever device is being powered.

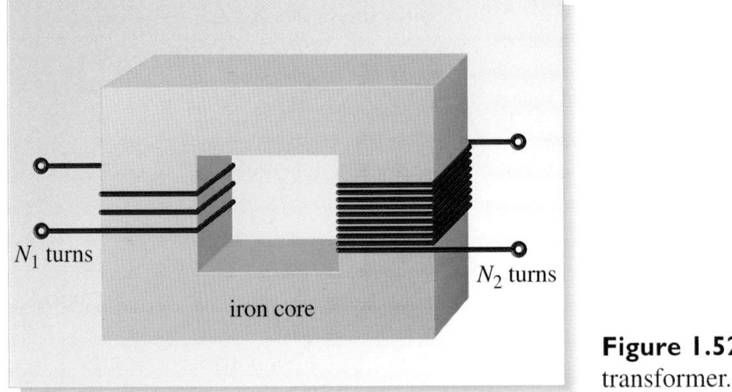

Figure 1.52 An ideal transformer.

To see how the characteristics of a transformer may be quantified, consider the consequences of the primary coil being connected to an AC source. The coil carries an alternating current, which sets up a periodically varying magnetic field. This field creates magnetic flux in both the primary and the secondary coils.

Earlier, in Section 2.8, you saw that the magnetic field inside a long cylindrical solenoid carrying a time-varying current is given by Equation 1.5. In this case, the magnitude of the magnetic field inside the primary coil can be written as

$$B(t) = \frac{\mu N_1 i_1(t)}{l} \tag{1.33}$$

where N_1 is the number of turns of the primary coil, l is its length, and $i_1(t)$ is the current flowing through it. Now the magnetic flux through each turn of a coil of area A is given by Equation 1.1a with $\theta = 0$:

$$\phi(t) = AB(t).$$

So the magnetic flux through *each turn* of the primary coil is

$$\phi_1(t) = \frac{\mu N_1 A}{l} i_1(t). \tag{1.34}$$

Assuming no loss of flux between the primary and the secondary coils, the magnetic flux through *each turn* of the secondary coil is the same as through each turn of the primary coil; that is

$$\phi_2(t) = \frac{\mu N_1 A}{l} i_1(t). \tag{1.35}$$

The total magnetic flux through the secondary coil is therefore given by

$$\Phi_2(t) = N_2\phi_2(t) = \frac{\mu N_1 N_2 A}{l} i_1(t) \tag{1.36}$$

where N_2 is the number of turns on the secondary. The magnitude of the rate of change of magnetic flux through the secondary coil is then given by

$$\left|\frac{d\Phi_2(t)}{dt}\right| = \frac{\mu N_1 N_2 A}{l}\left|\frac{di_1(t)}{dt}\right|. \tag{1.37}$$

Using Faraday's law, Equation 1.37 is also equal to the magnitude of the induced EMF in the secondary coil, so

$$|V_2(t)| = \frac{\mu N_1 N_2 A}{l}\left|\frac{di_1(t)}{dt}\right|. \tag{1.38}$$

By direct analogy with self-inductance, Equation 1.38 can be rewritten as

$$|V_2(t)| = M\left|\frac{di_1(t)}{dt}\right| \tag{1.39}$$

where the constant

$$M = \frac{\mu N_1 N_2 A}{l} \tag{1.40}$$

is the **coefficient of mutual inductance** of the coils. The SI unit is again the henry.

Notice that Equation 1.40 has been derived for the special case where both coils have the same area A. Although Equation 1.39 is a general expression that is true in all cases, it is not possible to derive a simple expression for the mutual inductance in most other situations.

What about the relationship between the EMF in the primary and secondary coils? One way of understanding this is as follows. First, recall the definition of self-inductance,

$$|V(t)| = L\left|\frac{di(t)}{dt}\right|. \tag{Eqn 1.9}$$

So the magnitude of the rate of change of current in the primary coil is

$$\left|\frac{di_1(t)}{dt}\right| = \frac{|V_1(t)|}{L_1}. \tag{1.41}$$

Substituting this into Equation 1.39 gives

$$|V_2(t)| = M\frac{|V_1(t)|}{L_1} \tag{1.42}$$

where the self-inductance of the primary coil is given by Equation 1.10 as

$$L_1 = \frac{\mu A N_1^2}{l}. \tag{1.43}$$

Now, substituting for M (from Equation 1.40) and L_1 (from Equation 1.43) into Equation 1.42 gives

$$|V_2(t)| = \frac{\mu N_1 N_2 A}{l} \times \frac{l\,|V_1(t)|}{\mu A N_1^2}.$$

On cancelling the various common terms,

$$|V_2(t)| = \frac{N_2}{N_1} \times |V_1(t)|. \tag{1.44}$$

Thus, to make the EMF in the secondary greater than the EMF in the primary, N_2/N_1 needs to be greater than 1. In other words, there must be more windings on the secondary than on the primary. This is the case illustrated in Figure 1.52, and is referred to as a *step-up* transformer. To make the EMF in the secondary less than the EMF in the primary, N_2/N_1 needs to be less than 1, so there must be more windings on the primary than on the secondary. This would give a *step-down* transformer. Remember that both $V_1(t)$ and $V_2(t)$ are sinusoidally varying EMFs, each of the form $V(t) = V_{max} \cos\omega t$, where V_{max} is a constant representing the maximum EMF during the cycle.

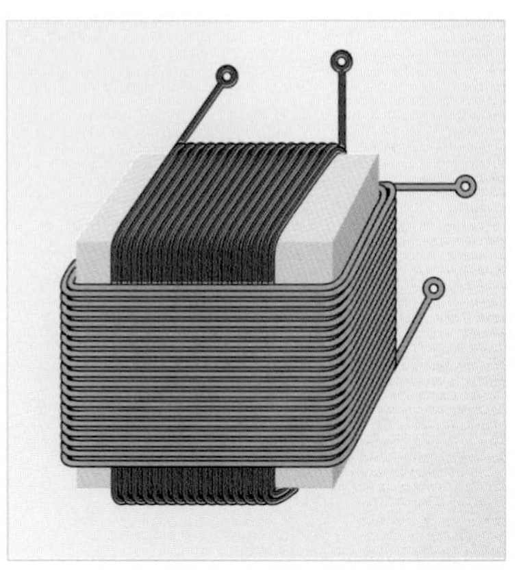

Figure 1.53 A more effective transformer winding arrangement. The windings are insulated from each other by shellac coatings.

Question 1.15 (a) An ideal transformer used to supply power to a food processor has 5000 turns on its primary winding. If the input EMF has a maximum value of 240 V, what must be the number of turns on the secondary winding to give an output EMF with a maximum value of 12 V? (b) The maximum rate of change of the current in the primary winding is 300 A s^{-1}. What is the mutual inductance of the transformer? ■

The practical theory of transformer design is rather more complicated than the situations described above; Figure 1.53 shows a more realistic design. With laminations, the eddy currents can be reduced to a tolerable level.

You are probably aware that electricity is distributed at high voltage. This is necessary to enable the current to be kept low so as to minimize heating losses. (Recall that the rate of heating loss is given by i^2R.) For example, the voltage in the transmission lines from the Hoover Dam to Los Angeles is set at 287 000 V. When the power reaches the main distribution centre in a large city, it must be stepped down, and this is usually done in stages. Figure 1.54 illustrates some stages in the distribution and stepping-down of voltage in the UK.

Figure 1.54 (a) High-voltage transmission lines; (b) stepping-down at a substation.

In Section 1.1 a *wireless lamp* was described — a device that causes an incandescent bulb to light up without being connected by wires to a power source. With the knowledge you have gained about induction, you should now be able to explain how this device works.

Look again at Figure 1.2. The solenoid in one box is connected to an AC power source, and that in the other is connected to a light bulb. The power source delivers alternating current to the solenoid it is attached to, and this creates a magnetic field. Because of the proximity of the two solenoids, the second solenoid (the one attached to the bulb) is also 'threaded' by this alternating magnetic field. The alternating field gives an alternating magnetic flux so, by Faraday's law, an alternating current is induced in the second circuit. This current lights the lamp.

A good 'magic trick' perhaps, but not really magic — just another application of the principles of electromagnetic induction.

4 Maxwell's equations

4.1 The field equations

Before Oersted's experiments showed that currents produce magnetic fields, electricity and magnetism were widely thought to be completely separate; Oersted showed that this could not be true. The work of Henry and Faraday on induction showed that the entanglement was closer yet, but it took Maxwell to gather together all the strands, add one more, and braid them together into a complete description of electromagnetism. Only part of the content of Maxwell's equations was discovered by Maxwell himself, but nevertheless it is almost impossible to overestimate the importance of the work that Maxwell did in extending the theory in such an elegant way.

Figure 1.55 James Clerk Maxwell.

James Clerk Maxwell (1831–1879)

Maxwell came from a well-to-do Scottish family. He attended Edinburgh Academy, and then the Universities of Edinburgh and Cambridge. His academic career began at Aberdeen University, but in 1868 the University was reorganized, and Maxwell was considered surplus to requirements and so declared redundant. Fortunately, he obtained an appointment at King's College London, and subsequently took the first Chair in Experimental Physics at Cambridge. His lasting memorial in Cambridge is the (old) Cavendish Laboratory, which he was instrumental in designing and getting built.

Maxwell's first scientific paper, discussing an aspect of geometry, was submitted when he was just 15. Apart from deriving the famous equations that take his name, Maxwell made significant contributions to kinetic theory and statistical mechanics, and was awarded the Adams Prize for his paper on the stability of Saturn's rings. He also did valuable work on colour perception and colour blindness, and found time to write influential textbooks on the theory of heat, matter and motion.

The first difficulty in appreciating Maxwell's theory of electromagnetism is that electromagnetic properties are best described by fields (this was not fully appreciated before Maxwell's time). The electromagnetic field in a region determines the forces that act on test charges and currents at each point in that region. It is a real thing, since it causes effects, yet any attempt to 'mechanize' it by defining a medium that it displaces is doomed to failure. In diagrams throughout this and previous books, arrows have been drawn to show you electric and magnetic fields in special and limited circumstances. This was done to help you visualize what is happening. But this vision is necessarily limited, for *six components* are needed (three for $\mathscr{E}$ and three for $\boldsymbol{B}$) to describe the electric and magnetic fields at *each four-dimensional point* in space–time. A generalized electromagnetic field is *impossible* to visualize.

The second difficulty is that Maxwell's equations are written in a mathematical language that is somewhat beyond the scope of this book. In fact, it is the use of very concise mathematical language that makes Maxwell's equations so powerful: they are able to say a great deal in only a few lines (Figure 1.56).

Despite these difficulties, Maxwell's equations will not be ignored here; they are much too important for that. Instead, the equations will be described in words and some of their consequences will be explained.

The first Maxwell equation is related to Coulomb's law, which states that electric fields are created by distributions of electric charges. The equation was first put into mathematical form by Gauss and can be expressed as:

(I) $\nabla \cdot \underset{\sim}{E} = \dfrac{\rho}{\varepsilon_0}$

(II) $\nabla \cdot \underset{\sim}{B} = 0$

(III) $\nabla \times \underset{\sim}{E} + \dfrac{\partial \underset{\sim}{B}}{\partial t} = 0$

(IV) $\nabla \times \underset{\sim}{B} - \dfrac{1}{c^2}\dfrac{\partial \underset{\sim}{E}}{\partial t} = \mu_0 \underset{\sim}{J}$

Figure 1.56 Maxwell's equations can be written in a very concise mathematical language.

Equation I

The electric flux due to an electric field through a closed surface is proportional to the charge contained inside it.

In contrast to this, the next equation says that matter *does not* have a magnetic quality equivalent to charge; instead, magnetic field lines are continuous, with neither a beginning nor an end, and loop from a north pole to a south pole. There are no such things as 'magnetic monopoles'. As noted for Question 1.4, the equation can be expressed as:

Equation II

The magnetic flux due to a magnetic field through a closed surface is zero.

The third equation is a combination of Faraday's law and Lenz's law, based on the fact that a changing magnetic flux causes an electric field, and the electric field opposes the change. It can be expressed as:

Equation III

The magnitude of the rate of change of the magnetic flux, due to a magnetic field, through a surface is equal to the magnitude of the EMF along the boundary of the surface. The direction of the EMF opposes the change causing it.

This third equation may be summarized more crudely by the statement:

A changing magnetic field causes an electric field.

Before Maxwell's time, it was known that a steady current caused a magnetic field. It's tempting to generalize this to a simple statement about magnetic fields being produced by currents in all cases. However, Maxwell noticed that such a statement violated the conservation of electric charge. The fact that this is so cannot be demonstrated without mathematics beyond the level of this book, so you are therefore asked to take Maxwell's word for it. However, you know that currents *do* create magnetic fields; indeed, this fact was used throughout the preceding sections. Faced with this inconsistency, Maxwell took the bold step of postulating that there was another term in the relation between magnetic fields and currents that was equal to zero in certain circumstances, but not in general. He found that the additional term was due to the fact that:

A changing electric field causes a magnetic field.

The statement above is important, and you should remember it. You should also see how it mirrors the earlier statement that a changing magnetic field causes an electric field.

Maxwell's fourth equation is perhaps the most difficult to state without resorting to symbols, but it has to do with the fact that magnetic fields can be produced by electric currents (as noted by Oersted) *and* by electric fields that change with time (as noted by Maxwell). It can be stated as:

Equation IV

The average magnetic field along the boundary of a surface depends on the current per unit area inside the surface, *and* on the rate of change of the electric field across the surface.

Written in words in this way, Maxwell's fourth equation looks somewhat clumsy, and of course, in writing out the equations in words, many injustices have been committed. Some subtleties have been ignored, and much of the glory has been lost. However, the significance of the equations themselves should not be doubted: taken together with the fundamental law that charge is conserved, and with the Lorentz force law, the equations provide an essentially complete description of electric and magnetic phenomena. Indeed, so complete is the description that it can be used as a basis for the development of electromagnetism rather than a summary of it. In some advanced treatments, the Maxwell equations and the Lorentz force law are postulated at the very beginning; all the other laws — Coulomb's law, Faraday's law, the laws governing the magnetic fields produced by various current distributions, and so on — can then be deduced from Maxwell's equations.

4.2 Electromagnetic radiation

The apparent symmetry in the following two statements has already been discussed.

A magnetic field that changes in time produces an electric field.

An electric field that changes in time produces a magnetic field.

Combining these two statements suggests an exciting possibility; perhaps changing electric and magnetic fields can be arranged in such a way that they cooperate in some sense, each creating and sustaining the other. Perhaps a changing electric field can produce a changing magnetic field, which in turn can produce a changing electric field, which in turn..., etc. Maxwell's mathematical investigations persuaded him that this was indeed a possibility, though the self-sustaining pattern of changing electric and magnetic fields would have to obey a number of constraints. One of these was that the pattern of fields would have to travel through space at a speed that was precisely determined by the Maxwell equations. Because the field pattern had to move away from its point of origin with a prescribed speed, it was said to radiate. What Maxwell had done was to predict the existence of *electromagnetic radiation* — a pattern of changing electric and magnetic fields with the ability to travel through a vacuum at a certain fixed speed.

From Maxwell's equations, it can be shown that this fixed speed is given by

$$c = \frac{1}{\sqrt{\varepsilon_0 \mu_0}} \tag{1.45}$$

where ε_0 and μ_0 are, respectively, the permittivity and permeability of free space.

Substituting known values for ε_0 and μ_0 gives the remarkable result that the speed is, to four significant figures, $2.998 \times 10^8\,\mathrm{m\,s^{-1}}$. This, as you are probably aware, is the speed of light in a vacuum. As Maxwell realized, his discovery of electromagnetic radiation was a major breakthrough. Richard Feynman put it as follows:

> Maxwell had made one of the great unifications of physics. Before his time, there was light, and there was electricity and magnetism. The latter two had been unified by the experimental work of Faraday, Oersted and Ampère. Then, all of a sudden, light was no longer 'something else', but was only electricity and magnetism in this new form — little pieces of electric and magnetic fields which propagate through space on their own.'
>
> R. P. Feynman, *et al.* (1964) *The Feynman Lectures on Physics, Vol.2*, Addison–Wesley.

The implication of this monumental discovery will be discussed in the next chapter.

5 Closing items

5.1 Chapter summary

1 The magnetic part of the Lorentz force ($\boldsymbol{F}_{\mathrm{mag}} = q\boldsymbol{v} \times \boldsymbol{B}$) may, under appropriate circumstances, give rise to an induced current in a conducting circuit moving through a magnetic field. A magnetic field that changes (in strength or direction), acting on a stationary conducting loop, may also induce a current in the loop.

2 More generally, induction requires consideration of the magnetic flux, ϕ, through the circuit. The magnetic flux through any flat surface is given by the product of the component of the average magnetic field perpendicular to the surface and the area of the surface: $\phi = AB\cos\theta$.

3 Faraday's law says that the magnitude of the EMF induced around the boundary of a surface is equal to the magnitude of the rate of change of the magnetic flux through that surface:

$$|V_{\mathrm{ind}}(t)| = \left|\frac{\mathrm{d}\phi(t)}{\mathrm{d}t}\right|.$$

4 The surface appearing in Faraday's law can be chosen at will. It does not have to be flat. Most usefully it can be a surface bounded by a conducting wire, in which case an induced EMF will cause a current to flow along the wire.

5 Lenz's law states that the direction of the induced EMF must be such as to oppose the change that caused it.

6 Induced currents, known as eddy currents, occur in solid blocks of ferromagnetic material such as iron, in which case they are usually unwanted, because they dissipate energy by resistive heating. Methods of reducing eddy currents are available, such as laminating the blocks into thin sheets of appropriate shape and direction.

7 Time-varying currents through coils may cause varying magnetic fluxes threading both the original coil and nearby coils. The resultant induced EMFs are proportional to the rate of change of the currents, and the proportionality constants are the coefficients of self- and mutual inductance, measured in henries:

$$|V_{\mathrm{ind}}(t)| = L\left|\frac{\mathrm{d}i(t)}{\mathrm{d}t}\right| \quad \text{and} \quad |V_2(t)| = M\left|\frac{\mathrm{d}i_1(t)}{\mathrm{d}t}\right|.$$

8 The coefficient of mutual inductance depends on the product of the number of turns of wire in each coil; the coefficient of self-inductance depends on the square of the number of turns in the relevant coil. For the specific cases of long cylindrical solenoids: $M = \mu AN_1N_2/l$ and $L = \mu AN^2/l$. Transformers rely on the phenomenon of mutual inductance to step-up or step-down varying EMFs. These are found in many places, from the power industry to the small transformers in home and office electronic equipment.

9 Rotating a loop of conducting wire in a uniform magnetic field at a constant angular speed results in an alternating current being induced in the loop. By using a pair of slip rings, this current may be tapped, thus providing a simple AC generator. By using a single split ring, properly orientated with respect to the phase of the current, the current may be rectified, thus providing a simple DC generator. By supplying appropriate currents to either of these simple devices, equally simple AC and DC electric motors may be constructed.

10 When a solenoid is attached to a battery in a circuit that also contains a resistor, the current in the circuit rises in accordance with an exponential law, approaching but never achieving a constant value. When the battery is disconnected, the current decays via another exponential law, approaching but never achieving a value of zero. A solenoid stores energy in its core magnetic field.

11 In circuits containing inductors and capacitors connected to an AC supply, the voltage and current are generally out of phase with each other. In a purely inductive circuit, the voltage leads the current by one-quarter of a cycle, whereas in a purely capacitive circuit, the voltage lags the current by one-quarter of a cycle.

12 *LC* circuits have a natural frequency, at which they are most responsive to an external driving signal. By varying either the capacitance or inductance of the circuit, it can be 'tuned' to a particular frequency, and such circuits provide the basis for radio transmitters and receivers, and, moreover, the whole telecommunications industry.

13 Maxwell's equations provide a unified theory of electric fields, magnetic fields and electromagnetic radiation. Electromagnetic radiation travels through a vacuum at the speed of light: $c = 1/\sqrt{\varepsilon_0\mu_0}$.

5.2 Achievements

Now that you have completed this chapter, you should be able to:

A1 Understand the meaning of all the newly defined (emboldened) terms introduced in the chapter.

A2 Calculate the magnitude of the magnetic flux and its rate of change with time in a variety of simple situations.

A3 Write down Faraday's law and Lenz's law, explaining their meaning and use, and then solve a variety of simple problems.

A4 Explain how the phenomena described by these laws underpin the operation of various devices that have been considered in this chapter, in particular, the telephone, the AC generator, the tuned circuit and the transformer.

A5 Determine the size and direction of the induced EMF or current in particular cases, such as loops moving through uniform fields, magnets moving through or past loops and self-induced currents in coils.

A6 Write down and work with the magnetic Lorentz force, particularly to find the direction of currents, and show that the result is in accordance with Lenz's law.

A7 Explain the origin of eddy currents, describe situations in which they are desirable and others where they are unwanted, and explain how their effect may be minimized.

A8 Calculate the self- (or mutual) inductance of a simple coil (or coils).

A9 Explain the origin of transient currents in circuits containing an inductor, and carry out simple calculations concerning rise and decay times and magnetic energy.

A10 Explain how phase lags between voltage and current arise in inductive and capacitive circuits, and how charge oscillations are sustained in an *LC* circuit.

A11 Calculate the natural frequency of oscillation of an *LC* circuit, and describe how such circuits form the basis for radio and other telecommunications.

A12 State Maxwell's equations in words, and explain the meaning of each.

5.3 End-of-chapter questions

Question 1.16 Consider again the wire loop being pulled through a uniform magnetic field as shown in Figure 1.8. Take the position of the wire to be that shown in the figure. The earlier analysis showed that a current will flow in the direction from W towards X. As an approximation to the true, and complicated, atomic processes inside the wire, assume that the current is carried by positive charges all flowing with the same speed u, moving clockwise from W to X. (The speeds are the same in a given arm, but differ in direction from arm to arm.) Find the magnetic Lorentz force on the charges moving with speed u in the arm from W to X, and relate this force to the external force pulling the loop. Explain how this agrees with Lenz's law.

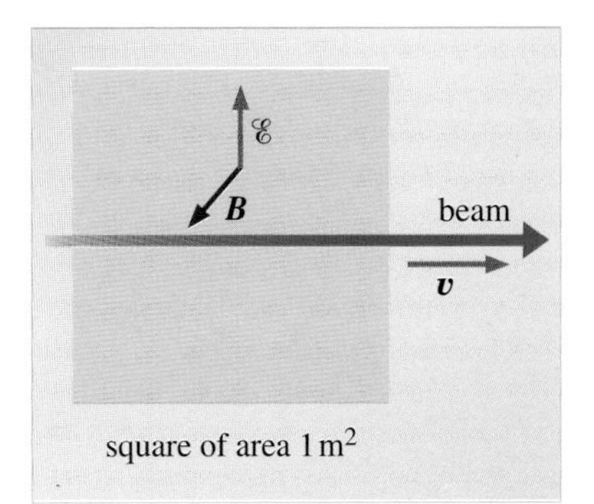

Figure 1.57 A beam of hydrogen ions passing through a region of crossed electric and magnetic fields.

Question 1.17 Figure 1.57 shows a beam of hydrogen ions passing perpendicularly through a surface whose area is $1\,\mathrm{m}^2$, and over which there is an electric field of strength $2 \times 10^2\,\mathrm{N\,C^{-1}}$. The ions have a velocity of $5 \times 10^3\,\mathrm{m\,s^{-1}}$ in the direction shown so that the electric and magnetic parts of the Lorentz forces are just balanced, and the ions move through the region without change of velocity. (a) Verify that the direction of the magnetic field is as shown in Figure 1.57, and determine its magnitude. (b) Determine the magnetic flux across the region.

Question 1.18 Imagine that a circular loop of wire, of resistance R and area A, is placed with its plane perpendicular to a magnetic field of magnitude $B(t)$ which always points in a certain fixed direction. Suppose that the magnetic field is uniform but not static; that is, at any particular instant, the magnetic field has the same strength at all points (uniform), but that the uniform field strength changes with time (non-static). In particular, suppose that the magnetic field strength at time t is given by $B(t) = B_0 + kt^2$, where B_0 and k are constants. Find an expression for the magnitude of the induced current in the loop of wire.

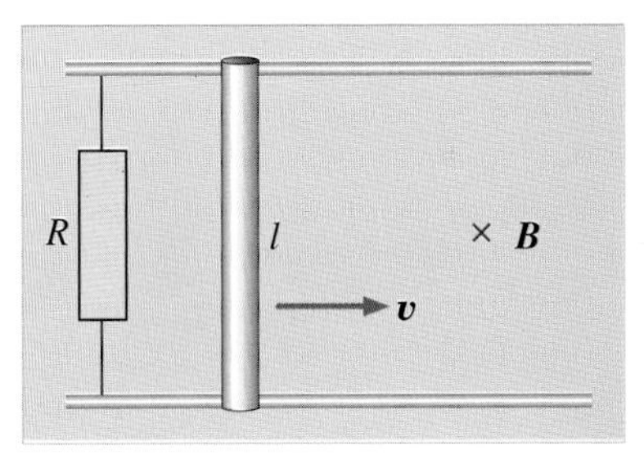

Figure 1.58 A conducting rod sliding along two fixed conducting rails, a distance l apart, linked by a resistor of resistance R. The rod has velocity v, and it moves at right-angles to a uniform static magnetic field of strength B directed into the page.

Question 1.19 (a) Figure 1.58 shows a conducting rod sliding along two fixed conducting rails separated by a distance l. The rod moves with constant speed v, and is travelling at right-angles to a uniform magnetic field of constant strength B pointing into the page. The rod and the rails are of negligible resistance, but the rails are linked by a resistance R, so at any instant the arrangement of rod, rails and resistor constitutes a circuit of total resistance R. Because the area of the circuit is changing with time, an induced current will flow around the circuit. What is the magnitude of the induced current? (*Hint* Consider the rate at which the area of the circuit increases with time.)

(b) Referring again to Figure 1.58, in order to induce a current of 1.5 A in a rod of length 10 cm, when the resistance between the rails is 3.0 Ω, with what speed would the rod need to travel through a magnetic field of strength 5.0 T?

Question 1.20 When a generator is supplying current to an external circuit, it is found that the force required to keep the coil turning is much greater than the force needed to overcome friction. What is the origin of the extra force that opposes the rotation of the coil?

Question 1.21 A radio receiver has a variable capacitor whose capacitance can be varied between 50 pF and 450 pF. When set up to receive medium-wave broadcasts, it uses a solenoid with 100 turns of wire and can tune in to stations with frequencies between 0.5 MHz and 1.5 MHz. (Remember 1 pF = 10^{-12} F.)

The radio is now adjusted to receive long-wave broadcasts, with frequencies between 100 kHz and 300 kHz, by winding more turns of wire on the solenoid. Assuming that other properties of the solenoid and circuit remain unchanged, how many turns of wire must there be on the solenoid to receive these broadcasts? ■

Chapter 2 Waves and electromagnetic radiation

1 Butterflies and oil spills

Consider the beautiful colour of the wings of the *Morpho rhetenor* butterfly shown in Figure 2.1a. You may think that the coloration is caused by a pigment, but research indicates otherwise. It is now thought that the coloration is caused by a thin-film *interference effect*, an effect very similar to the one that causes the spectrum of colours seen in a thin layer of oil on top of a puddle (Figure 2.1b). Interference is a phenomenon that is associated with all sorts of waves and, like a number of other apparently unrelated physical phenomena, can be explained by considering the properties of waves. The aim of this chapter is to describe various wave properties, and hence to explain these phenomena.

Figure 2.1 Thin-film interference of light waves can give spectacular colours: (a) wing colour in the *Morpho rhetenor* butterfly; (b) the effect of a thin layer of oil on a puddle.

This chapter begins by considering how waves may be described physically and mathematically. It moves on to look at two categories of waves known as *travelling waves* and *standing waves*, and includes examples of each in terms of sound and musical instruments. Sections 5 and 6 then consider light and related electromagnetic radiation, and in particular how it *propagates* through different media, how it can undergo *reflection* or *refraction* at interfaces, and how *diffraction* and *interference* effects arise. The chapter ends with a discussion of some more examples of these phenomena in the natural world.

2 Describing waves

2.1 What is a wave?

Most people probably have some idea of what constitutes a wave. For instance, a stone dropped into the centre of a pond generates waves on the surface of the water; the waves travel outwards and may eventually cause a cork at the edge of the pond to bob up and down with a regular motion. Similarly, an earthquake — a sudden motion of part of the Earth's crust — generates seismic waves, which travel through the Earth, and may cause damage to buildings some distance away on the surface. Another image that the word 'wave' often conjures up is that of water waves travelling across the sea, or breaking on a beach (Figure 2.2).

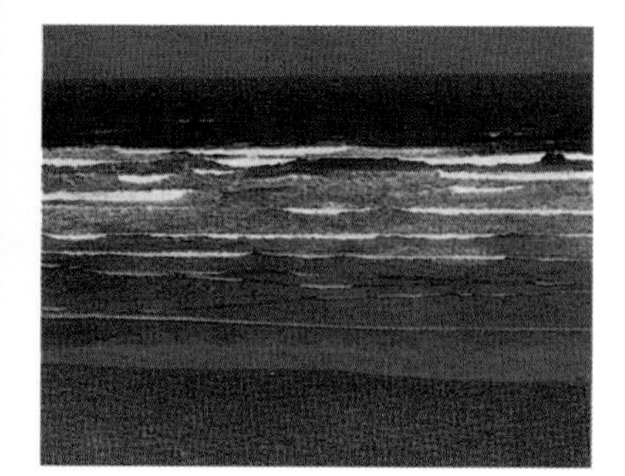

Figure 2.2 Water waves travelling across the sea and breaking on a beach.

Describing waves physically is not an easy task; even the most basic features are challenging. For example, when a wave approaches a beach, what is it that is actually travelling? It can't be the water itself, or the entire ocean would gradually accumulate on the beach! Indeed if you look at small boats or other floating objects you will see that their main response to a passing wave is to move up and down, so it is pretty clear that the movement of the water is essentially vertical — at right-angles to the direction in which the wave is travelling. (The water motion is actually slightly more complicated than this, but up-and-down motion is a reasonable first approximation.)

In the case of ocean waves, a medium (i.e. the water) is essential to the existence of the wave, but it is clear that the wave does not consist of an overall motion of the medium in the direction of the wave. Moreover, there are waves, notably the electromagnetic waves introduced in Chapter 1 and further discussed later in this chapter, that can travel through a vacuum and which therefore exist even in the absence of a medium. So if a wave does not consist of the overall motion of a medium, what is it that does constitute a wave?

Returning to the example of boats bobbing up and down on the sea, note that it takes *energy* to set the boats in motion. This energy comes from some distant source, perhaps a passing ship, and is delivered to the boats by the wave. A characteristic feature of many waves is their ability to transport energy from one place to another, without any permanent change in the medium through which they are passing. Sound waves carry energy from a vibrating source such as vocal chords or a musical instrument. The energy from the sound wave eventually produces vibrations in the eardrum, so the sound is heard. Light waves, radio waves and other forms of electromagnetic radiation (discussed later) are essentially periodic fluctuations of electric and magnetic fields, again carrying energy from place to place. Radio waves are produced by vibrations of electrical charge in a transmitting antenna. When the energy from the wave reaches the receiving antenna of a radio set, similar charge vibrations are established. Much more impressive demonstrations of energy carried by waves are shown in Figure 2.3.

In many cases a **wave** can be described as a *periodic*, or regularly repeating, 'disturbance' that conveys *energy* from one point to another.

Figure 2.3 Energy carried by waves: (a) wreckage in Crescent City, California from the 1964 Good Friday tsunami ('tidal wave'); (b) surfing.

It is appropriate to discuss 'waves' along with 'dynamic fields' in this book, since a field is *always* necessary in order to have a wave. In the case of sound waves the field may be a pressure field or density field in a particular medium; in the case of a wave on a string the field may be a strain field in the string. Electromagnetic waves rely on electric and magnetic fields, as will be discussed later.

2.2 Wave properties

Later, the mathematical description of waves will be considered, but first let's look at some more qualitative ways in which the properties of waves may be described.

While standing on a beach, you will have seen or heard waves break onto the shore with a fairly regular time interval between one 'crash' and the next. Each crash represents one wave crest breaking onto the shore and the *time interval* between two of them is known as the **period** of the wave, usually represented by the symbol T. In general, the period of a wave may be defined as the time between one part of the wave (say the crest) passing a fixed point in space and the *next* identical part of the wave (the next crest) passing the same fixed point. In the case of ocean waves, a convenient fixed 'point' is the shoreline.

A quantity related to the period of a wave is its **frequency**. This is the rate at which waves pass a fixed point. The frequency is therefore simply the reciprocal of the period of a wave, $f = 1/T$. If the period is measured in the SI unit of seconds (s), the frequency will have the SI unit of hertz (Hz), where $1\ \text{Hz} \equiv 1\ \text{s}^{-1}$.

As well as being periodic with time, a wave is also periodic in space. The word 'wave' is often used to describe a single crash onto the beach, but it really refers to the entire sequence of crests and troughs, stretching away into the distance. The form of that spatial pattern of crests and troughs at any instant of time is called the *wave profile*. The *distance* between one wave crest and the next is known as the **wavelength** of the wave, and is usually represented by the Greek letter lambda λ. In general, the wavelength of a wave is defined as the distance between one part of the wave profile, at a particular instant in time, and the next identical part of the wave profile at the same instant of time. Two adjacent crests of the wave are a convenient pair of locations to use for this definition, although any pair of equivalent points will do as long as the wave has the same 'height' and is changing in the same way at both points.

Question 2.1 Try the following **thought experiment**. Imagine that you are standing on a beach and watching wave crests break on the seashore.

(a) If the wavelength suddenly becomes smaller (i.e. the wave crests are closer together than before), but the wave continues travelling at the same speed, what happens to the period of the wave? What happens to the frequency?

(b) If, instead, the waves suddenly begin travelling more slowly across the sea, but with the same wavelength as before, what happens to the period of the wave? What happens to the frequency? ■

You will probably have noticed that the definitions for the period and wavelength of a wave are rather similar. The period is a time interval and refers to instants separated in *time*; the wavelength is a distance and refers to points separated in *space*. The relationship between them is clearly related to the *speed* at which the wave is moving, and that should come as no surprise since speed is the usual way of relating distances and times. Later in this chapter the relationship between the period, wavelength and speed of a wave will be quantified.

2.3 Transverse and longitudinal waves

The **direction of propagation** of a wave is the direction in which energy is transported. As you have seen, the particles of the medium through which the wave propagates are not themselves permanently transposed from place to place; rather they vibrate periodically about a fixed position. In fact a wave can be thought of as the result of correlated oscillations occurring at every point along the path of the wave. It is the direction of these oscillations which determines whether a particular wave is described as *transverse* or *longitudinal*.

If the individual oscillations are at right-angles to the direction of propagation of the wave itself, the wave is described as a **transverse wave**; an example is shown in Figure 2.4. Here a wave travels along a string that has been jerked up and down at one end. If one small segment of the string is marked by bright colouring, it can be seen that the marked segment simply moves up and down: it doesn't travel along with the wave.

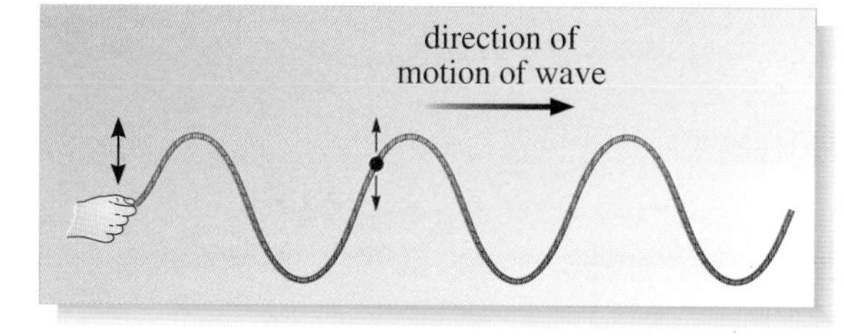

Figure 2.4 A representation of a transverse wave on a string. The coloured segment moves up and down at right-angles to the direction in which the wave travels.

In contrast to this, a sound wave travelling through air is an example of a **longitudinal wave**. In this case the particles move back and forth parallel to the direction of propagation of the wave, although the mean position of the particles remains unchanged. Longitudinal waves can also be set up in a long spring (a 'slinky') by moving one end back and forth along its length. A sound wave travelling through air is similar to this, as shown in Figure 2.5.

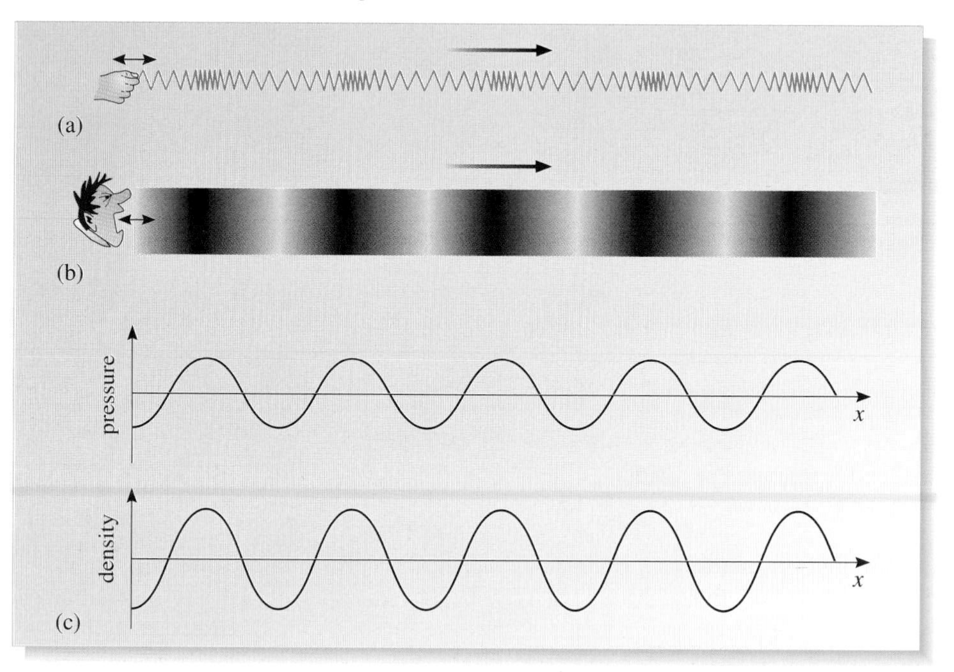

Figure 2.5 (a) Longitudinal waves on a spring are analogous to (b) sound waves, which are longitudinal waves in air. (c) The sound waves can be represented graphically by plotting the variation of the air pressure or density against position.

When the end of the spring vibrates back and forth, it generates periodic compressions and extensions of the spring, and these travel along the spring with a uniform speed. In exactly the same way, when human vocal chords vibrate back and forth, they generate periodic compressions and rarefactions in the air, and these travel away from the diaphragm with a uniform speed. In the regions of compression, the pressure and density of the air are higher than average, whereas in the regions of rarefaction the pressure and density are lower than average. These variations of pressure and density can be represented graphically in the way shown in Figure 2.5c; you should recognize the similarity between these graphs and the representation shown in Figure 2.4, which is a snapshot of a transverse wave on a string.

2.4 Examples of transverse and longitudinal waves

Sound waves are longitudinal waves that can travel through a gas, liquid or solid. As a sound wave passes through air, the pressure fluctuations are only of the order of 1 pascal — very small when compared with normal atmospheric pressure, which is about 10^5 Pa (the pascal is the SI unit of pressure, defined as $1\,\mathrm{Pa} \equiv 1\,\mathrm{N\,m^{-2}}$). The range of frequencies audible to human beings is about 20 Hz to 20 kHz, although the ability to hear at the upper end of the range declines quite markedly with age. Frequencies higher than about 20 kHz are termed *ultrasonic*, and sound above this frequency is referred to as **ultrasound**. Animals such as dogs and bats can hear frequencies in the ultrasound region; these frequencies are also widely used in medicine (see Figure 2.6) and materials testing.

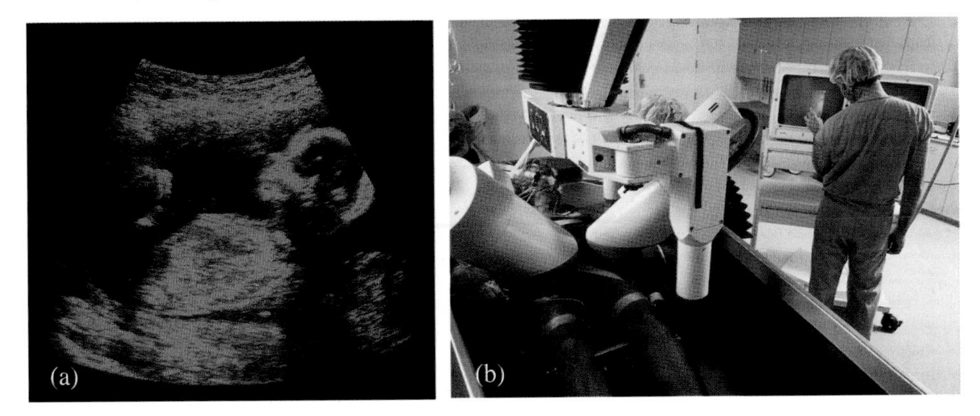

Figure 2.6 Medical uses of ultrasound include (a) monitoring foetal development in pregnancy and (b) shattering kidney stones. The foetal imaging relies on the fact that different tissue types reflect and absorb ultrasound to different extents. The shattering of kidney stones is a type of non-invasive surgery, which relies on the energy carried by the ultrasound wave being transferred to the kidney stone.

Animals such as bats, whales and dolphins use the principle of **echo location** to detect objects using sound. Sound pulses are sent out in particular directions and their reflections are detected. From the time taken for the pulses to return, the positions of intervening objects can be determined. Humans have copied this technique in a number of ways. Sonar (*so*und *n*avigation *a*nd *r*anging) can be used to determine the depth of the sea or the location of shoals of fish, for instance. In seismic surveying, a loud impact on the ground, created by an explosive charge or by dropping a heavy weight, generates sound waves that travel into the Earth and are reflected back from interfaces between the underground layers (Figure 2.7). The sound waves are detected by an array of instruments at the surface, and the data can be processed and displayed by computer. From the arrangement of the rock layers, geophysicists can estimate the chances of finding particular minerals or a reservoir of trapped oil or gas.

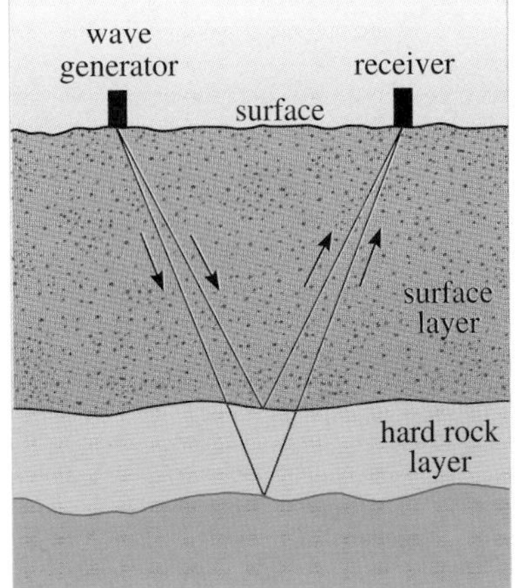

Figure 2.7 Seismic surveying: waves will be reflected back to some extent by any boundary between different rock layers.

Question 2.2 A dolphin emits a pulse of sound towards a shoal of fish. If the speed of sound in water is about $1500\,\mathrm{m\,s^{-1}}$, and the pulse returns after 0.20 s, how far away is the shoal of fish? ■

The waves that travel through the Earth following an earthquake are known as **seismic waves**. Seismic waves can either travel within the Earth or at its surface; those within the solid Earth can either be longitudinal (known as *P-waves* or *primary waves*) or transverse (*S-waves* or *secondary waves*), as shown in Figure 2.8.

However, away from the surface, only longitudinal waves can travel through fluids, since a transverse motion of particles in the body of a fluid is not passed on from particle to particle. The fact that S-waves cannot be detected everywhere following an earthquake gave early support to the idea that beneath the mantle the Earth has a liquid outer core, as shown in Figure 2.9.

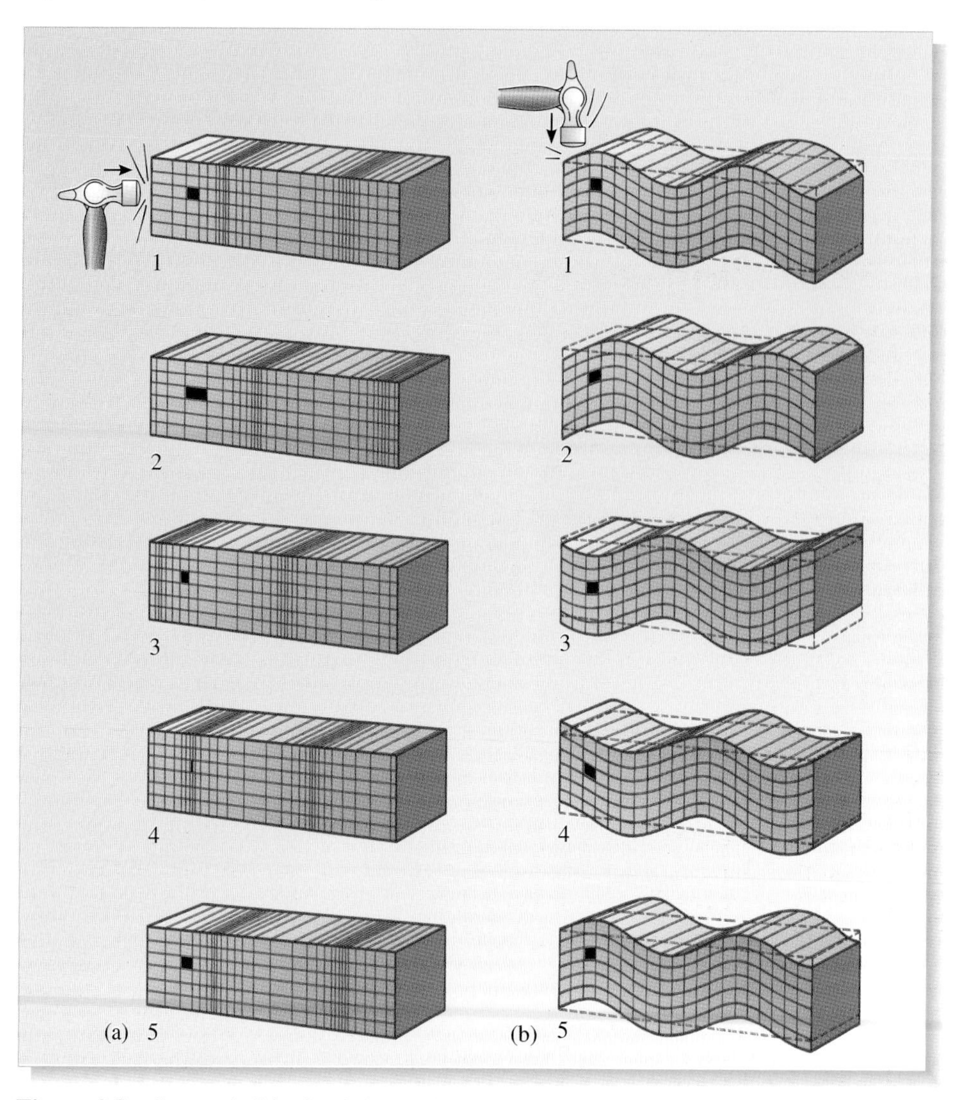

Figure 2.8 Stages 1–5 in the deformation of a block with the passage of (a) a longitudinal P-wave and (b) a transverse S-wave.

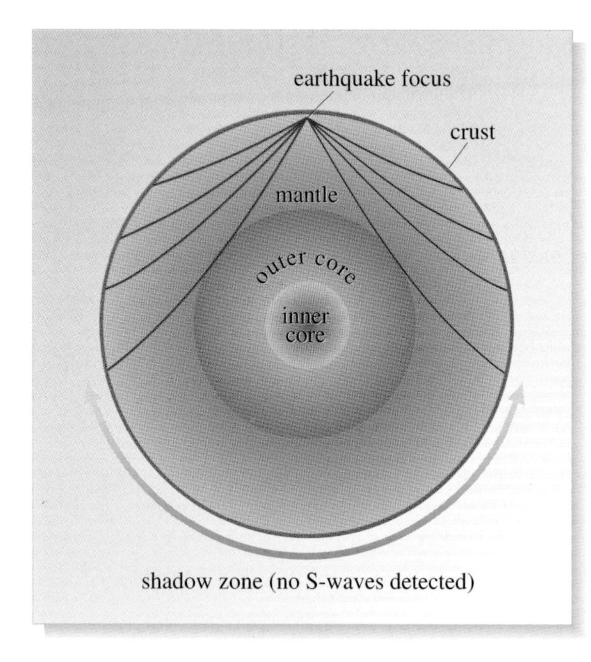

Figure 2.9 S-waves cannot pass through the Earth's liquid outer core, and so a 'shadow zone' exists where no S-waves are detected at the surface. The bending of the path of the seismic waves is the result of *refraction*, which is discussed in Section 6.3.

3 Travelling waves

To progress in this discussion of waves and wave properties, a rather more precise use of terminology must be introduced. The waves considered so far are termed **travelling waves**; these are simply waves that propagate (i.e. travel) from one place to another. (Later, Section 4 will consider another type of wave known as a *standing wave*.)

Having seen some examples of travelling waves, both transverse and longitudinal, a more quantitative description of travelling waves is now presented.

Open University students should leave the text at this point, and use the multimedia package *Waves*. The activity will occupy about one hour. You should return to the text when you have completed this activity.

3.1 The mathematical description of travelling waves

Figure 2.10 shows some graphical depictions of travelling waves, similar to those you saw in Section 2. You can see that the displacement of a particle from its mean position varies sinusoidally with distance along the wave. The **amplitude** of the wave is defined as the maximum magnitude of the displacement from the equilibrium position (*not* the full height from crest to trough). The amplitude is labelled as A on Figure 2.10, where the wavelength λ of the wave is also indicated. The wavelength is the distance, measured parallel to the direction in which the wave is travelling, between successive points that are oscillating in identical ways (i.e. points that have the same instantaneous displacement and move with the same velocity). As noted earlier, wavelength is often thought of, for convenience, as the distance between two successive wave crests.

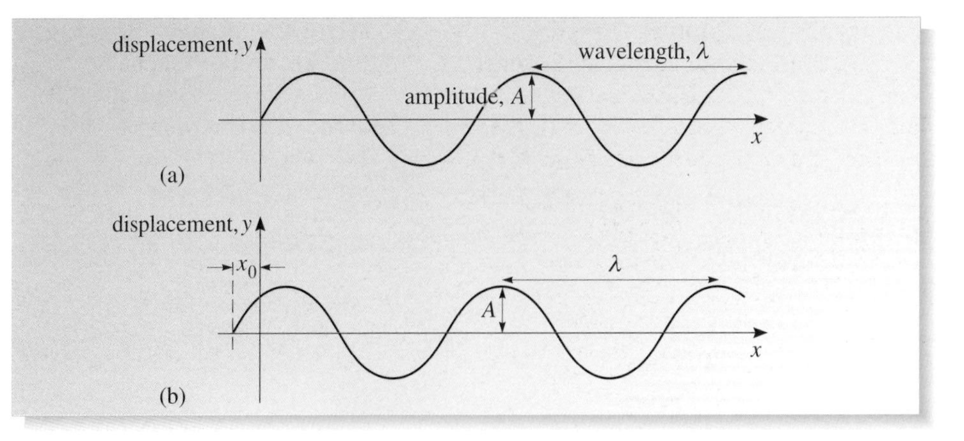

Figure 2.10 The displacement due to a travelling wave at a given instant in time. (a) In this example the displacement at $x = 0$ is taken to be $y = 0$ for simplicity. (b) Here the wave profile has exactly the same shape but it is shifted a distance x_0 along the horizontal axis; that is, there is a *phase difference* between the graph in (b) and the graph in (a).

In Figure 2.10a, the wave profile is pictured at a moment when the displacement y happens to be zero at the point $x = 0$. The wave profile at that particular instant can therefore be represented by the equation

$$y(x) = A \sin kx \tag{2.1a}$$

where the quantity $k = 2\pi/\lambda$ is called the **angular wavenumber**. By contrast, the wave profile shown in Figure 2.10b is shifted to the left by a small amount x_0 with respect to that in Figure 2.10a. The expression representing this shifted profile is

$$y(x) = A \sin(kx + kx_0).$$

It is usual to replace kx_0 with the symbol ϕ (the Greek letter phi). So the general expression for a wave profile is

$$y(x) = A \sin(kx + \phi). \tag{2.1b}$$

Equation 2.1a is therefore a special case of Equation 2.1b with $\phi = 0$.

The difference between the two wave profiles in Figure 2.10 is described by saying that there is a **phase difference** between them. In view of the mathematical descriptions given in Equations 2.1a and 2.1b, the phase difference is described by the quantity ϕ, and the profile in Figure 2.10b (Equation 2.1b) is said to 'lead' that in Figure 2.10a (Equation 2.1a) by ϕ.

● In what SI unit should the angular wavenumber be expressed?

❍ 2π is dimensionless, and in the SI system λ is measured in metres, so the SI unit of k is m^{-1}. ■

Question 2.3 Consider the variation of pressure with position at a given instant of time for a sound wave of wavelength 1.2 m and amplitude 1 Pa. To verify that this wave can be modelled by Equation 2.1 (with y representing pressure and x representing the position along the direction of propagation from some chosen origin), calculate values of y for nine values of x at 0.1 m intervals from $x = 0.1$ m to $x = 0.9$ m. Hence, sketch a graph to show the variation of y with x, assuming that $y = 0$ when $x = 0$. ■

Note that the wave profiles of Figure 2.10 and Equation 2.1 represent a travelling wave at an instant in time; the term 'snapshot' is sometimes used to emphasize this point. However, the displacement y at any given value of x will actually vary with time. To appreciate the effect of this, first think about the variation of the displacement at one particular point, $x = X$ say, as shown in Figure 2.11a.

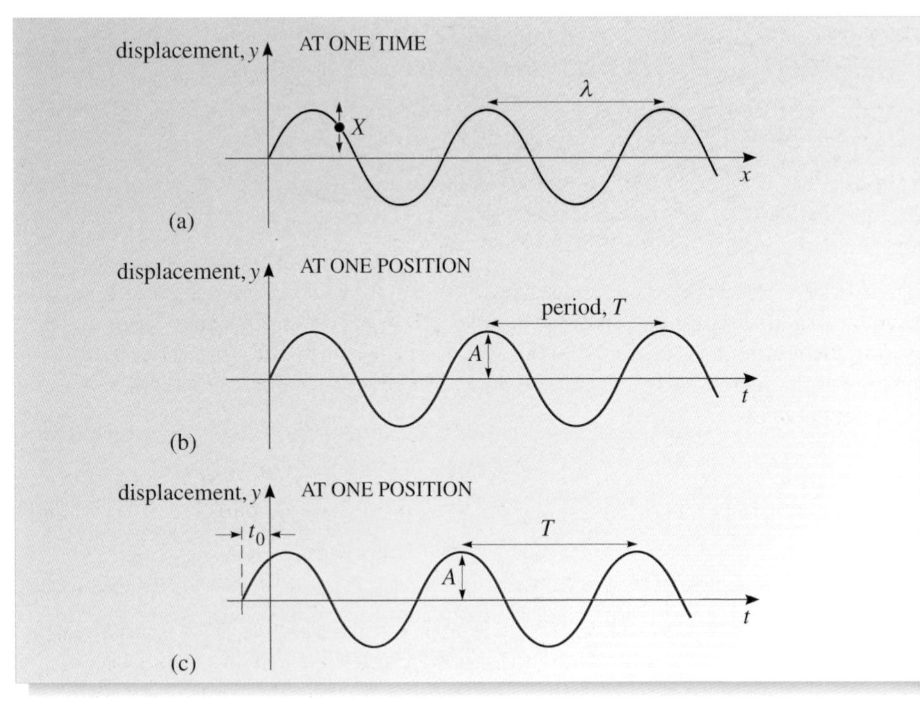

Figure 2.11 (a) A snapshot of a transverse wave at an instant of time; (b) the variation of the displacement of one particle (for example, point X) with time, assuming that the displacement $y = 0$ at a time $t = 0$; (c) another displacement versus time graph, assuming that $y \neq 0$ at $t = 0$; this graph is shifted along the horizontal axis by an amount t_0 with respect to that in (b).

The variation of the displacement y with time t is shown in Figure 2.11b. It can be represented by the following equation

$$y(t) = A \sin \omega t. \qquad (2.2a)$$

Here, ω is the **angular frequency**, given by $\omega = 2\pi/T$, where T is the period (or equivalently $\omega = 2\pi f$, where f is the frequency). A more general graph of y against t is shown in Figure 2.11c. This has been shifted to the left by t_0 compared with Figure 2.11b, and may be described by the equation

$$y(t) = A \sin (\omega t + \omega t_0).$$

Again it is usual to replace ωt_0 with the symbol ϕ giving

$$y(t) = A \sin (\omega t + \phi). \qquad (2.2b)$$

Equations 2.1 and 2.2 both describe particular aspects of the wave — one just in terms of position at a particular time, and the other just in terms of time at a fixed position. These equations can be combined to give a complete mathematical description of the wave specifying the value of y at position x *and* time t. This general equation is

$$y(x, t) = A \sin (kx - \omega t + \phi) \qquad (2.3a)$$

$$\text{where } k = \frac{2\pi}{\lambda} \text{ and } \omega = \frac{2\pi}{T}.$$

By analogy with the terminology used to describe oscillations in *Describing motion*, the quantity $(kx - \omega t + \phi)$ that appears on the right-hand side of Equation 2.3a is called the **phase** of the wave. This quantity determines the stage in its cycle of values that the wave has reached at x and t. The quantity ϕ, the constant part of the phase, is called the **phase constant**, or sometimes the initial phase since it determines the phase of the wave at $x = 0$ and $t = 0$.

● What is the SI unit of the phase constant ϕ?

❍ ϕ must be measured in the same unit as kx and ωt, since it is added to these terms. As kx and ωt are both dimensionless numbers, ϕ must also be a dimensionless number. It therefore has no units, SI or otherwise. Despite this you will often see it treated as an angle, in which case its value may be quoted in radians or degrees. ■

It is interesting to note that, in contrast to a one-dimensional oscillation, a wave travelling in one dimension requires a function of *two* variables (x and t) for its complete description. The need for at least two variables is a fundamental characteristic of waves.

You may wonder why Equation 2.3a includes a minus sign and why the general equation is not given as:

$$y(x, t) = A \sin(kx + \omega t + \phi). \qquad (2.3b)$$

In fact, Equation 2.3a represents a wave moving from left to right as time progresses, whereas Equation 2.3b represents a wave moving from right to left. Also note that, in many cases, it is possible to choose the points at which $x = 0$ and $t = 0$, so that the phase constant ϕ is zero, thus simplifying the description of the wave.

Question 2.4 A sinusoidal wave travelling along a string has wavelength 10 cm, period 0.05 s, and amplitude 2 cm. Sketch graphs (with an indication of scale) showing:

(a) the displacement versus position at a certain instant;

(b) the displacement versus time at a certain position.

State any additional assumptions you had to make in drawing the graphs.

Question 2.5 Consider the wave described by $y(z, t) = A \sin(kz - \omega t)$, where $A = 2\,\text{m}$, $k = 3\,\text{m}^{-1}$ and $\omega = 4\,\text{s}^{-1}$.

(a) What is the amplitude of the wave?

(b) What is the wavelength of the wave?

(c) What is the period of the wave?

(d) What is the phase constant of the wave? ■

3.2 The speed of waves

Question 2.1 demonstrated that the speed of a travelling wave is related to its wavelength and period, and the mathematical relationship between these quantities will now be derived. Consider for simplicity a wave on a string. The only movement of the individual parts of the string is up and down, but the wave as a whole can be considered to move along the string, as illustrated in Figure 2.12.

If you compare part (a) of Figure 2.12 with part (i) you will see that the wave has travelled exactly one wavelength between the two, and that the time interval between these two diagrams is the period T. Thus, the wave has travelled a distance of one wavelength in a time of one period, so its speed υ must be

$$\upsilon = \text{distance/time} = \text{wavelength/period}$$

$$\upsilon = \frac{\lambda}{T}. \qquad (2.4)$$

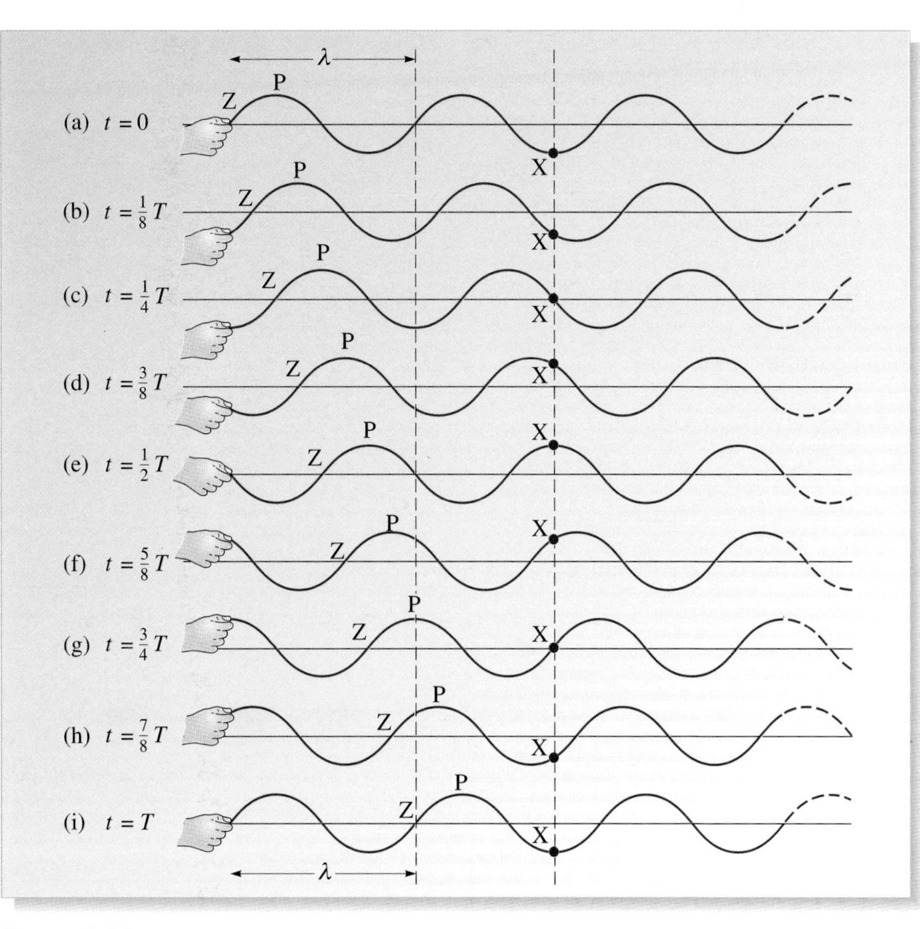

Figure 2.12 Movement of a wave along a string. Each segment of string, for example the red segment marked X, does not move along the string, but the wave as a whole travels along the string. The position of zero displacement Z travels steadily down the string chasing the peak P.

Since $f = 1/T$, this may be rewritten as

$$v = f\lambda. \tag{2.5}$$

Equation 2.5 is the key equation that describes the relationship between wave speed, wavelength and frequency for any travelling wave. Note that it is a direct consequence of the definitions of λ and f.

- Write down an expression for wave speed in terms of angular wavenumber k and angular frequency ω.

○ Since $k = 2\pi/\lambda$ and $\omega = 2\pi f$, substituting into Equation 2.5 gives

$$v = \frac{\omega}{2\pi} \times \frac{2\pi}{k} = \frac{\omega}{k}. \quad \blacksquare$$

The speed of sound waves, ultrasound and seismic waves depends to a great extent on the medium through which the waves are passing (see Table 2.1). As the sound wave propagates, the particles in the medium are forced into longitudinal vibrations. The speed at which this is accomplished depends on how easy it is to compress and expand the medium and this in turn is described by a property of each medium, known as the *axial modulus*. The speed of sound also depends on the density of the

Table 2.1 The speed of sound in various materials.

Material	Speed/m s^{-1}
carbon dioxide	267
air	330
helium	972
hydrogen	1315
water	1480
glycerol	1860
steel	5121

medium through which it is travelling, and it can be shown that in general the speed of longitudinal waves is given by:

$$v = \sqrt{\frac{\text{axial modulus}}{\text{density}}}\,. \tag{2.6}$$

A metal such as steel has a higher density than air but it also has a very much higher axial modulus, so sound travels much faster through steel than it does through air. As an illustration of this, you may have noticed that you can often hear a noise from the metal rails as a train approaches long before you hear the sound of the train propagated through the air.

Question 2.6 (a) A typical granite just below the surface of the Earth has a density of $2.7 \times 10^3\,\text{kg m}^{-3}$, and an axial modulus of $8.5 \times 10^{10}\,\text{N m}^{-2}$. Calculate the speed of seismic P-waves in the granite.

(b) For a fixed density, what happens to the speed of P-waves if the axial modulus increases by factors of 2, 3 and 4?

(c) For a fixed axial modulus, what happens to the speed of P-waves if the density increases by factors of 2, 3 and 4? ■

Even within one medium, the speed of sound can vary considerably if other factors vary. The speed of sound in a gas or liquid depends on its temperature. At the time of writing (1999) the Acoustic Thermometry of Ocean Climate (ATOC) experiment is using this variation as a way of measuring the temperature of an ocean. Measurements are being recorded of the time taken for sound to travel from a loudspeaker off the coast of California to a series of microphones across the Pacific Ocean. It may be possible to use the results of experiments like these as a sensitive indicator of global warming.

The speed of a transverse wave travelling along a string is given by the expression

$$v = \sqrt{\frac{F_\text{T}}{\mu}} \tag{2.7}$$

where F_T is the tension in the string, and μ is its mass per unit length.

Equation 2.7 shows that the wave speed will increase when the tension is increased. This is reasonable, as increased tension means that each segment of the string pulls more strongly on the following segment, and therefore makes its displacement change more rapidly. On the other hand, the wave speed decreases when the mass per unit length of the string is increased. This is because increasing the mass per unit length of the string also increases the inertia of the string, which makes it respond more slowly to the forces that are causing its displacement to change.

- Show that the unit of $\sqrt{(F_T / \mu)}$ is the same as the unit of speed.

- ❍ The unit of F_T is the newton and the unit of μ is that of mass/length — that is, kg m^{-1}. But $1\,\text{N} = 1\,\text{kg m s}^{-2}$ (remember from Newton's second law that $F = ma$). So the unit of F_T/μ is $\text{kg m s}^{-2}/\text{kg m}^{-1} = \text{m}^2\,\text{s}^{-2}$. The unit of $\sqrt{(F_T / \mu)}$ is therefore $\sqrt{\text{m}^2\,\text{s}^{-2}} = \text{m s}^{-1}$, which is, of course, the unit of speed. ■

Question 2.7 When one end of a very long string is shaken from side to side with a period of 0.5 s, waves are generated with a speed of $2.0\,\mathrm{m\,s^{-1}}$ along the string. If the tension in the string is doubled, and the period of vibration remains the same, what will be the speed, frequency and wavelength of the waves that are generated? ■

3.3 Plane waves and spherical waves

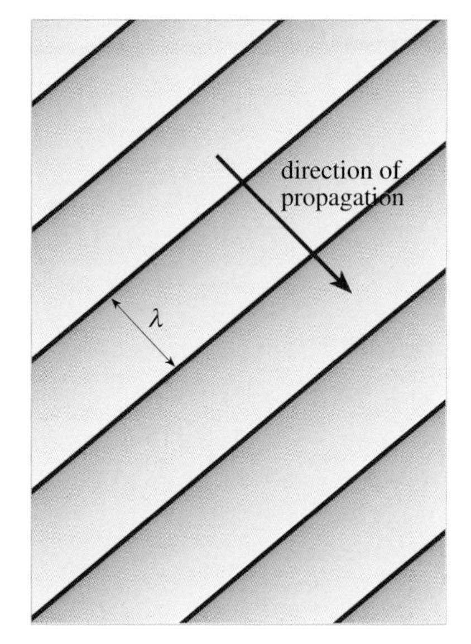

Figure 2.13 The wavefronts of a two-dimensional wave represent lines where the phase of the wave has the same value. For a two-dimensional plane wave, the wavefronts are parallel lines, separated by a distance equal to the wavelength of the wave.

The water waves discussed earlier were an example of what is essentially a two-dimensional wave, since their propagation is confined to the surface of the ocean. One way of representing such a wave is to draw lines along the crests of the wave, as shown in Figure 2.13. The parallel lines in this figure represent the wavefronts of the wave; along each wavefront the phase of the wave has the same value. Clearly, the wavefronts could have been drawn along the troughs of the wave or along any other series of locations with the same phase, but the crests are a convenient location to use. Wherever they are drawn, the wavefronts are separated by a distance equal to the wavelength of the wave. Notice also that the wavefront at any point is at right-angles to the direction of propagation of the wave; this is a property of wavefronts in general.

The simplest kind of two-dimensional or three-dimensional wave is a **plane wave**. The defining characteristic of such a wave is that all the points on a plane that is perpendicular to the direction of propagation will have the same value of phase. Consequently, for a plane wave, the wavefronts are *parallel lines* (in two dimensions) and parallel planes (in three dimensions), as seen in Figure 2.13.

● How would the wavefronts of the waves resulting from a stone dropped into a pond differ from those described above?

❍ The wavefronts would take the form of concentric circles, as shown in Figure 2.14. The wavefront is still perpendicular to the direction of propagation at any point. ■

Figure 2.14 For a two-dimensional circular wave, the wavefronts are concentric circles.

A simple generalization of the (two-dimensional) circular waves from the previous question is to look at what happens in three dimensions. When sound waves are generated by a loudspeaker, for example, they are not constrained to move in one direction like a wave on a string, nor in two dimensions like those resulting from a stone dropped into a pond. Consequently, sound waves do not travel in a narrow band as was implied by Figure 2.5; instead they spread out in space as indicated in Figure 2.15. Waves that spread out equally in all directions are called **spherical waves**. The wavefronts corresponding to these spherical waves are concentric spherical surfaces, and once again are perpendicular to the direction of propagation at any point.

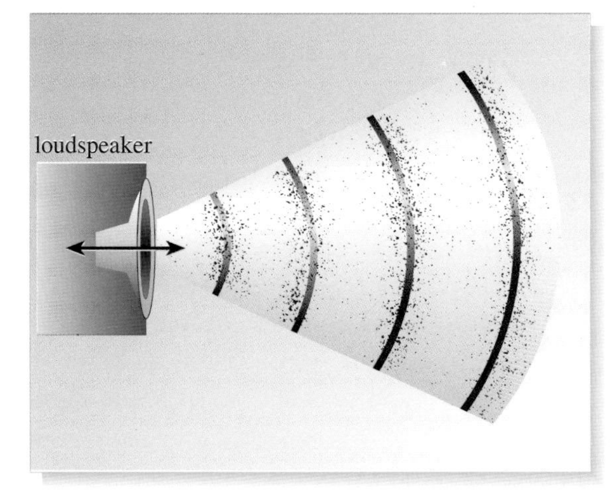

Figure 2.15 Sound waves spreading out in three dimensions from a loudspeaker are an example of spherical waves.

- At a large distance from the source of the sound, what would be the shape of the wavefronts?

- Sufficiently far from the source of the sound, the curvature of the wavefronts would be negligible in any localized region; consequently the wavefronts would approximate to a set of parallel planes. The original spherical wave would effectively be converted into a (three-dimensional) plane wave. ■

To summarize:

> **Wavefronts** are lines (in two dimensions) or surfaces (in three dimensions) that connect points in a wave which have the same phase. They are perpendicular to the direction of propagation of the wave at every point.

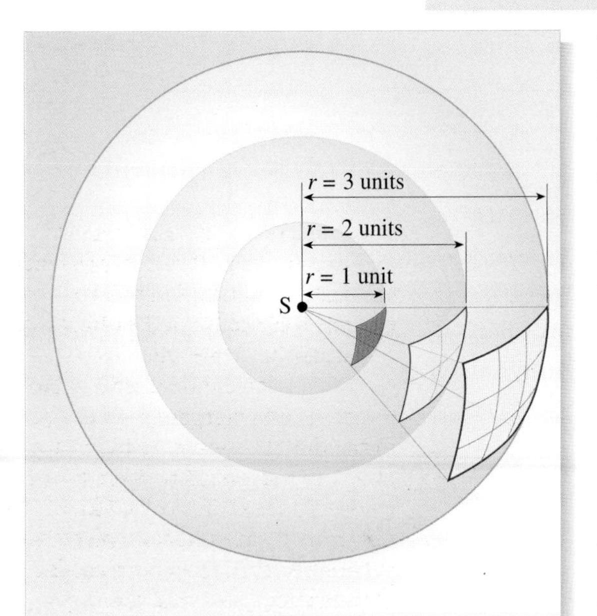

Figure 2.16 The energy emitted per second by the source S passes through the surface of a sphere of area $4\pi r^2$. The energy that passes through the shaded 'square' at $r = 1$ unit then passes through *four* 'squares' of the same area at $r = 2$ units, and through *nine* squares of the same area at $r = 3$ units. The energy per second passing through each of the three shaded squares will therefore be in the ratio $1 : \frac{1}{4} : \frac{1}{9}$. Hence the energy per second reaching a detector of fixed area will decrease as $1/r^2$, where r is the distance of the detector from the source.

When dealing with spherical waves, an important consequence of the fact that waves spread out is that the amplitude of the waves decreases as the distance from the source increases. To understand why this happens, the energy carried by the waves must be considered.

Since spherical waves travel and spread out, it is not very useful to talk about the total energy of the waves. So, to describe the strength of a wave, the amount of energy carried by the wave in unit time across unit area perpendicular to the direction of propagation is used instead. This quantity is known as the **intensity** I of the wave, and its SI unit is the joule per square metre per second, $\text{J m}^{-2}\,\text{s}^{-1}$, or the watt per square metre, W m^{-2}.

If the source radiates its energy equally in all directions (such a source is described as being *isotropic*), then the radiated energy must pass evenly through the surface of an imaginary sphere whose centre is located at the source (see Figure 2.16). Furthermore, as the energy propagates away from the source, it continues to fill the expanding surface area of the ever-increasing sphere. The surface area of a sphere is given by the formula, area = $4\pi r^2$, where r is the radius of the sphere. The energy flowing through the sphere per second is $I \times 4\pi r^2$, and this must be true at all distances if energy is not being absorbed. So $I \times 4\pi r^2$ = constant, and therefore

$$I = \frac{\text{constant}}{4\pi r^2} \propto \frac{1}{r^2}.$$

Thus, for a spherical wave, intensity I is related to the distance from the source r by an *inverse square law*:

$$I \propto \frac{1}{r^2}. \tag{2.8a}$$

For a plane wave, there is no decrease in intensity with distance from the source, and so:

$$I = \text{constant}. \tag{2.8b}$$

In some situations this allows us to find a relationship for the variation of amplitude with distance from the source. Recall from *Predicting motion* that the total energy in a vibrating system is given by

$$E = \tfrac{1}{2}kA^2. \qquad (PM, \text{Eqn } 2.44)$$

In other words the total energy (comprising the sum of the kinetic and potential energies at a particular point) is proportional to (amplitude)^2. In a similar way, the intensity of a wave at a certain position is proportional to (amplitude)^2 at that position

$$I \propto A^2. \qquad (2.9)$$

Combining Equations 2.8a and 2.9 gives

$$A^2 \propto \frac{1}{r^2}$$

and hence an amplitude

$$A \propto \frac{1}{r} \qquad (2.10a)$$

for a spherical wave.

Thus, the amplitude of a spherical wave is inversely proportional to the distance from the source. This is only true if the wave is not obstructed, or absorbed, and if the distance from the isotropic source is much greater than the size of the source.

Similarly, for a plane wave, combining Equations 2.8b and 2.9 yields:

$$A = \text{constant}. \qquad (2.10b)$$

Question 2.8 A loud shout has an intensity of $8 \times 10^{-5}\,\text{W m}^{-2}$ at a distance of 1 m from the source. (a) Given that the threshold of human hearing is about $10^{-12}\,\text{W m}^{-2}$ at voice frequencies, and that sound spreads out evenly in all directions, how far away could such a shout be heard in open country? (b) What is the ratio of the amplitude of the sound wave at this distance to the amplitude 1 m from the source? ■

3.4 The Doppler effect

Another property of travelling waves is that the observed frequency (and wavelength) of a wave depends on the relative motion between the source of the wave and the observer. The first person to explain this effect was Christian Doppler (Figure 2.17), and the phenomenon is now known by his name — the **Doppler effect**.

Most people have had direct experience of the Doppler effect. Perhaps its commonest manifestation is in the sound of a passing vehicle, such as a train or a motorcycle. The most obvious feature of the sound of a passing vehicle is that it gets louder as the vehicle approaches, and fainter as it gets further away, which has nothing to do with the Doppler effect. However, if you ignore the changing loudness of the sound and listen instead to its pitch (i.e. its frequency), then you will notice the Doppler effect. The pitch is higher when the vehicle is approaching than when it is receding, and the effect is even more noticeable if the vehicle is sounding a siren. It is this dependence of the observed pitch on the motion of the vehicle that constitutes the Doppler effect. Similar effects arise when an observer moves towards or away from a fixed source of sound.

Figure 2.17 Christian Doppler (1805–53) was an Austrian physicist and mathematician. In 1842 he explained the cause of the effect that now bears his name whereby a relative movement between a source of sound and an observer gives rise to a shift in the observed frequency of that sound. His ideas were tested a few years later in Holland in a bizarre experiment that involved a locomotive pulling an open carriage of trumpeters past an observing team consisting of musicians with a good ear for pitch. Note that it was only when train technology had developed sufficiently that high enough speeds were attained for the Doppler effect to be observed.

A moving source of sound

The cause of the Doppler effect in the case where the source of sound is moving is illustrated in Figure 2.18. An observer is stationary and the source of sound is moving *towards* the observer with a speed V. The source emits sound waves with a frequency f_{em}. The concentric circles in Figure 2.18 represent the surfaces of maximum pressure in four successive compressions of the sound wave at a particular time (i.e. four wavefronts). Notice that, since the source is travelling in the direction shown, by the time that compression number 2 is emitted, the source has moved to catch up slightly with compression number 1. A similar effect exists for each of the four compressions shown. The net result is that the observer receives compressions that appear to be closer together than the emitted wavelength λ_{em}. This is equivalent to observing an increase in the frequency — that is, a higher-pitched sound.

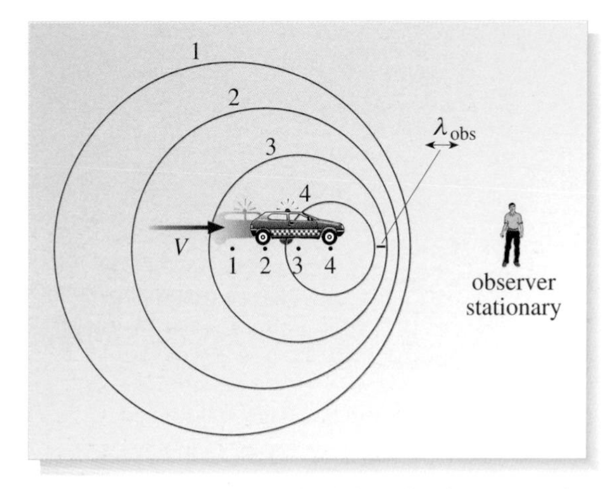

Figure 2.18 The Doppler effect: moving source of sound and stationary observer.

To see exactly how much higher the frequency is, consider the following argument. During one period T_{em} of the emitted sound wave, the source moves a distance $d = V \times T_{em} = V/f_{em}$ towards the observer. So the apparent wavelength measured by the observer is $\lambda_{obs} = \lambda_{em} - d$. Hence

$$\lambda_{obs} = \lambda_{em} - \frac{V}{f_{em}}. \qquad (2.11)$$

Equation 2.11 can be expressed in terms of frequencies, since $\lambda_{obs} = v/f_{obs}$ for the observed wavelength and frequency, and $\lambda_{em} = v/f_{em}$ for the emitted wavelength and frequency, where v is the speed of sound (the same in both cases). Therefore

$$\frac{v}{f_{obs}} = \frac{v - V}{f_{em}}$$

and rearranging this equation, the observed frequency of the sound is

$$f_{obs} = f_{em}\left(\frac{v}{v - V}\right). \qquad (2.12)$$

The change in frequency, the so-called Doppler shift, is therefore $\Delta f = f_{obs} - f_{em}$, and so

$$\Delta f = f_{em}\left(\frac{v}{v - V} - 1\right)$$

$$= f_{em}\left(\frac{V}{v - V}\right). \qquad (2.13)$$

As long as the speed of the source (V) is less than the speed of sound (v), the part of Equation 2.13 in brackets is positive. The shift in frequency is therefore positive and so $f_{obs} > f_{em}$, as stated earlier.

- What do you suppose happens to the observed frequency if the source of sound is moving *away* from the observer?

○ The relative motion will now *increase* the wavelength according to the equation

$$\lambda_{obs} = \lambda_{em} + \frac{V}{f_{em}}$$

This leads to the result:

$$f_{obs} = f_{em}\left(\frac{v}{v+V}\right).$$

The Doppler shift is therefore

$$\Delta f = f_{em}\left(\frac{-V}{v+V}\right)$$

and so in this case $f_{obs} < f_{em}$. The same result is obtained more simply by changing the sign of V in Equation 2.13. ■

A moving observer

Figure 2.19 shows the situation where a stationary source is emitting sound waves of frequency f_{em} (and wavelength λ_{em}) and an observer is approaching the source at a speed V. Four successive compressions are once again indicated, centred on the source. In this case, the observer is receiving, or crossing, compressions more often than if the observer were stationary, and so the observed wavelength is shorter. Equivalently, the observed frequency is higher.

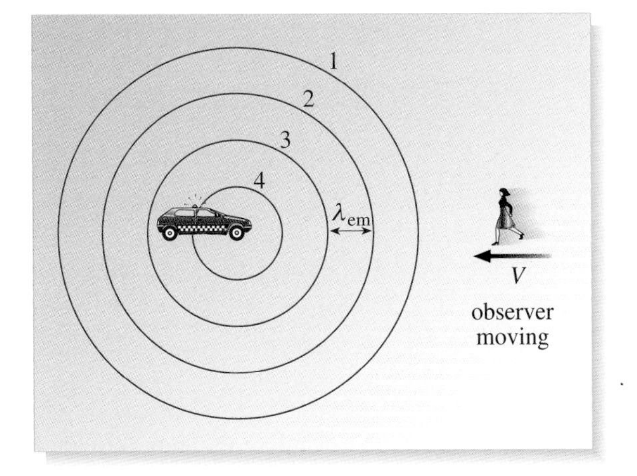

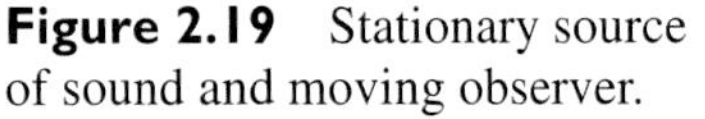

Figure 2.19 Stationary source of sound and moving observer.

The procedure to calculate this frequency shift is slightly different from the case of a moving source encountered above. As the observer moves towards the source of sound, the relative speed between the observer and the compressions is simply $(v + V)$. Since the compressions are a distance λ_{em} apart, the time taken between crossing one and then crossing the next is

$$T_{obs} = \frac{\lambda_{em}}{v+V}. \qquad (2.14)$$

As before, what is required is a relationship involving frequencies, so there is still some work to do. The observed period can be rewritten in terms of the observed frequency using $T_{obs} = 1/f_{obs}$. The wavelength emitted by the source can also be rewritten as $\lambda_{em} = v/f_{em}$. Making these substitutions, Equation 2.14 becomes

$$\frac{1}{f_{obs}} = \frac{v/f_{em}}{v+V}$$

This may in turn be rearranged to give the observed frequency of sound as

$$f_{obs} = f_{em}\left(\frac{v+V}{v}\right) \tag{2.15}$$

and the Doppler shift, $\Delta f = f_{obs} - f_{em}$, is therefore

$$\Delta f = f_{em}\left(\frac{v+V}{v} - 1\right)$$

or

$$\Delta f = f_{em}\left(\frac{V}{v}\right). \tag{2.16}$$

The shift in frequency is therefore positive and so $f_{obs} > f_{em}$, as stated earlier. Motion of the source towards the observer, or motion of the observer towards the source, both cause an increase in the observed frequency.

● What do you suppose happens to the observed frequency if the observer is moving *away* from the source of sound?

❍ Following a similar derivation to that above, but replacing the relative speed $v + V$ by $v - V$, we obtain

$$f_{obs} = f_{em}\left(\frac{v-V}{v}\right)$$

The Doppler shift is therefore

$$\Delta f = f_{em}\left(-\frac{V}{v}\right)$$

and so $f_{obs} < f_{em}$. ■

Motion of the source away from the observer, or motion of the observer away from the source, both cause a decrease in the observed frequency. These results *only* apply to waves supported by a medium. Neither Equation 2.13 nor Equation 2.16 applies to electromagnetic waves, such as light waves. The formulae appropriate in the case of electromagnetic waves are presented in Chapter 4.

Doppler measurement of blood flow

As mentioned earlier, ultrasound offers a safe way of investigating the internal structure and organs of animals, including humans. Ultrasound is transmitted into the body and is reflected at boundaries between different types of tissue. The position of organs can then be determined by a simple extension of the echo location method described earlier. The familiar two-dimensional displays of organs and foetuses (for example, Figure 2.6a) are created by transmitting ultrasound into the body in a range of different directions, and combining the information gathered from each direction.

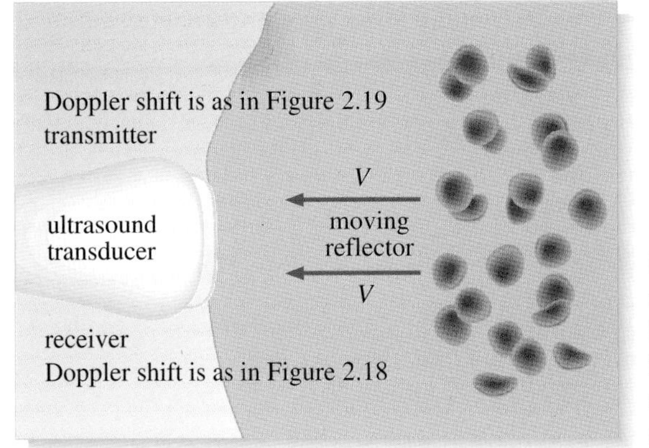

Figure 2.20 The reflection of ultrasound from particles in the blood gives a Doppler shift, which is a combination of both a moving source of sound and a moving observer.

Ultrasound is now also routinely used to measure continuous movement of fluids in the body, such as the flow of blood, in a technique that relies on the Doppler effect. If a fluid such as blood is flowing with speed V directly towards a stationary ultrasound beam, then particles in the blood act as moving reflectors as shown in Figure 2.20. The transducer acts as both source and receiver of the ultrasound in this case. The total Doppler shift is therefore made up of two components:

1 The shift as the result of the reflecting surface (as *observer*) moving towards the ultrasound transducer (the source) with speed V. From Equation 2.16

$$\Delta f_1 = f_{\text{em}}\left(\frac{V}{v}\right).$$

2 The shift as a result of the reflecting surface acting as a *source* of ultrasound waves, and moving towards the receiver with speed V. In this case $v \gg V$, so Equation 2.13 can be simplified to give

$$\Delta f_2 = f_{\text{em}}\left(\frac{V}{v}\right).$$

Combining these two Doppler shifts gives an overall Doppler shift of

$$\Delta f = \Delta f_1 + \Delta f_2 = 2f_{\text{em}}\left(\frac{V}{v}\right). \tag{2.17}$$

In many cases, however, the blood is travelling at an angle to the direction of the ultrasound beam. If the angle between the ultrasound beam and the direction of blood flow is θ, then the Doppler shift is simply

$$\Delta f = 2f_{\text{em}}\left(\frac{V}{v}\right)\cos\theta. \tag{2.18}$$

Question 2.9 In a blood flow investigation, a source of 3.0 MHz ultrasound is directed at 45° to a blood vessel. A Doppler shift of 280 Hz is detected. Assuming that the ultrasound beam travels at a speed of $1500\,\text{m}\,\text{s}^{-1}$ in the intervening tissue, what is the speed of the blood flow? ■

In practice, the blood in a blood vessel does not all flow at the same rate, so the analysis of Doppler shift measurements is quite complex. However, any constrictions in a blood vessel can be rapidly identified by the change that they produce in the normal signal.

4 Standing waves

So far, travelling waves of various kinds have been considered, but they have always been remote from other waves. What happens when two waves meet? In this section, one outcome will be considered, namely the setting up of *standing waves*. Interference between waves will be considered more generally in Section 6.

4.1 Reflection of travelling waves at a boundary

Firstly, think about what happens when a wave is incident at a boundary. Consider a wave propagating along a string and eventually reaching a position at which the string is attached to something else. The resulting behaviour of the wave depends on what happens physically at the point of attachment. This is characterized mathematically by the *boundary conditions*, which are the requirements imposed on the wave displacement y (and possibly also its derivatives) at the boundary.

For simplicity, a single pulse will be considered first, rather than a continuous wave. Imagine that the string is attached to a fixed object, such as a wall, as shown in Figure 2.21. In this case the boundary condition implies that the transverse displacement at the wall must be zero. When the pulse travelling in the $+x$-direction reaches the wall, the moving string will exert an upwards force on the wall. However, the wall is rigid and immovable, so it remains motionless. Consequently, no mechanical energy will be transferred to the wall. In accordance with Newton's third law, the wall will exert an equal downward reaction force on the string and this will cause a reflected pulse to travel away from the wall in the $-x$-direction. This reflected pulse will carry away all the incident energy, so it must have the same amplitude as the incident pulse. However, its displacement will be in the opposite direction to that of the original pulse.

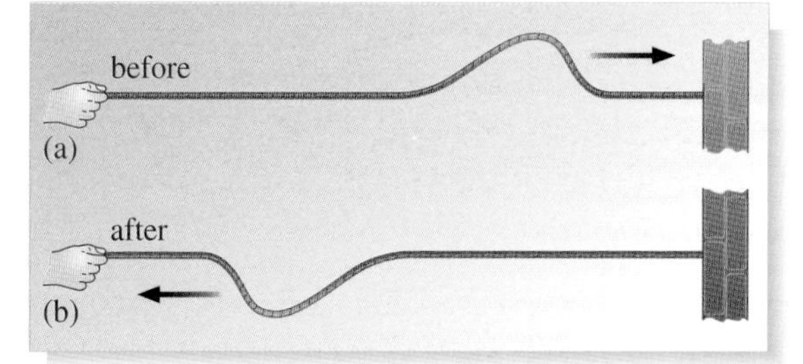

Figure 2.21 (a) A wave pulse travels in the $+x$-direction towards a wall; (b) after reflection, a wave pulse travels back in the $-x$-direction, with the same shape and amplitude, but opposite displacement.

If, in place of a rigid wall, the string were attached to a massless, frictionless ring on a pole, it would have a free end at the boundary rather than a fixed end. Under these conditions there will still be a reflected wave but its nature will be different from that of a reflection from a wall. This is indicated in Figure 2.22, where you can see that the transverse displacement of the reflected pulse does not suffer the sign reversal that occurred in the case of a rigid wall. This is because the ring tends to carry on upwards, since it has nothing to the right to pull it back.

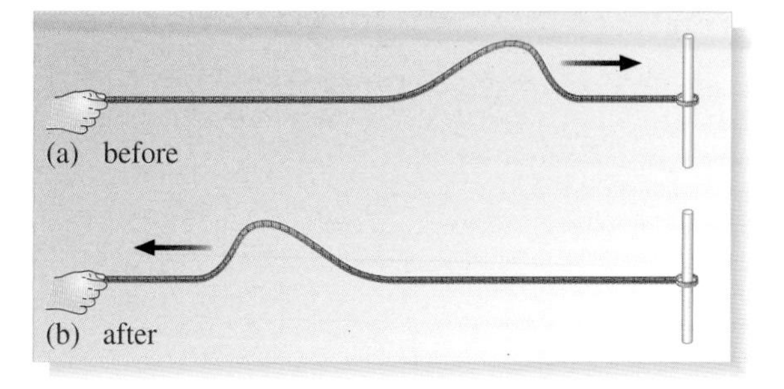

Figure 2.22 (a) A wave pulse travelling towards a freely moving boundary; (b) the reflected pulse, with displacement in the same direction as the incident pulse.

4.2 The principle of superposition

What would happen if the reflected pulse in Figure 2.21 or Figure 2.22 met a second pulse travelling in the $+x$-direction? The outcome is predicted by the principle of superposition, which may be stated as follows.

> The **principle of superposition** states that if two or more waves meet at a point in space, then at each instant of time the net disturbance at that point is given by the sum of the disturbances created by each of the waves individually.

For the specific case of a wave on a string, the 'disturbance' is simply the displacement of the string. The rather general wording of the definition above is to allow for other situations that you will meet later in this chapter.

So, as two pulses meet, the resultant displacement at each instant can be predicted simply by adding the displacements of the individual pulses, as illustrated in Figure 2.23. Note that after the pulses have passed each other, they both continue on their way unchanged.

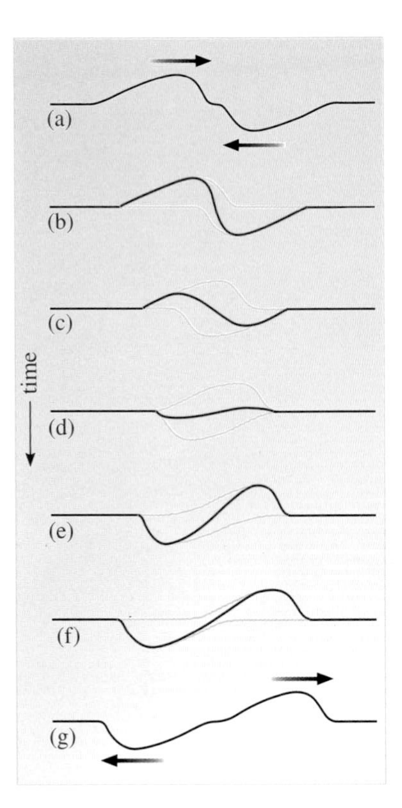

Figure 2.23 Using the principle of superposition to predict the resultant displacement as two wave pulses, travelling in opposite directions, meet and continue on their way. The sketches illustrate several instants of time. The sum of the displacements from the individual pulses is shown in red.

Question 2.10 (a) What would be the outcome if a symmetrical pulse, as shown in Figure 2.24 was reflected at a solid boundary, and then met another similar pulse travelling towards the boundary?

(b) Sketch the transverse wave that would result if a symmetrical pulse, as shown in Figure 2.24, was reflected at a *free* boundary and then met another similar pulse travelling towards the boundary. ■

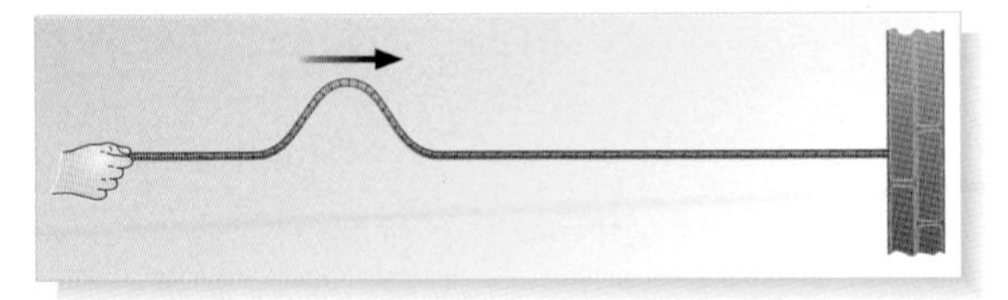

Figure 2.24 A symmetrical transverse pulse.

4.3 The superposition of two travelling waves

The next thing to consider is the combined effect of two periodic sinusoidal waves, travelling in opposite directions along a string, as shown in Figure 2.25 (overleaf). The wave travelling from left to right is shown in red and the wave travelling from right to left is shown in blue, but note that the waves have identical amplitude, frequency and wavelength. One could be the result of the other being reflected. The principle of superposition applies to this situation in exactly the same way as before.

Question 2.11 Use the principle of superposition to sketch the combined displacement due to the two waves at the instants shown in Figure 2.25b–d. Stages (b) to (d) represent the passage of time as the red wave passes the blue one. ■

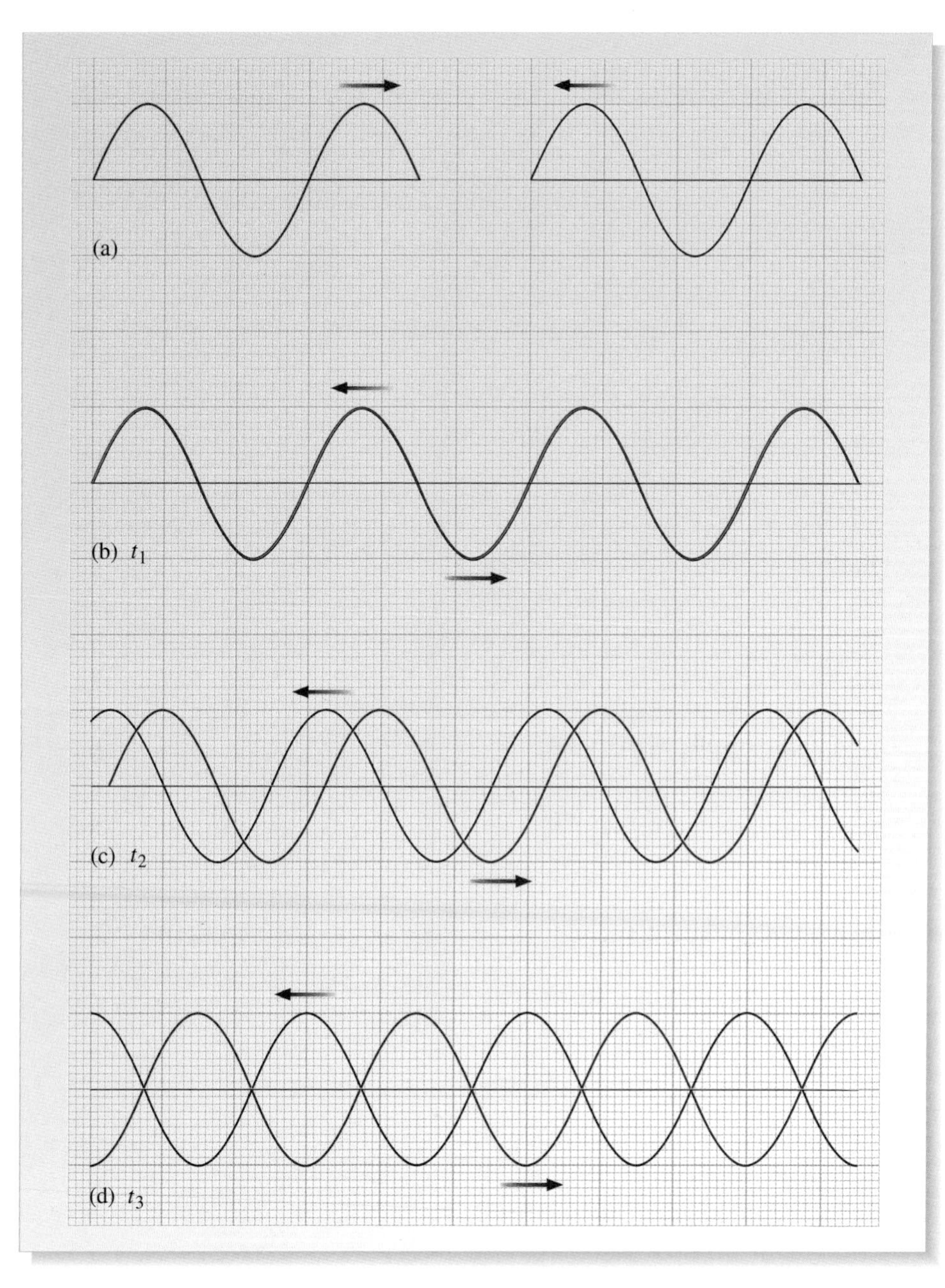

Figure 2.25 Identical travelling waves, travelling in opposite directions along a string. (a) The wave shown in red is travelling from left to right, and the wave shown in blue is travelling from right to left. (b)–(d) show the position of the waves at later times.

● How does the resultant displacement (from your answers to Question 2.11) vary as time progresses?

❍ The resultant maximum displacement is initially twice that of the red and blue waves. Then it falls, and (when the two travelling waves have a phase difference of π) it is zero everywhere. ■

If you had continued to plot the resultant displacement as the two travelling waves moved on, you would have obtained the repeating cycle shown in Figure 2.26.

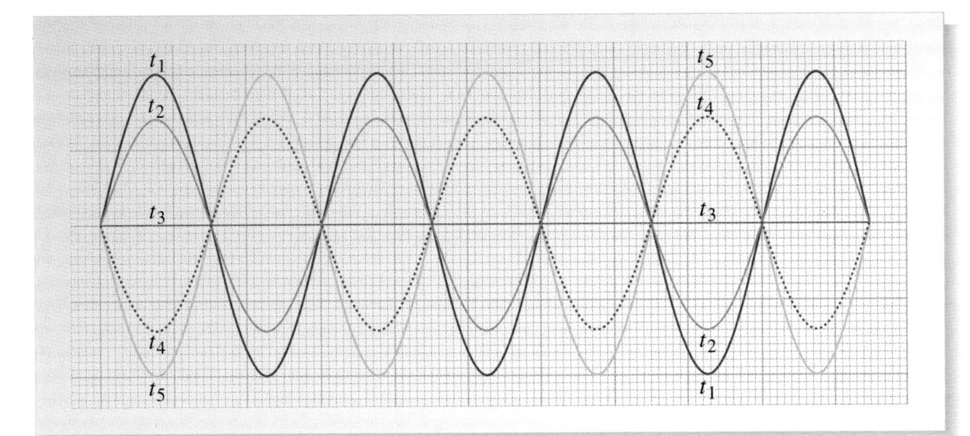

Figure 2.26 The standing wave produced by the superposition of the two travelling waves shown in Figure 2.25. t_1 comes before t_2, which comes before t_3. The cycle would continue to t_4 and t_5, before returning via t_4, t_3 and t_2 to t_1, and then starting again.

This is called a **standing wave**: it has the property that there are fixed positions, every half wavelength, where the displacement is always zero. These positions are called **nodes** (remember: *no d*isplacement). There are then points of maximum vibration amplitude, midway between these nodes, called **antinodes**.

Box 2.1 The mathematical description of a standing wave

The principle of superposition has been used to obtain a standing wave by sketching the resultant displacements when two travelling waves pass each other. There is a more mathematical approach, which you are not required to remember, but it gives the same result:

The displacement of a wave travelling from left to right is given by $y_- = A\sin(kx - \omega t)$. Similarly the displacement of the wave travelling from right to left is given by $y_+ = A\sin(kx + \omega t)$. (In both cases, assume that $y = 0$ at $x = 0$ and $t = 0$, so that the phase constant $\phi = 0$.) From the principle of superposition the resultant is

$$y = y_- + y_+ = A\sin(kx - \omega t) + A\sin(kx + \omega t).$$

There is a mathematical relationship which states that, for any variables X and Y,

$$\sin X + \sin Y = 2\sin\left(\frac{X+Y}{2}\right)\cos\left(\frac{X-Y}{2}\right)$$

In this case, with $X = kx - \omega t$ and $Y = kx + \omega t$, and recalling that $\cos -x = \cos x$, the combination of the two sine waves simplifies to

$$y(x,t) = 2A\cos\omega t\sin kx.$$

This is the mathematical description of a standing wave. Note that in this case the function $y(x,t)$ is a *product* of a cosine function involving time and a sine function involving position. One of the outcomes of this is that $y = 0$ for $kx = 0$, π, 2π, 3π, etc., whatever the value of t. In other words the nodes of the standing wave will be at $x = 0$, π/k, $2\pi/k$, $3\pi/k$, etc. — that is, at $x = 0$, $\lambda/2$, λ, $3\lambda/2$, etc. — just as you might expect.

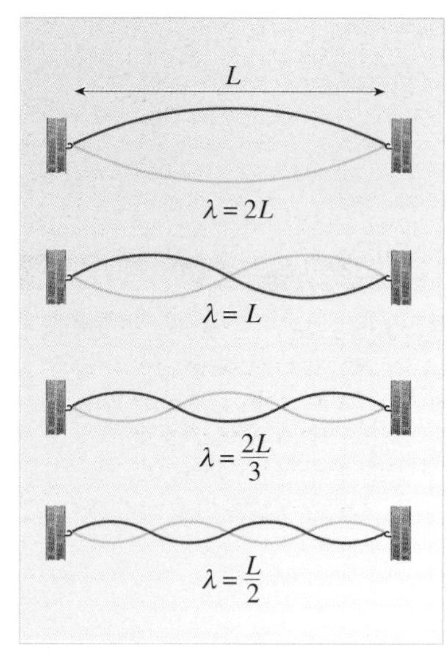

Figure 2.27 The first four modes of vibration in a string fixed at both ends.

4.4 Bounded waves

You have seen how a standing wave can be created on a string but what has not been properly investigated is what happens at the *ends* of the string. You probably noticed that the box about the mathematical description of a standing wave assumed that $y = 0$ at $x = 0$ and $t = 0$. This implies that at least one end of the string on which the standing wave exists was fixed. In fact, it is frequently the case that the string is fixed at *both* ends, in other words there are **boundary conditions**. It turns out that *any* boundary conditions that constrain a standing wave result in a restriction on the allowed frequencies or wavelengths that can be sustained. This result has profound implications, which you will meet when you use quantum mechanics to describe the behaviour of electrons in atoms, but for now let's return to the simple example of a string fixed at each end.

In this case the boundary conditions mean that the displacement y of the standing wave must be zero at each end of the string; that is, there must be a node at each end of the string. Thus, there are only certain allowed *modes* of vibration, as indicated in Figure 2.27.

The **fundamental** mode of vibration, which is also described as the first **harmonic**, is the simplest mode, with the longest possible wavelength. It has a node at each fixed end, and an antinode in the middle. The length of the string is L, so the fundamental wavelength λ is equal to $2L$. The other allowed standing waves, known as the second harmonic, third harmonic, fourth harmonic, etc., have $L = \lambda$, $3\lambda/2$, 2λ, etc. Thus, $\lambda = L$, $2L/3$, $L/2$, etc.

Question 2.12 (a) Sketch the shape of the next two allowed standing waves after those shown in Figure 2.27.

(b) If the string shown in Figure 2.27 has a length of 3.0 m, what are the wavelengths of the fifth and sixth harmonics? ■

In general, the length of the string must be a whole number of half wavelengths; that is

$$L = n\left(\frac{\lambda}{2}\right) \tag{2.19}$$

where n is a positive integer. Note that $n = 1$ for the first harmonic, $n = 2$ for the second harmonic, etc. Equation 2.19 can be rearranged to give

$$\lambda = \frac{2L}{n}. \tag{2.20}$$

To find an expression for the frequency of the nth harmonic, Equation 2.20 can be combined with Equations 2.5 and 2.7. From Equation 2.5

$$f = \frac{v}{\lambda}.$$

Substituting for v from Equation 2.7 and λ from Equation 2.20 gives

$$f = \frac{n}{2L}\sqrt{\frac{F_{\mathrm{T}}}{\mu}} \tag{2.21}$$

where F_{T} is the tension in the string, and μ is its mass per unit length.

Question 2.13 The E (highest) string on the violin has a length of 0.32 m, a mass per unit length of 0.38 g m^{-1}, and is under a tension of 68 N. What are the wavelength and frequency of the fundamental and the second harmonic? ■

The allowed modes of vibration are sometimes said to be *resonant* standing waves. Attempts can be made to set the string oscillating at other frequencies. Waves of this frequency will then move along the string and be reflected at the end. However, it is only when the frequency of the oscillation (the forcing frequency) is the same as that of one of the harmonics that the string will vibrate with substantial amplitude and duration; that is, at these frequencies resonance occurs. For other frequencies the vibration of the string will only have a very small amplitude and will die out rapidly.

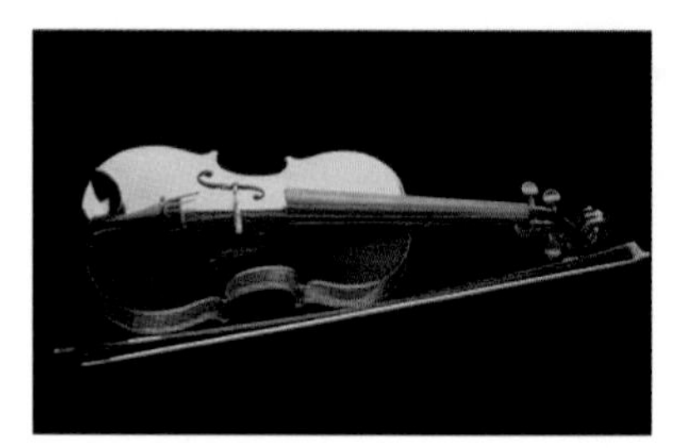

Figure 2.28 A violin. Vibrations in the strings cause the front and back of the violin, and the air between them to vibrate in resonance. The tension in the strings is altered by turning the tuning pegs.

4.5 Standing waves in musical instruments

An understanding of the previous section immediately leads to an appreciation of the physics of stringed musical instruments such as guitars, violins, harps and pianos. In all these cases, a mechanical impulse (a finger, pick, bow or hammer) creates a wave on the string, and the only components of the wave that persist on the string are those that represent allowed standing waves. Since the instruments to which the strings are attached are not perfectly rigid, energy is transferred from the strings to the body of the instrument, which resonates and produces the actual sound associated with the musical instrument (Figure 2.28).

Equation 2.21 shows that frequencies of the allowed modes of vibration of the string depend on the length of the string, the tension and the mass per unit length. These quantities are varied in different ways in different instruments. The strings in a violin all have the same length, so the fundamental mode in each case has the same wavelength. However the fundamental frequency varies from string to string because each of the strings has a different mass per unit length, and they are all under a slightly different tension, as shown in Table 2.2.

Table 2.2 Typical characteristics of violin strings.

String	G	D	A	E
length/cm	32	32	32	32
mass per unit length/g.m^{-1}	3.0	1.5	0.61	0.38
tension/N	47	53	48	68
frequency/Hz	196	293	440	660

Of course, a violin is not limited to producing just four notes. A violinist can produce a wide range of notes by placing a finger at different points on the strings, thereby altering the string's effective length.

● What are the wavelength and frequency of the fundamental note produced by the E string when the violinist places a finger 18cm from the bridge on the neck of the violin in Figure 2.28?

❍ The finger fixes the string and therefore shortens the length on which waves can be set up to 18 cm. So, assuming the tension remains the same, the fundamental wavelength will be twice this, 36 cm. The fundamental frequency will be

$$f = 660\ \text{Hz} \times \frac{32\ \text{cm}}{18\ \text{cm}} = 1173\ \text{Hz}\cdot$$

(This is the frequency of high D, two octaves above the violin's open D string.) ■

Figure 2.29 (a) Strings of differing length and mass per unit length are used to produce notes of different pitch (frequency) in different members of the violin family of instruments. (b) The harp and (c) piano also produce notes of different frequency through the use of strings with differing length and mass per unit length.

The other members of the violin family are the viola, cello and double bass (Figure 2.29a). These instruments have longer strings with a larger mass per unit length, so they produce *lower* frequencies than the violin. The harp and piano (Figure 2.29b and c) also use strings of different length and mass per unit length to produce the different notes. In general, the length of the string and its mass per unit length are predetermined by the construction of the instrument, but the tension can be adjusted, and this is precisely how a musical instrument is *tuned* to produce the right note.

The application of standing waves to musical instruments is not limited to instruments with strings. Instruments as diverse as flutes, trumpets and organs all make essential use of vibrating columns of air. The air column is set in motion by, for example, the player blowing across the mouthpiece of a flute, or setting the reed of a clarinet in motion with their lips and breath. In the flute or recorder (as well as the so-called 'open' organ pipes), the tube of the instrument is essentially open at both ends (the mouthpiece is one 'open end'). This means that there is an antinode rather than a node at each end (the antinode is actually slightly beyond the end of the tube, but details such as this need not bother us here), so standing waves are set up as shown in Figure 2.30a. The clarinet (and the so-called 'closed' organ pipes) on the other hand, are open at one end and closed at the other. So there is a node at one end and an antinode at the other; standing waves are set up as shown in Figure 2.30b.

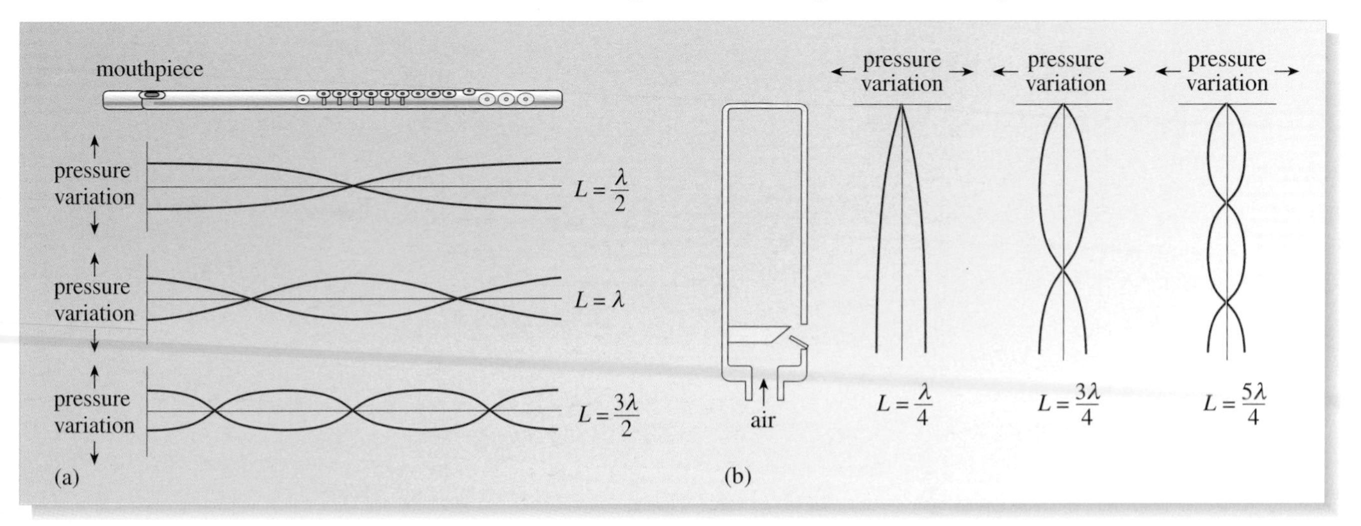

Figure 2.30 Representations of standing waves set up in (a) a flute (b) a 'closed' organ pipe. In (a) there is an antinode at each end and $f = nv/2L$, where L is the length of the closed tube, v is the speed of sound in air and $n = 1, 2, 3$, etc. In (b) there is a node at one end and an antinode at the other, and $f = nv/4L$. In this case n must be odd; that is, $n = 1, 3, 5$, etc.

The equations for the allowed frequencies on each occasion are given in the caption to Figure 2.30. You don't need to remember these equations, but they show clearly that the fundamental frequency increases as length decreases. Thus, to get a higher-pitched note (i.e. a note of higher frequency), the length of the air column must be decreased. Figure 2.31 shows various ways in which this can be achieved.

Figure 2.31 Various ways of altering the length of the vibrating column of air in a musical instrument: (a) a somewhat draconian and irreversible approach! (b) in a real clarinet, uncovering the holes shortens the effective length of the air column and raises the pitch (frequency) of the note; (c) brass instruments are long metal tubes, bent into coils to make them easier to hold (the trombone has a sliding section that can be moved out to lengthen the tube); (d) pipe organs have many pipes of differing lengths.

Why does one musical instrument sound so different from another? The previous discussion of sound has been confined to simple sinusoidal waves that can be characterized by a single wavelength (or single frequency). In practice the pressure variations in the sound wave coming from a musical instrument are rather more complicated. Even a single note has a more complicated waveform. For example, when a violinist plays the A string, which has a fundamental frequency of 440 Hz, the pressure at a nearby point might vary with time in the way shown in Figure 2.32a. The fundamental period is 1/440 Hz = 2.27 ms, but the wave is far from being sinusoidal. It is a *superposition* of many of the higher-frequency standing waves on the string, as well as the fundamental, and the relative contributions will vary depending on the source of the sound. This is partly what gives each musical instrument its characteristic richness and uniqueness of sound.

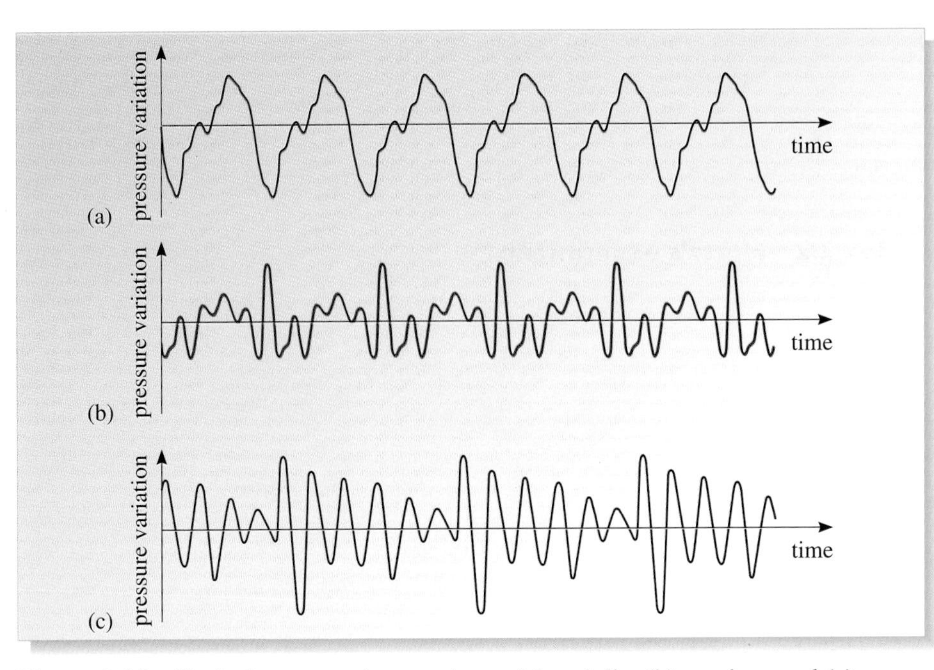

Figure 2.32 Typical pressure changes due to (a) a violin, (b) an oboe, and (c) a French horn; the different combinations of harmonics produced by different instruments are partly responsible for their unique sound.

5 Electromagnetic radiation

So far in this chapter you have seen how travelling waves and standing waves can be described mathematically, and you have seen various examples of each, often involving sound. The remainder of this chapter concerns another important phenomenon that involves waves, namely *electromagnetic radiation*. In Chapter 1 you saw that the crowning achievement of Maxwell's equations was the prediction of electromagnetic radiation that travelled, in a vacuum, at the speed of light. From this it seems reasonable to conclude that light itself is a form of electromagnetic radiation. Other forms of electromagnetic radiation are now known to include **radio waves**, **microwaves**, **infrared radiation**, **ultraviolet radiation**, **X-rays** and **gamma rays**.

5.1 The wave model of electromagnetic radiation

So what exactly *is* electromagnetic radiation? One of the early theories of light suggested that particles were emitted *from* the eye when an object was seen. Isaac Newton was among those who realized that something had to enter the eye to enable vision to take place, but he still thought in terms of particles, or 'corpuscles', of light emitted from luminous objects. In 1680, Christiaan Huygens proposed that light travelled from place to place by means of a wave motion, but it was not until experimental evidence was provided by Thomas Young early in the nineteenth century that this idea was widely accepted. Later still, in the early twentieth century, quantum theory suggested that light interacts with matter as though it is a stream of particles (known as photons). As you will see in later books of *The Physical World*, this so-called 'wave–particle duality' of light lies at the heart of an understanding of the world: light *propagates* like a wave, but is *emitted* or *absorbed* like a particle. For the rest of this chapter, only the propagation of light and other electromagnetic radiation is considered, so the wave model is what is needed.

Maxwell's equations actually predicted that electromagnetic radiation takes the form of waves, but when Maxwell's work was published in 1867 many physicists were reluctant to accept it. The experiments that conclusively demonstrated both the existence of electromagnetic radiation and its wave-like nature were performed by Heinrich Hertz (Figure 2.33) between 1885 and 1889.

Box 2.2 Hertz's experiment

Hertz used an electrical oscillator made of two brass conductors, each connected to an induction coil, and separated by a tiny gap across which sparks could leap. Hertz reasoned that charge would oscillate rapidly back and forth, and, if Maxwell's predictions were correct, electromagnetic waves would be transmitted. To confirm this, he made a simple receiver from a loop of wire with a small brass sphere at one end and a point at the other. When the point was held very close to the sphere, the presence of electromagnetic waves was revealed by sparks crossing the gap between the point and the sphere. Hertz found he was able to detect electromagnetic waves several tens of metres from the oscillator. He went on to prove the presence of waves by showing that they exhibited various properties only associated with wave motion.

The story of Heinrich Hertz has an interesting postscript. He was aware that his work had given evidence for the electromagnetic waves predicted by Maxwell, but he did not see that his experiment had any other useful purpose. When asked what use the waves might be, Hertz is reputed to have replied:

'It is of no use whatsoever. This is just an experiment that proves Maestro Maxwell was right. We just have these mysterious electromagnetic waves that we cannot see with the naked eye. But they are there.'

'So what's next?' asked a student.

With the benefit of hindsight, Hertz's reply 'Nothing, I guess' has a certain irony. Hertz's waves were radio waves. Guglielmo Marconi (1874–1937) read of Hertz's work and used radio waves for signalling.

Hertz also discovered the photoelectric effect, further discussed in *Quantum physics: an introduction*, although again he did not appreciate the significance of his discovery. He died of blood poisoning at the age of 37.

Figure 2.33 Heinrich Hertz (1857–94) studied under Hermann von Helmholtz at the University of Berlin, but he was professor of physics at Karlsruhe Polytechnic when he demonstrated the production of electromagnetic waves.

All forms of electromagnetic radiation travel at the same speed in a vacuum — the speed of light. This **speed of light** is normally given the symbol c, and since 1983 it has been *defined* to have a value of 299 792 458 m s^{-1}. For most purposes it is adequate to use the value 3.00×10^8 m s^{-1} to three significant figures. Equation 2.5 ($v = f\lambda$) is true for all waves, but for electromagnetic waves it is usually stated as

$$c = f\lambda. \tag{2.22}$$

The value of c is perhaps one of the most widely known physical constants. Do remember, though, that electromagnetic waves only travel with this speed when they are travelling through a vacuum. When light travels through a material rather than through a vacuum, its speed v is less than c. For air the difference is only 0.03%, but the speed of light in water and in glass is substantially smaller, as you can see in Table 2.3.

Table 2.3 The speed of light in various materials

Material	Speed/m s^{-1}
air	3.00×10^8
water	2.25×10^8
olive oil	2.05×10^8
crown glass	1.97×10^8
quartz	1.94×10^8
flint glass	1.85×10^8
diamond	1.24×10^8

Figure 2.34 Sound needs a medium to travel through but light does not. The bell in the evacuated glass jar can be seen but not heard.

Electromagnetic waves are more difficult to visualize than any waves discussed so far. The motion of waves on strings and on the surface of water can actually be seen; sound waves in air, though not visible, are analogous to longitudinal waves in springs, and so the longitudinal vibrations of the air can be visualized quite easily. But electromagnetic waves can travel through a vacuum (as illustrated in Figure 2.34), so they cannot involve any material vibrations; instead they involve oscillations of both the electric and magnetic field. It is these fields that vary during the passage of an electromagnetic wave, in much the same way that the pressure of air varies during the passage of a sound wave.

Maxwell predicted that the electric and magnetic fields in an electromagnetic wave are always at right-angles to each other, and both fields are at right angles to the direction of propagation. A snapshot of such a wave is shown in Figure 2.35. In this case we are supposing that the electric field is confined to the vertical plane, and that its direction and magnitude at each point along the propagation axis is indicated by the orange curve. The magnetic field at each point along the propagation axis is then indicated by the green curve. Both fields vary with time in such a way that the whole wave moves with speed c. Furthermore, if you look along the direction that the wave is travelling, a clockwise rotation through 90° is always needed to go from the electric field direction to the magnetic field direction.

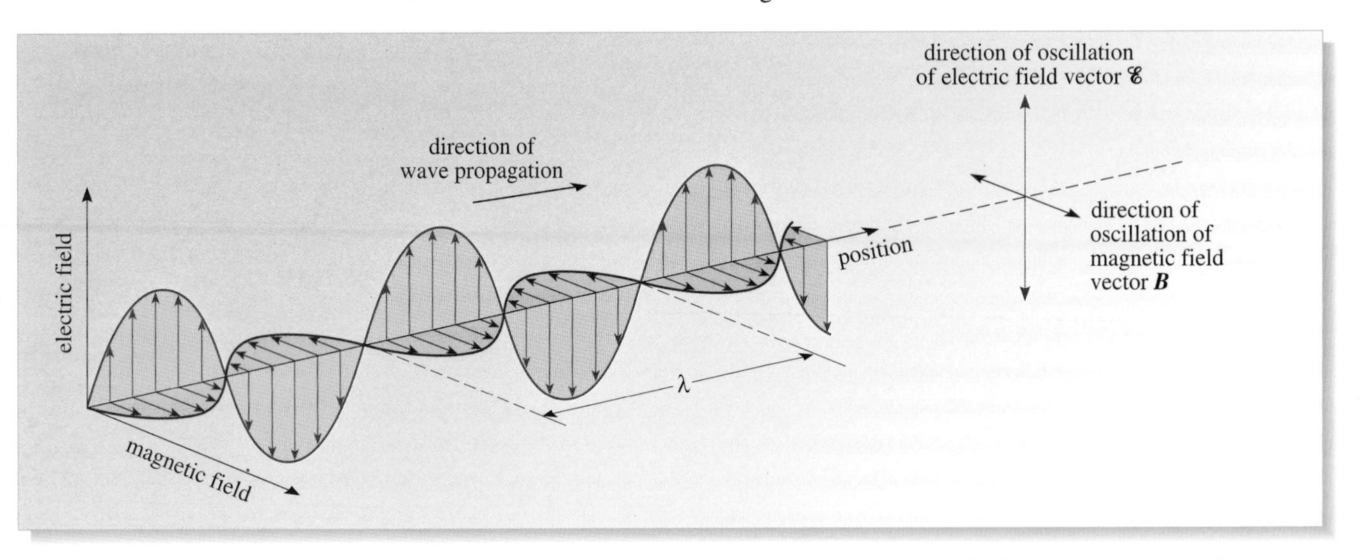

Figure 2.35 The variations of electric and magnetic fields in an electromagnetic wave.

5.2 Polarization and the applications of polarized light

Light was at first thought to be a longitudinal wave: the realization that it is in fact transverse came from a study of its **polarization**. Polarization occurs when there is a restriction placed on the direction in which the vibrations in a transverse wave can take place. For a transverse wave, the vibrations *must* be at right-angles to the direction of propagation (by definition), but that still leaves them free, in principle, to take up any orientation. Two such orientations for the electric field of an electromagnetic wave are shown in Figure 2.36, but there are, in principle, an infinite number of other orientations that are intermediate between these extremes. Figure 2.36a shows the electric field vibrating in the vertical plane, and in Figure 2.36b it vibrates in the horizontal plane. In each case the wave is said to be **plane** (or *linearly*) **polarized** along the direction of oscillation.

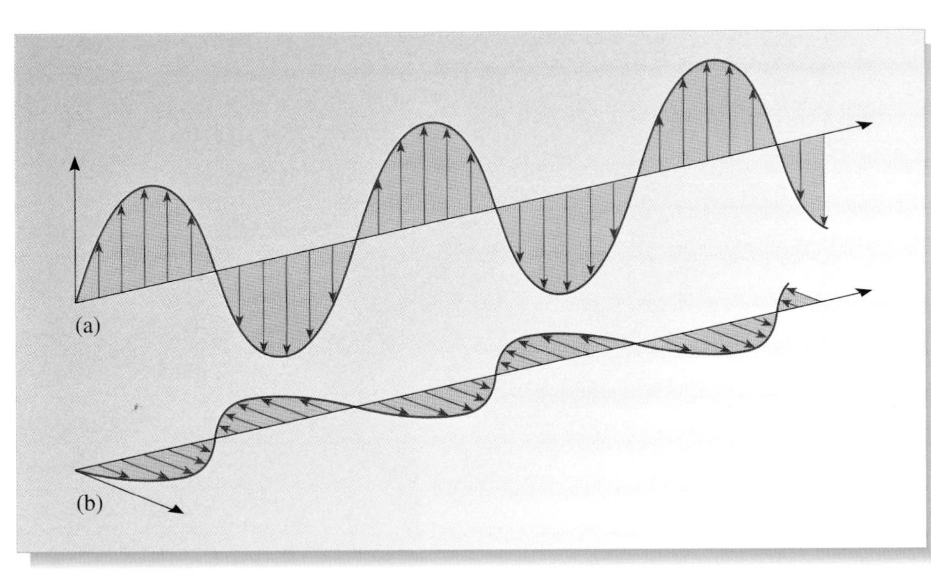

Figure 2.36 A plane polarized wave: (a) in a vertical plane; (b) in a horizontal plane.

Light is generally unpolarized; in other words it is a superposition of a large number of polarized waves which have electric fields in every possible direction perpendicular to the direction of propagation. This is because atoms generally emit light quite independently of the surrounding atoms, so even though the light emitted from an individual atom is polarized, when the light from a vast number of atoms is combined, there is no preferred polarization direction. Radio waves, on the other hand, are usually polarized because they are generated by currents in a single transmitting antenna with a definite orientation. None the less, visible light *can* be polarized, as can all other forms of electromagnetic radiation.

● What does Figure 2.37 tell you about the orientation of the electric and magnetic fields in the television signal broadcast in the area where the photograph was taken?

❍ The television antennae are all *horizontal*, so that they can receive an electric field that is polarized in a *horizontal* direction (i.e. the direction of the current, or electron, flow). The magnetic field direction must therefore be vertical. ■

Figure 2.37 Television antennae.

Although light is generally unpolarized, it is possible to polarize it by various means; and polarized light has several applications. Probably the best-known method of polarizing light is to pass it through a polarizing filter (known commercially as 'Polaroid'). When light, initially with many electric field orientations, passes through a polarizing filter, it emerges with only one allowed electric field orientation, as shown in Figure 2.38. The polarizing direction of the filter is established during manufacture by the orientation of the molecules of the polarizing material (which are at *right-angles to* the allowed electric field orientation). If the polarized light then meets a second polarizing filter, with its polarizing direction at right-angles to that of the first filter, no light at all can pass. Two such Polaroid sheets are said to be *crossed*.

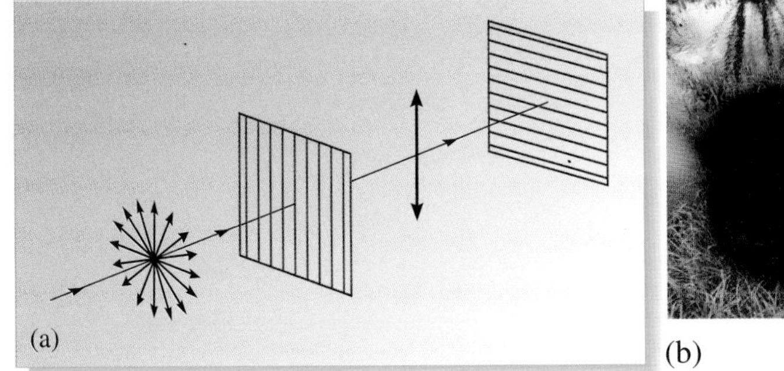

(b)

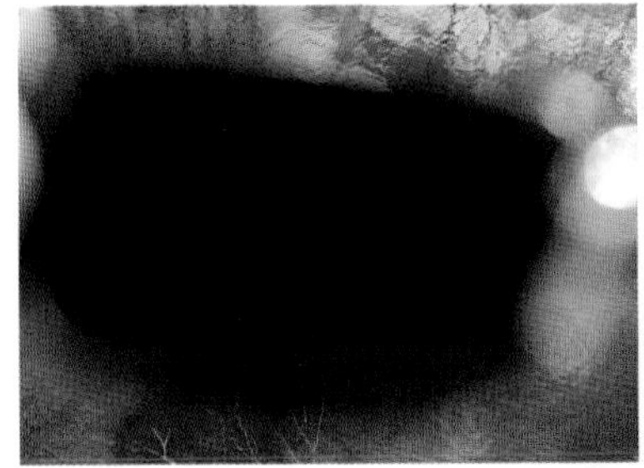

(c)

Figure 2.38 (a) When light passes through one polarizing filter, it is plane polarized. When another filter at right-angles to the first is encountered, no light can pass. (b) The view through a single Polaroid filter. (c) The view through two crossed Polaroids

Figure 2.39 The phenomenon of birefringence in calcite.

Some substances, including some minerals, are said to be *birefringent*, which means that light entering them is split into two different components polarized in different directions (see Figure 2.39). Some materials are also able to rotate the plane of polarization of polarized light. When a piece of one of these materials is placed between crossed polarizing filters, some light will now pass (see Figure 2.40a). This property can be used to identify some minerals and rocks, such as the peridotite shown in Figure 2.40b.

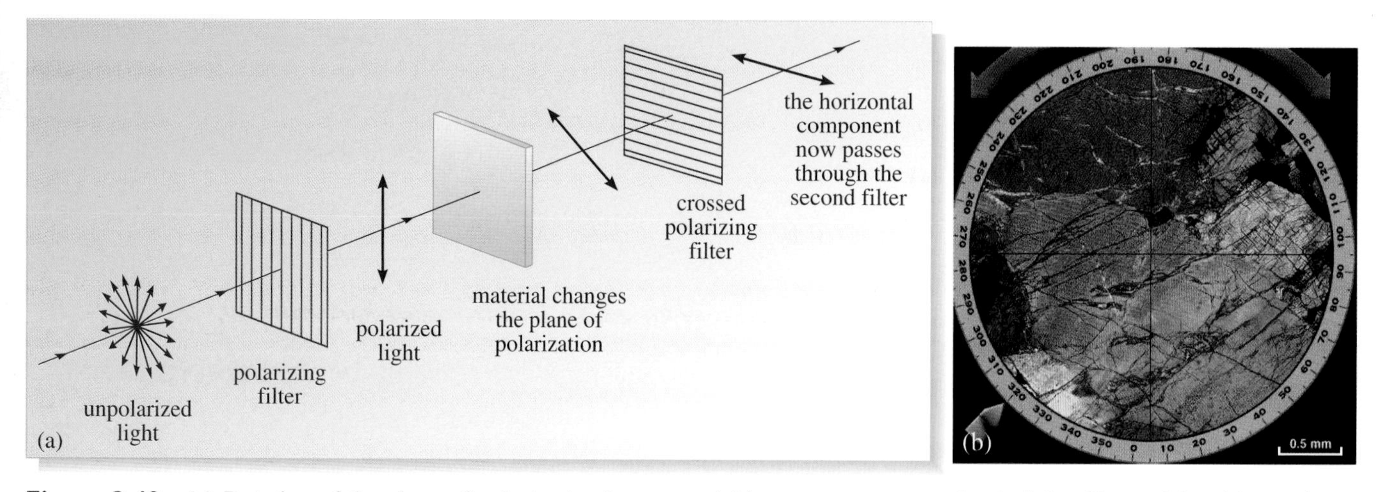

Figure 2.40 (a) Rotation of the plane of polarization by a material between two crossed polarizing filters; (b) a thin section of peridotite viewed by transmission of light between crossed polarizing filters.

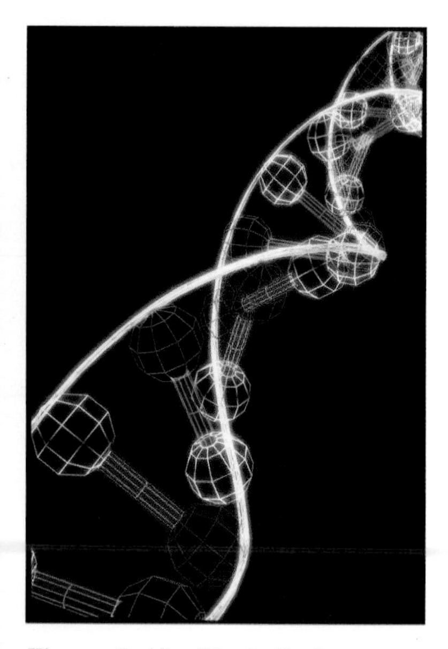

Figure 2.41 The helical structure of the DNA molecule rotates the plane of polarization of light.

The rotation of the polarization of light as it passes through certain solutions is due to the structure of the molecules. The helical structure of DNA is one structure that leads to this effect (Figure 2.41). Furthermore, some molecules are *chiral* ('handed'), which means that they exist in two different forms which are mirror images of each other (Figure 2.42). These two forms of the same molecule will rotate the plane of polarization by a particular angle but in opposite directions.

As well as the polarization that occurs when light passes through certain materials, reflected light also tends to be plane polarized. Light reflected from a wet road or swimming pool is chiefly polarized in a horizontal direction. Only vertically polarized light can pass through Polaroid sunglasses, thus reducing unwanted reflections and glare, as shown in Figure 2.43.

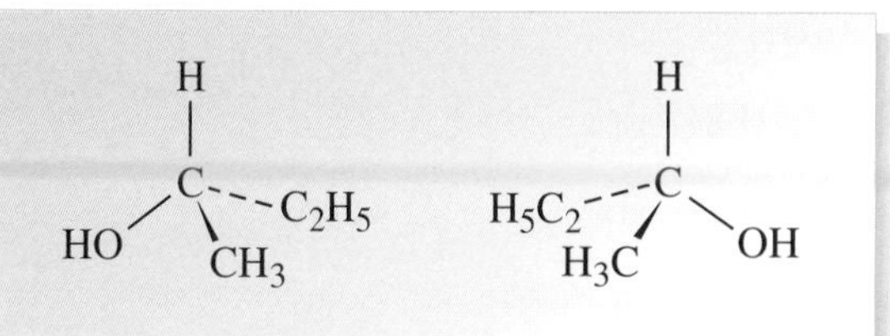

Figure 2.42 Two forms of butan-2-ol. The two molecules are mirror images of each other. Solutions of one or the other form of the molecule rotate plane-polarized light in opposite directions. This can be used as a test for the presence of one or the other type of molecule.

Sunlight scattered off molecules in the atmosphere may also be polarized, which is why, on a clear day, Polaroid sunglasses make certain regions of the sky appear darker than others. It is thought that insects, including bees and ants, may use this polarization of sunlight to navigate. Experiments have shown that the direction of travel of ants can be altered by holding a polarizing filter between them and the Sun.

Question 2.14 Two sheets of Polaroid are placed close together in front of a light bulb so that no light passes through them. Describe what will be seen as the second sheet of Polaroid is rotated through 180°, and explain why this effect is seen. ■

(a)

(b)

Figure 2.43 (a) Reflections from a glass window; (b) a photograph of the same glass window taken through Polaroid sunglasses.

5.3 The electromagnetic spectrum

Visible light has now been identified as a type of electromagnetic radiation, and all electromagnetic radiation can be modelled as transverse electromagnetic waves travelling at the speed of light. Why is it then that all electromagnetic radiation does not have identical properties? The explanation lies in the fact that the different kinds of radiation in the **electromagnetic spectrum** (Figure 2.44) correspond to waves of very different frequencies, and therefore different wavelengths. The frequencies range from about 10^{24} Hz for gamma rays (γ-rays) to around 10^5 Hz for some radio waves (wavelengths of less than 10^{-15} m to more than 10^3 m). Visible light, to which human eyes are sensitive, ranges in wavelength from around 400 nm to 700 nm (where 1 nm = 10^{-9} m). On a diagram like Figure 2.44 it appears quite insignificant! The reason why visible light is of great importance to us — out of all proportion to the extent of its wavelength range in the electromagnetic spectrum — is that the spectrum of radiation emitted by the Sun has its peak in the visible region. As a consequence, life on Earth has evolved to make use of this particular kind of radiation.

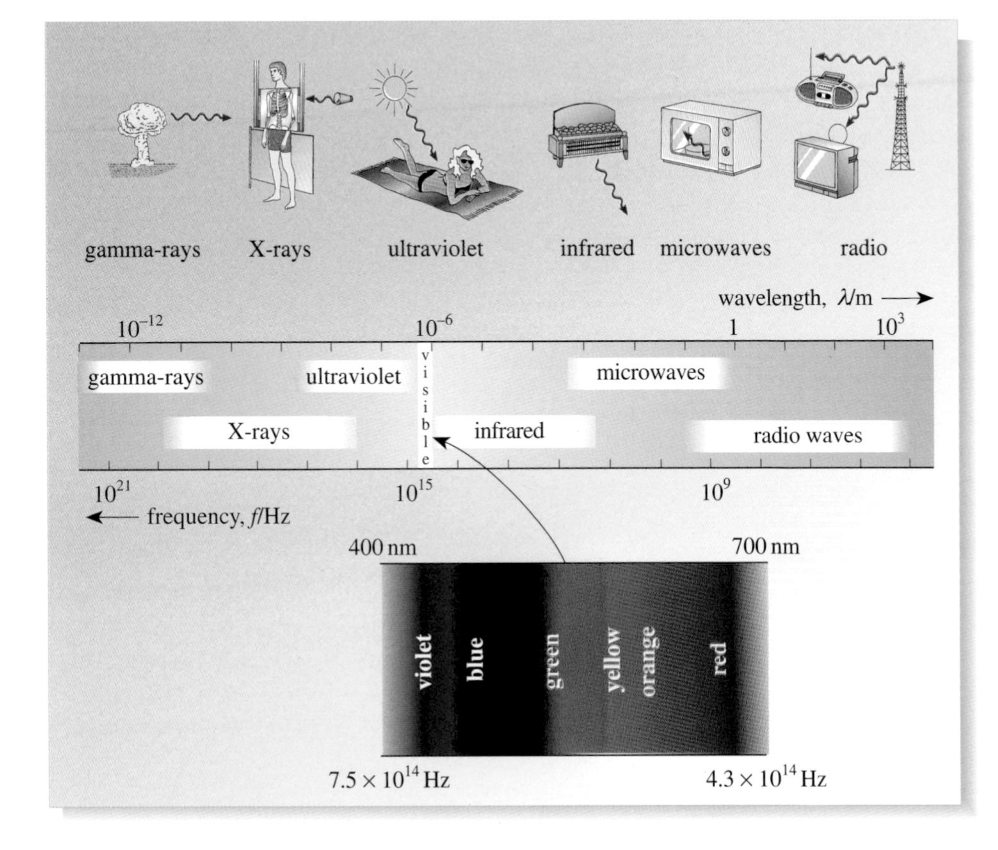

Figure 2.44 The electromagnetic spectrum; the inset shows an expanded version of the visible region.

Certain sections of the electromagnetic spectrum have been given labels such as 'infrared', 'ultraviolet', etc., but you should note that the spectrum has a *continuous* variation of wavelengths from one end to the other, and there is some overlap between different sections; the names are merely conventional labels assigned to radiation in different regions.

Radio waves and microwaves

Radio waves are at the long-wavelength end of the electromagnetic spectrum (with typical wavelengths in the range 15 cm to 2 km), and are so named because they are used to carry radio and television signals. Radio waves are not readily absorbed by most sorts of matter, and can travel great distances through space; it is for this reason that they have been used successfully in radio astronomy to make some remarkable observations, including the discovery of *pulsars* — tiny collapsed stars rotating at up to several hundred times a second and emitting intense beams of radio waves. Some of the most spectacular objects discovered to date are *radio galaxies*, which emit huge jets of matter from their cores, believed to harbour massive black holes (Figure 2.45).

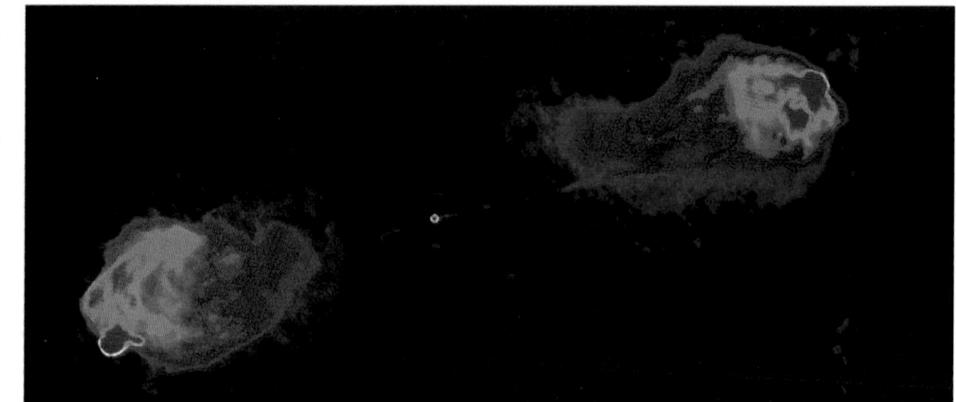

Figure 2.45 This false-colour map of the radio galaxy Cygnus A shows two radio-wave-emitting jets emerging from the centre of the galaxy, and ending in huge lobes of radio-wave-emitting gas. Each jet is of the order of 10^5 light years long. The colours represent different intensities of radio-wave emission.

Figure 2.46 An antenna used as part of the UK mobile telephone network which relies on microwave signals.

Microwaves have wavelengths from about 1 mm to about 15 cm, although the distinction between microwaves and radio waves is not well defined. Like radio waves, microwaves are used in communications and one of the implications of their shorter wavelength is that a beam of microwaves can be sent in a more precise direction than a beam of radio waves. In the UK, mobile phones operate in the microwave range of the electromagnetic spectrum (Figure 2.46). Both radio waves and microwaves are used in radar, which operates in a similar way to the echo location discussed earlier, but using electromagnetic radiation. The microwaves used to heat food in microwave ovens have a wavelength of about 12 cm.

Infrared radiation

When an object is heated, it will emit electromagnetic waves with a range of wavelengths depending on the object's temperature. If the temperature is similar to that found on the surface of the Earth, infrared (IR) waves with wavelengths in the region of 10^{-5} m are mainly emitted. Devices that are sensitive to infrared radiation are used for a variety of detection and imaging purposes. One such example is satellite-based remote sensing. Sensors on satellites as far as 36 000 km from the Earth gather information throughout the visible and infrared parts of the electromagnetic spectrum. This means, for example, that two objects that may both appear green to us, would be clearly distinguishable because they reflect different amounts of infrared radiation. The data can be recorded digitally and can be displayed using artificially assigned colours. Some uses of remote sensing are in meteorology, geology, oceanography, pollution control and the study of marine ecosystems (Figure 2.47).

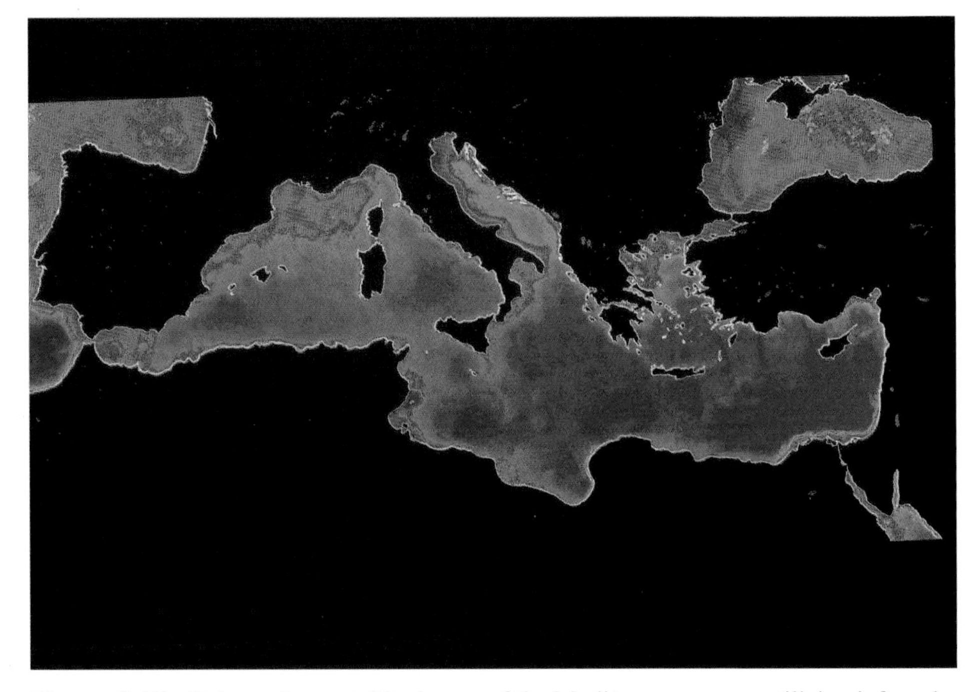

Figure 2.47 False-colour satellite image of the Mediterranean area utilizing infrared reflectance to observe the distribution of phytoplankton (microscopic aquatic plant life) in surface waters. Colours represent varying phytoplankton densities from red (most dense) to blue (least dense). Land areas are black. The Mediterranean, which may be affected by pollution, is fairly barren compared with the plankton-rich waters of the northern Atlantic (top left) and Black Sea (upper right).

Infrared radiation is also increasingly used in thermal imaging devices used to 'see in the dark' when conditions preclude the use of visible light. For example, rescue workers use thermal imaging devices to detect people in smoke-filled buildings (Figure 2.48).

Figure 2.48 An infrared image of a person in a smoke filled room (left), whereas using visible light, nothing can be seen (right).

Ultraviolet radiation

Most of the ultraviolet radiation (radiation with wavelengths from about 10 nm to 400 nm) found on Earth originates in the Sun. UV radiation is readily absorbed by objects such as the human body and interacts with body chemicals. This is responsible for tanning — an effect that can be reproduced using UV-emitting fluorescent tubes (Figure 2.49a). In small doses, this can be harmless or even beneficial; for example, UV radiation is responsible for production of vitamin D by the body. A deficiency of vitamin D causes the condition known as *rickets*, in which the bones of growing children are malformed and fail to harden. In fact, the earliest life forms on this planet may have started because of the ability of UV radiation from the Sun to facilitate reactions between ammonia, methane, carbon dioxide and water. (These reactions form amino acids, the basic building blocks of proteins, which are, in turn, the essential ingredients of all life forms.) However, exposure to shorter-wavelength UV waves can be dangerous, leading to melanoma (skin cancer; see Figure 2.49b) and to eye cataracts.

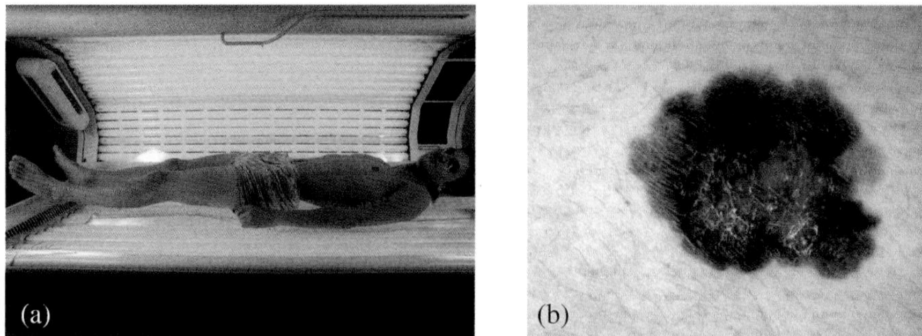

Figure 2.49 (a) A desirable effect of exposure to UV radiation is the tanning of skin; (b) a deleterious effect is the production of skin cancers.

X-rays and γ-rays

X-rays have wavelengths from about 10^{-12} m to 10^{-8} m, whereas γ-rays have wavelengths extending from less than 10^{-14} m to about 10^{-11} m. There is no sharp dividing line between the two, and the distinction is usually made with reference to the origin of the radiation, rather than its wavelength. In Earth-bound laboratories, X-rays are produced by bombarding metal plates with electrons, and γ-rays are obtained from radioactive sources. X-rays have numerous uses, from medical imaging and radiation therapy in hospitals to luggage examination at airports. They are also used to detect sub-surface defects in metals, and the technique of X-ray diffraction (discussed in more detail in Section 6) has been used to study the internal structure of matter on the atomic scale.

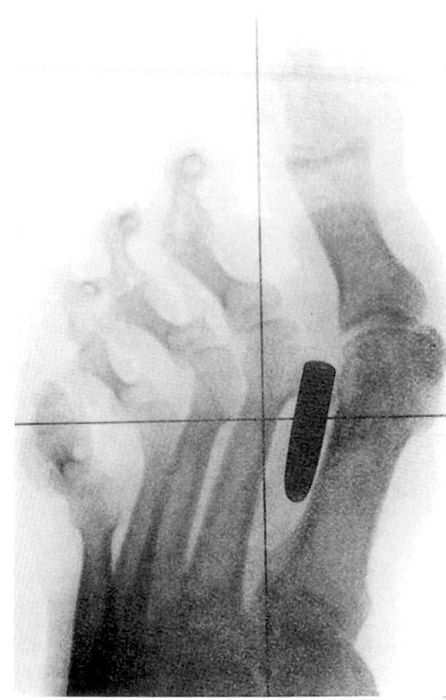

Figure 2.50 An early X-ray photograph showing a bullet in a foot.

X-rays were discovered by Wilhelm Röntgen in 1895, and the usefulness of X-rays in diagnostic medicine was quickly realized. Unfortunately, early X-ray machines were unshielded and poorly controlled, so both patients and hospital staff were exposed to large doses of radiation. In the early days the use of X-rays was limited to obtaining an image of an X-ray shadow on a photographic plate (see Figure 2.50), but CT (computer tomography) scanning can now be used to build up a detailed three-dimensional picture by producing images of 'slices' of the body, and even of soft tissue such as the brain (Figure 2.51).

The principle of the gamma camera is rather different. In this case the γ-rays being detected are from a radioactive tracer ingested by the patient. The progress of the tracer through a particular organ can be monitored.

Question 2.15 (a) BBC Radio 5 is broadcast using a carrier wave with a frequency of 909 kHz. What is the wavelength of the carrier wave? (b) X-rays used in a CT scanner have a wavelength of 0.021 nm. What is the frequency of these X-rays? (Assume that the speed of light $c = 3.00 \times 10^8\ \mathrm{m\,s^{-1}}$.) ■

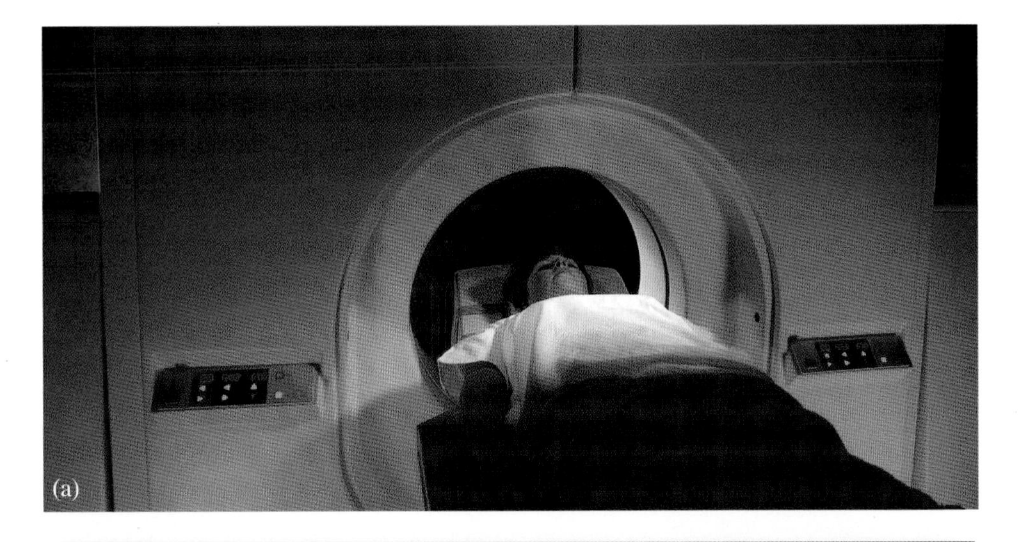

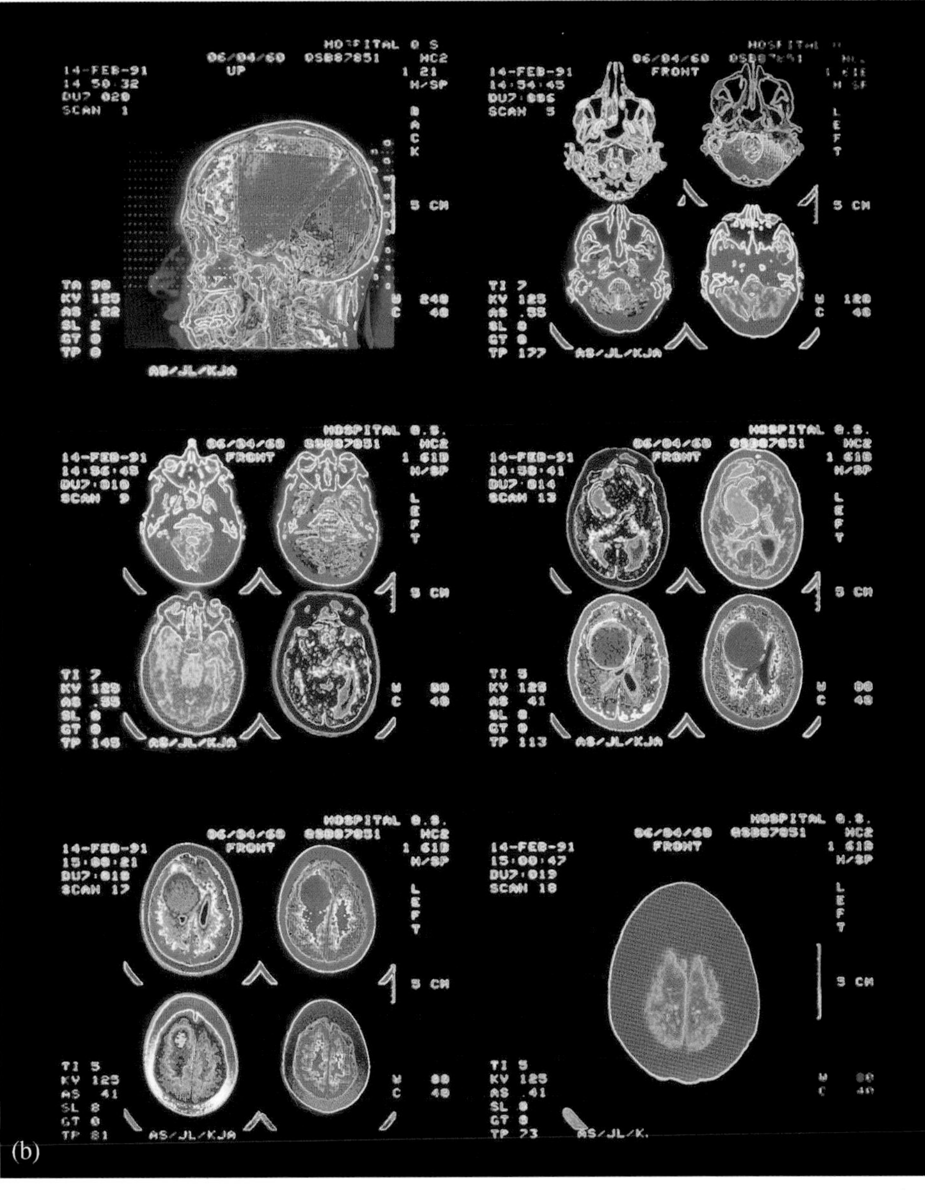

Figure 2.51 (a) A modern CT scanner; (b) a series of axial slices through the brain as obtained from a CT scanner.

6 The propagation of waves

This section will discuss in more detail the ways in which waves propagate, especially when they encounter a boundary, move from one medium to another, meet some sort of obstacle, or meet another wave. Reflection, refraction, diffraction and interference (the terms will be explained shortly) are phenomena exhibited by all waves, and it is possible to explain them mathematically in terms of wave motion. However, the full theory is complicated and rather cumbersome. Fortunately, it is possible to obtain a good deal of insight into these phenomena by means of a much simpler approach introduced in 1678 by the Dutch physicist Christiaan Huygens (Figure 2.52). Throughout this section, *light waves* will be used as the example, because they allow the phenomena to be visualized more easily. However, similar effects occur with other types of waves too, such as other forms of electromagnetic radiation and sound waves.

Figure 2.52 Christiaan Huygens (1629–95) was a contemporary of Newton, and had similarly wide-ranging scientific interests. Among other things he discovered the true shape of the rings of Saturn, developed the pendulum clock and lectured to the Royal Society on his own theory of gravitation. However, he is best remembered for propounding the wave theory of light.

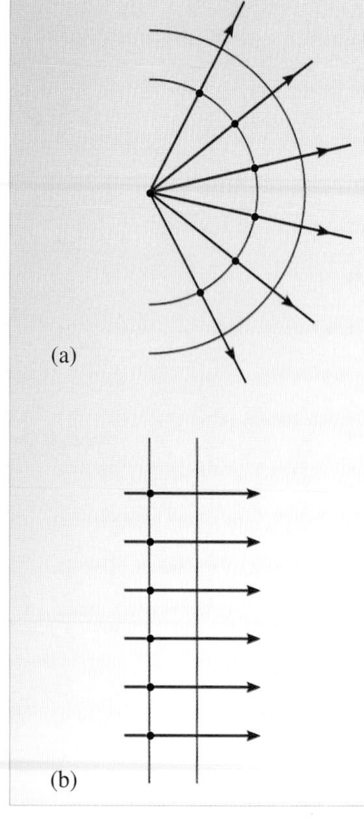

Figure 2.53 Rays are lines drawn at right-angles to wavefronts, so in (a) the rays are straight lines spreading radially from the source, and in (b) the rays are straight parallel lines perpendicular to the wavefronts.

6.1 The Huygens principle

Two important concepts when discussing the propagation of waves are *wavefronts* (which you met earlier in Section 3.3) and *rays*. In the absence of any obstructions, the wavefront through any point will be perpendicular to the direction of propagation of the wave at that point. To indicate the direction of propagation, it is convenient to draw lines with arrowheads on them. As shown in Figure 2.53, they are drawn at right-angles to propagating wavefronts, leading to the following definition:

Rays are lines showing the direction of propagation of a wave. They are perpendicular to the wavefronts.

It is important to note that light paths, as indicated by rays, are *reversible*. That is to say, if a ray describes the path taken by light in propagating from A to B, then the *opposite* ray will describe the path taken by light in propagating from B to A. This effect is known as the **reversibility of light paths**.

Huygens realized that it is possible to predict how a given wavefront will advance by means of geometrical constructions similar to those of Figure 2.53a and 2.53b. These constructions are based on the following principle:

> **The Huygens principle**
>
> Each point on a wavefront can be regarded as a small secondary source of waves; in three dimensions these secondary waves are spherical; in two dimensions they are circular.
>
> The secondary waves from each point spread out in all directions with the wave speed v, just as ripples spread out across the surface of water when a stone is dropped into it.
>
> During an interval Δt, the individual secondary waves travel a distance $v\Delta t$ from each point on the wavefront. The new position of the wavefront at the end of the interval is the *envelope* of these individual waves — a curve drawn tangentially through the individual secondary wavefronts.
>
> This new wavefront acts in turn as a further source of secondary waves.

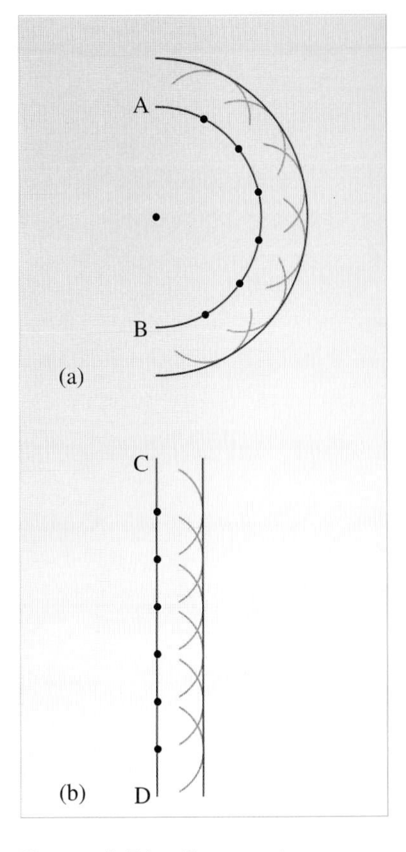

Figure 2.54 Constructions applying the Huygens principle for an unrestricted wavefront in the case of (a) a circular wavefront and (b) a plane wavefront.

Figure 2.54a shows how the Huygens principle works for a wave expanding from a point source when the wavefront is unrestricted. For clarity only six point sources on the wavefront AB have been included, but many more could have been shown. In practice it is only necessary to select as sources a set of points that are separated by about one wavelength. After a short period of time (typically about the period of the wave) the new wavefront can be constructed by drawing a smooth line that just touches each secondary wavefront.

Figure 2.54a shows the wavefront near the original point source. The time interval for the construction is the wave period, so a circular wavefront (this would be spherical in three dimensions) creates a new circular wavefront that has expanded by one wavelength. Figure 2.54b shows the situation far away from the original point source; now the radii of the wavefronts are so large that any part of the wavefront will appear to be the straight line CD (a flat plane in three dimensions), and will give rise to other plane wavefronts as it advances.

6.2 Reflection

Study note *Open University students will not be required to remember the details of the Huygens derivations of the laws of reflection and refraction presented in this and the next section. However, you should be able to follow the basic reasoning, and you must be able to recall and apply the conclusions of that reasoning (i.e. the laws of reflection and refraction).*

You may be wondering why, in the Huygens construction, only the expanding secondary waves travelling in the direction of the original wave are considered, rather than those in the opposite direction as well. If so you deserve congratulations! This is one of the flaws in the Huygens approach, and it needs Maxwell's theory to resolve it.

In discussing echo location in Section 2, the word 'reflection' was used without specifying the details of the process. In order to put the record straight, the Huygens principle will now be applied to one of simplest types of reflection, that of a *plane* wavefront by a *plane* mirror as shown in Figure 2.55.

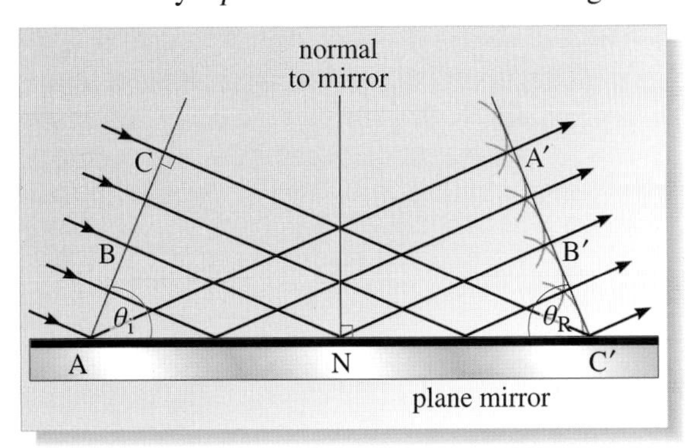

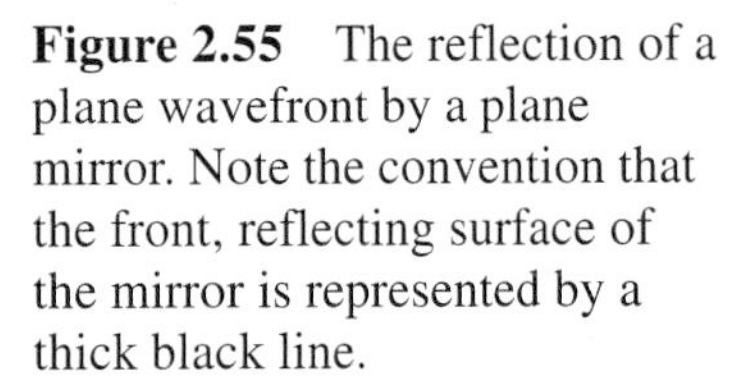
Figure 2.55 The reflection of a plane wavefront by a plane mirror. Note the convention that the front, reflecting surface of the mirror is represented by a thick black line.

The line ABC is part of a plane wavefront moving in the direction shown by the arrows (on the rays) and beginning to arrive at the plane mirror at an angle of θ_i. Point A on the wavefront is just arriving at the mirror, whereas the parts of the wavefront corresponding to B and C will arrive later, at points N and C′, respectively, on the mirror. (The line perpendicular to the mirror at point N is referred to as the **normal** to the mirror.) Secondary waves produced at the same time from points A, B and C will expand outwards from these points, all travelling at the same speed. In the time taken for the secondary wave produced at C to reach the mirror at point C′, the secondary wave from A will have travelled an equal distance and will be arriving at A′, and that from B will be arriving at B′ having covered the same distance (BN + NB′). The resulting wavefront, after reflection at the mirror, will be the line A′B′C′, which is tangential to each of the secondary waves. Now the right-angled triangles ACC′ and AA′C′ are identical, so the lines ABC and A′B′C′ make the same angle with the mirror surface. Therefore:

$$\theta_i = \theta_R. \tag{2.23a}$$

In other words, the angle at which the incident wavefront strikes the mirror (the **angle of incidence** θ_i) is equal to the angle at which the reflected wavefront leaves the mirror (the **angle of reflection** θ_R).

It is more usual, and certainly much easier to see 'the wood for the trees', if the wavefronts in Figure 2.55 are dispensed with altogether and only the *rays* are drawn. In fact, only one key ray (and its reflection) is usually needed, as shown in Figure 2.56. As shown in this figure, the angles of incidence and reflection are expressed respectively as the angle between the incident ray and the normal to the mirror surface, and the angle between the reflected ray and the normal to the mirror surface. This is perfectly valid, since the angle between the wavefront and the mirror surface has to be the same as the angle between the *normal* to the wavefront (the ray) and the *normal* to the mirror surface. Hence, with reference to Figures 2.55 and 2.56, $i = \theta_i$ and $R = \theta_R$. (Make sure you're convinced!) So the **law of reflection** can be simply expressed as:

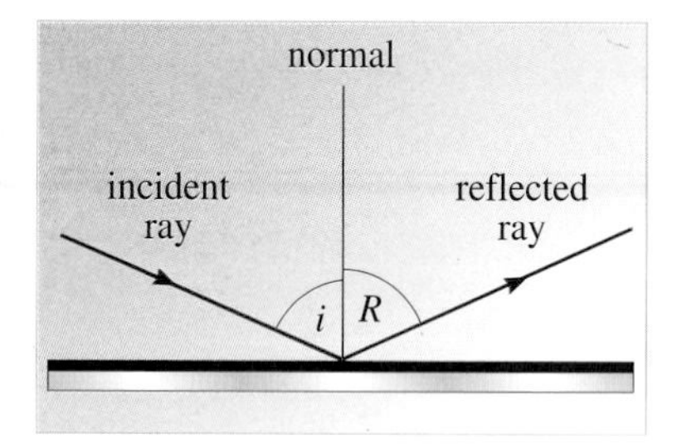

Figure 2.56 The ray representation of the law of reflection.

$$i = R. \tag{2.23b}$$

One final point is that, although all the representations of reflection here have, of necessity, been drawn two-dimensionally, the Huygens construction applies in *three* dimensions. What this implies is that nature also imposes the further restriction that the incident and reflected rays and the normal to the mirror surface must all lie *in the same plane* (the plane of the paper in Figures 2.55 and 2.56). Summarizing:

The law of reflection

- The incident ray, the reflected ray and the normal all lie in the same plane.
- The angle of reflection is equal to the angle of incidence.

● At what angle is a parallel beam of light reflected if it strikes a flat mirror from a direction exactly perpendicular to the mirror's surface?

❍ If the parallel beam strikes the mirror at an angle of 90° to the mirror surface, the angle of incidence (i.e. the angle between the normal to the mirror surface and the direction of the beam) must be equal to *zero degrees*. Hence, by the law of reflection, the angle of reflection must also equal zero degrees; that is, the beam is reflected back along its own path. ■

Note that, although the justification here has been developed in terms of a plane wavefront striking a flat reflecting surface, the law of reflection could have been deduced by considering any wavefront striking any surface. And since it only invokes the Huygens principle, it also applies to all types of waves, irrespective of type or wavelength. If parallel light rays from a distant source fall on a large parabolic mirror (Figure 2.57), they are reflected and converge at a single point. This is the principle of reflecting telescopes and is further discussed in the next chapter of this book. For now, this will simply be used as an example of the universal nature of the law of reflection: for each ray incident on the mirror, the angle of incidence equals the angle of reflection. Also, the use of parabolas to bring beams to a focus is not restricted to light waves. Figure 2.58 illustrates the use of parabolic reflectors to focus radio waves, microwaves and sound waves from a distance.

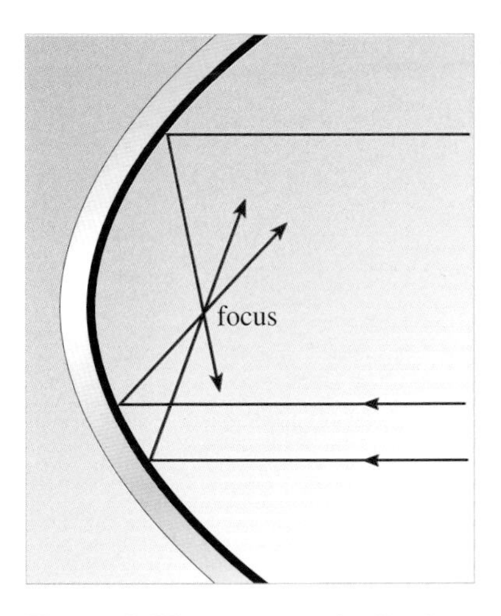

Figure 2.57 The law of reflection in operation in a parabolic reflector.

(a)

Figure 2.58 Parabolic reflectors can be used to focus (a) radio waves, (b) infrared radiation (detected as heat), (c) sound waves.

(c)

6.3 Refraction

All the arguments in the previous section assumed a *perfectly* reflecting plane surface. But what happens when the surface with which the wave interacts is *not* 100 per cent reflecting? Such a case arises in many everyday situations involving transparent media. For instance, if you place a straw into a glass of water, you will observe that the straw appears to be bent at the surface of the water (Figure 2.59). This effect is due to the **refraction** (change of direction) of light rays at the air–water surface. As with reflection, all types of waves can be refracted; examples with

Figure 2.59 A straw placed in a glass of water appears to be bent.

light waves are usually chosen simply because they are so easy to visualize. Refraction is caused by a change of wave speed as the wave moves from one medium to another.

Can the Huygens principle provide a model for understanding what happens when a substantial percentage of the wave passes *through* the boundary from one medium to another? As you might expect, the answer is yes, as shown in Figure 2.60. The line ABC shows the position of a wavefront at the instant that point A strikes the boundary, whereas the arcs touching A′ and B′ show the position of the spherical waves (originating from A and B″, respectively) at the instant, a time t later, when the point C on the wavefront has reached the boundary at C′.

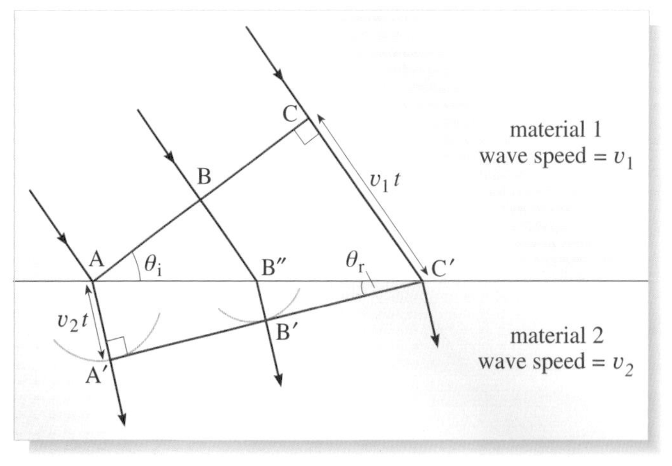

Figure 2.60 The refraction of a wavefront (reflection not shown) at the interface between two media.

What distinguishes this case from that of reflection is that there are two different media involved here, and the speed of a wave depends on the particular medium in which that wave is travelling (see Table 2.3). If, as here, the speed v_2 in the second medium is less than the speed v_1 in the first, the secondary spherical waves spreading out from A will not have travelled as far in the second medium as they would have in the first. In other words, the distance AA′ (equal to $v_2 t$) will be less than the distance CC′ (equal to $v_1 t$). Intermediate points on the wave — points like B — will spend part of their time travelling at speed v_1 and part at v_2. In the case of B, exactly midway between A and C, the distance BB″ will be $\frac{1}{2}v_1 t$ and the radius B″B′ of the secondary spherical wave emanating from B″ will be $\frac{1}{2}v_2 t$. When the envelope of all the secondary spherical wavefronts is drawn (i.e. the line A′B′C′), it is clear that the overall effect of slowing the wave down as it enters the second medium is to cause it to slew around and move in a different direction.

The quantitative relationship between the incident direction of the wavefront and its direction after refraction at the boundary can now be derived. From triangle ACC′:

$$\sin\theta_i = \frac{CC'}{AC'} = \frac{v_1 t}{AC'} \qquad (2.24)$$

and from triangle AA′C′:

$$\sin\theta_r = \frac{AA'}{AC'} = \frac{v_2 t}{AC'}. \qquad (2.25)$$

Dividing Equation 2.24 by Equation 2.25 gives

$$\frac{\sin\theta_i}{\sin\theta_r} = \frac{v_1}{v_2} \qquad (2.26a)$$

where the t and AC′ terms have cancelled.

As in the case of reflection, the diagrammatic representation of refraction can be simplified by drawing just the key rays and the normal to the surface: this is shown in Figure 2.61. Also, as before, the angle of incidence i can be measured as the angle between the incident ray and the normal to the surface, and the **angle of refraction** r (note the lower-case r here, compared with the capital R for reflection) as the angle between the refracted ray and the normal. Since $i = \theta_i$ and $r = \theta_r$ (the angle between the wavefront and the surface must be the same as the angle between the *normal* to the wavefront and the *normal* to the surface):

$$\frac{\sin i}{\sin r} = \frac{v_1}{v_2}. \tag{2.26b}$$

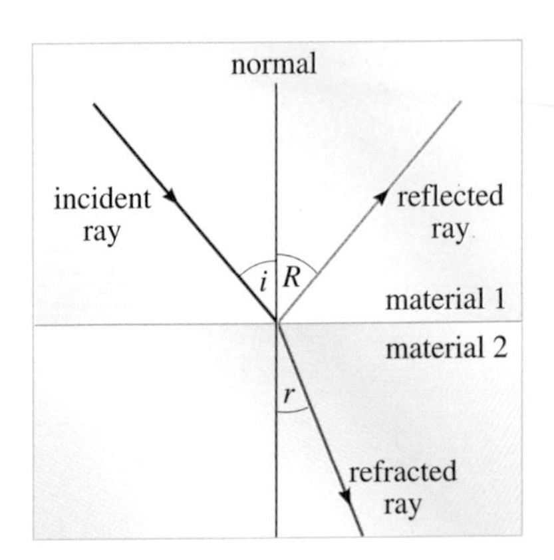

Figure 2.61 The reflection and refraction of a light ray at the boundary between two transparent materials.

Question 2.16 A beam of parallel light strikes an air–glass interface with an angle of incidence of 45.0°. The speed of light in air is $3.00 \times 10^8\,\text{m}\,\text{s}^{-1}$, whereas in the glass the light has a speed of $1.88 \times 10^8\,\text{m}\,\text{s}^{-1}$. What is the angle of refraction of the beam as it passes into the glass? ■

In the case of visible (and near-visible, e.g. ultraviolet and infrared) radiation, it is common practice to express the speed of the wave in a particular medium as a fraction of c, the speed of light in a vacuum. Hence, for the two media shown in Figure 2.61,

$$v_1 = \frac{c}{n_1} \quad \text{and} \quad v_2 = \frac{c}{n_2} \tag{2.27}$$

where the numbers n_1 and n_2 are known as the **refractive indices** of medium 1 and medium 2, respectively. Note that, since v is always less than or equal to c, all refractive indices must be greater than or equal to 1. To illustrate the range of values that occur, the refractive indices of the media listed in Table 2.3 are given in Table 2.4.

Table 2.4 The refractive indices of various materials.

Material	Refractive index, n
air	1.00
water	1.33
olive oil	1.46
crown glass	1.52
quartz	1.55
flint glass	1.62
diamond	2.42

● The speed of a particular beam of green light in a certain kind of glass is $0.65c$. What is the refractive index of this glass for this green light?

○ Since $v = c/n$ (Equation 2.27), the refractive index is

$$n = c/v = c/0.65c = 1.54.$$ ■

Equation 2.26b can now be rewritten using Equation 2.27. Thus

$$\frac{\sin i}{\sin r} = \frac{n_2}{n_1}. \tag{2.28}$$

This is frequently written in the more convenient and compact form

$$\frac{\sin i}{\sin r} = n_{21} \tag{2.29}$$

where $n_{21} = n_2/n_1$ is the ratio of the two *absolute* refractive indices; in other words, it is the *relative* refractive index of the two media. Since, to a very good approximation, the speed of light in air is c (the defined speed of light in a vacuum), it follows that the refractive index of air is very close to unity (the refractive index of a vacuum). Hence, if the incident wave in Figure 2.61 is a light wave travelling in air, $n_1 \approx 1.0$ and n_{21} in Equation 2.29 is effectively equal to n_2. In such a situation, where we only have to be concerned with the refractive index of the *second* medium, it is common practice to dispense altogether with the subscript on n in Equation 2.29.

Equation 2.29, together with the restriction that the incident ray, the refracted ray, and the normal to the boundary interface must all lie in the same plane, constitute what is known as **Snell's law of refraction**. Note that when a wave passes from a 'higher speed' medium into a 'lower speed' medium, the ray is refracted *towards* the normal; when passing from a lower- to a higher-speed medium, the ray is refracted *away* from the normal.

Snell's law of refraction

- The incident ray, the refracted ray and the normal all lie in the same plane.
- The angle of incidence and the angle of refraction are related by

$$\frac{\sin i}{\sin r} = \frac{v_1}{v_2} = \frac{n_2}{n_1} = n_{21} \qquad \text{(Eqns 2.26–2.29)}$$

where v_1 is the speed of the waves in the first medium, v_2 is the speed of the waves in the second medium, n_1 is the refractive index of the first medium, n_2 is the refractive index of the second medium, and n_{21} is the relative refractive index of the two media.

Refraction of light, together with the fact that your brain likes to assume that light entering the eye has travelled in a straight line, explains why the elusive fish never seems to be in exactly the position that you think it is (Figure 2.62a). An expert spear-fisher (or heron) makes allowances for refraction and aims beyond the apparent position of a fish. Refraction of light is widely used in all optical systems that rely on lenses or prisms. These are further discussed in Chapter 3.

Figure 2.62 (a) Refraction at a water–air interface causes an object to appear to be in a different position; (b) a mirage.

Sound waves in the atmosphere are refracted as they pass through layers of air at different temperatures, since the speed of sound in air depends on temperature. That's why sound seems to travel much further in the evening after a hot day; refraction bends the waves that would have disappeared into the atmosphere back down to Earth. Similarly, seismic waves from earthquakes are refracted as the changing density and axial modulus of the rocks inside the Earth cause a change in the speed of the waves (see Figure 2.9). Another example of refraction in nature is the mirage (Figure 2.62b).

6.4 Dispersion

There is one very important difference between the law of refraction as derived here and the law of reflection developed earlier: it has to do with the frequency dependence of the two phenomena. As stated previously, the law of reflection ($i = R$) is *independent* of the type, or frequency, of the wave being reflected; an X-ray wave, a sound wave, red light and blue light are all reflected in such a way that the angle of incidence always equals the angle of reflection. But refraction does not work in quite the same way. As you have just seen, refraction results, in essence, from the fact that the speed of a wave is different in different materials. But that's not the whole story. *The speed of a light wave in any particular material also depends on the frequency of the wave.* For instance, the speed of blue light in water is $2.242 \times 10^8\,\mathrm{m\,s^{-1}}$, whereas the speed of red light in water is $2.252 \times 10^8\,\mathrm{m\,s^{-1}}$. In general, higher-frequency light waves travel more slowly than lower-frequency light waves in the same medium. Consequently, the value of v_1/v_2 in Equation 2.26 will change as the frequency of the radiation changes, and the refractive index of a material is therefore a function of frequency.

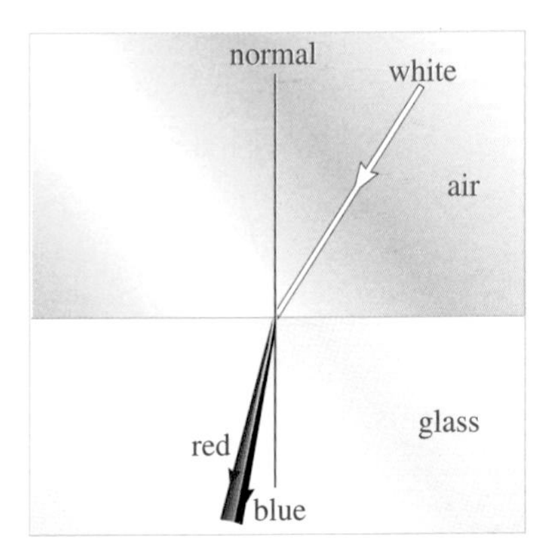

Figure 2.63 White light is dispersed into its constituent colours as it crosses an interface such as air–glass. The speed of light in the medium depends on the frequency, so the angle of refraction is different for each colour.

An important overall consequence of this is that the angle through which a ray of visible light is bent as it crosses an interface will be different for different colours. This phenomenon is known as **dispersion**, so called because if the incident beam were white light (a mixture of all the colours in the visible spectrum), the refraction phenomenon would cause the different colours to be separated out — dispersed — through a range of different angles, as shown in Figure 2.63.

6.5 Total internal reflection

Although reflection and refraction have been treated as if they were completely independent phenomena, the likelihood is that, when a wave strikes an interface, both reflection *and* refraction will occur. Even at an air–glass interface, where you might expect all the light to be refracted into the glass, up to about 10 per cent of the incident energy may be reflected back into the original medium (the air). You have almost certainly witnessed this phenomenon for yourself by observing your own reflection in, for example, a bus or train window.

But there is one particular circumstance where the *refracted* energy at an interface — which you might normally expect to comprise in excess of 90 per cent of the total energy — can be *completely suppressed*. It occurs when a wave is passing from a medium of high refractive index (low wave speed) to one of lower refractive index (higher wave speed) — say, for example, from glass into air. This is illustrated in Figure 2.64. In (a), the angle of refraction is about 45°, and the refracted beam might be expected to carry most of the energy from the incident beam. As the angle of incidence is increased, however, the situation will eventually occur where the angle of refraction r equals exactly 90°, as shown in Figure 2.64b. The value of i corresponding to $r = 90°$ is the largest possible value of *i for which refraction can still take place*. This particular value of the angle of incidence is called the **critical angle**. Its value can be obtained by substituting the value $\sin r = \sin 90° = 1$ into any form of the refraction equations. For example, using Equations 2.26 and 2.28,

$$\sin i_{\mathrm{crit}} = \frac{v_1}{v_2} = \frac{n_2}{n_1}. \qquad (2.30)$$

Values of critical angles with respect to air for some materials are given in Table 2.5.

Table 2.5 The critical angles of various materials with respect to air.

Material	Critical angle for material/air interface
water	48.8°
olive oil	43.2°
crown glass	41.1°
quartz	40.2°
flint glass	38.1°
diamond	24.4°

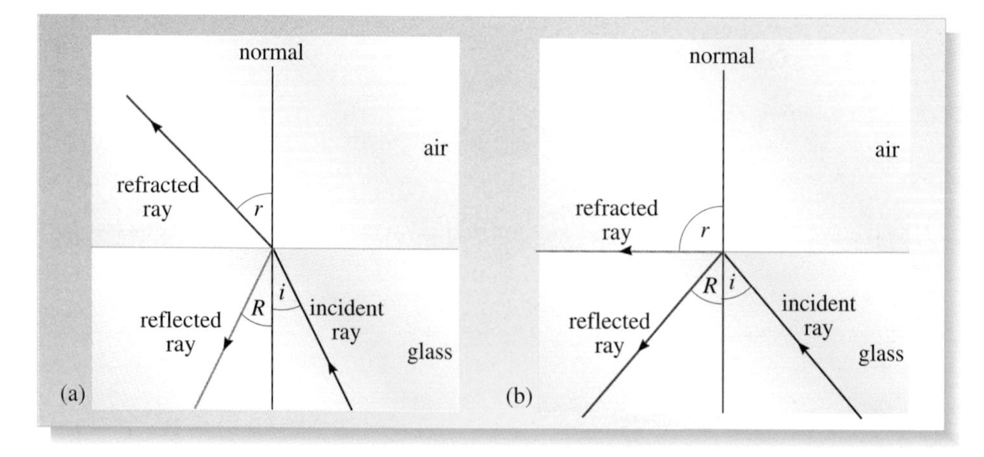

Figure 2.64 Reflection *and* refraction at a glass–air interface: (a) when the angle of refraction is about 45° and the refracted beam carries most of the incident energy; (b) when the angle of incidence is such that the angle of refraction is 90°, and the *reflected* beam carries virtually all the incident energy. The particular angle of incidence shown in (b) is known as the *critical angle*.

- Calculate the critical angle for light passing from a vacuum into a material of refractive index 1.50.

❍ There is no critical angle for this situation! When a wave passes from a medium of high speed to one of low speed, the angle of refraction is always *less* than the angle of incidence. ■

Of course, there is nothing to prevent the angle of incidence being increased to values greater than the critical angle. What do you think would happen to the *refracted* ray under these circumstances? Clearly, the angle of refraction cannot be increased beyond 90°; in fact, when the angle of incidence is greater than the critical angle, no light at all is refracted into the second medium. The energy, which has to go somewhere, is now *all reflected* back into the original medium. Such behaviour is called **total internal reflection**.

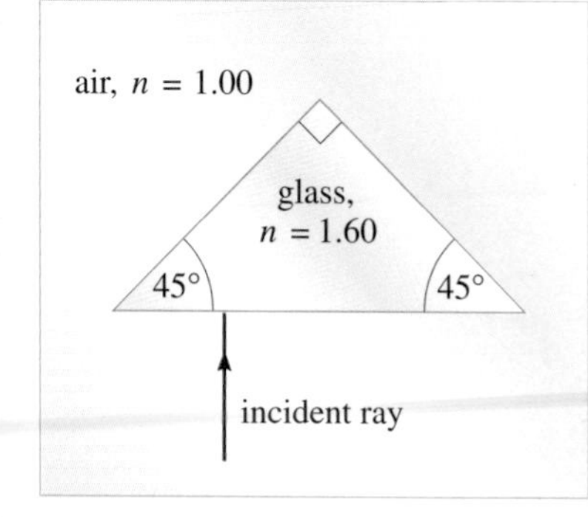

Figure 2.65 For use with Question 2.17.

Question 2.17 Figure 2.65 shows a triangular glass prism, in which the apex angle of the triangle is 90° and the other two angles are 45°. A ray of light, travelling in the plane of the paper, is incident normally on the 'base face' of the triangle, as shown. Given that the refractive index of the glass is 1.60, and that the refractive index of air is 1.00, sketch on the diagram the subsequent progress of the incident ray. ■

The basic principle illustrated by Question 2.17 has important practical applications. When the idea is extended into three dimensions, the triangular-shaped piece of glass is transformed into a *corner-cube* — that is, a piece of glass shaped like the sliced-off corner of a cube. This device then has the optical property of reflecting an incident ray of light — no matter from what direction it is incident — back in the same direction. The corner-cube is the basis of the rear reflectors used on cars or bicycles, and in 'cat's eyes' on roads; and an array of 100 of these corner-cube reflectors, placed on the Moon's surface by the Apollo 11 astronauts, has enabled scientists to use laser lunar-ranging techniques to obtain accurate measurements of the distance between the Earth and the Moon.

Box 2.3 Optical fibres

In our new 'digital age', information is increasingly encoded into a stream of binary digits (bits), and then transmitted as a sequence of on–off pulses (representing the ones and the zeros). The limiting rate at which the information can be transmitted is determined by the frequency of the 'waveform' used to carry the data. For radio waves, travelling down coaxial cables, this is usually in the region of tens of megahertz, or tens of megabits per seconds. But this can be increased by several orders of magnitude if the signal is converted from electrical pulses to pulses of light and the signal then transmitted down an optical fibre. Infrared radiation is usually used in such high-speed optical systems. The typical frequency of this radiation is in the region of 10^{14} Hz, which in theory would permit data rates approaching 10^{14} bits per second. In practice, however, although optical pulses can be *generated* almost as fast as this (with the latest in sub-picosecond laser technology), optical pulses this short cannot yet be *detected*. But the potential is there!

The optical fibre that carries the infrared signal is made from a cylindical core of glass, typically a few millimetres in diameter, surrounded by a cladding of *less* dense glass (see Figure 2.66). The light is injected into the glass core through the end of the fibre (and therefore at a small angle relative to the fibre's axis). The result of this is that the light is trapped within the core, because the angle of incidence of the light beam at the interface between the core and the cladding will be greater than the critical angle. Consequently, the beam will suffer a multiplicity of total internal reflections, thereby enabling the optical signal to propagate down the fibre.

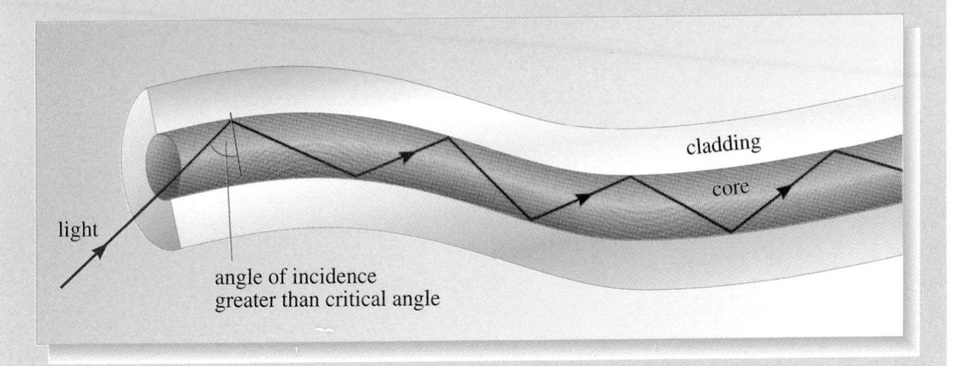

Figure 2.66 Light travels along an optical fibre by multiple total internal reflections at the interface between the fibre's core and its cladding.

Apart from the high data rates, there are a couple of other advantages that optical fibres have over 'electrical' cables. Firstly, because the signals are optical, they are not subject to electrical interference. And secondly, because light waves do not create an external magnetic field (unlike an electric current flowing down a wire), they are far less susceptible to external surveillance, and thus represent a much more suitable option for handling 'secure' exchanges of information.

Optical fibre cables now carry the bulk of undersea transatlantic telephone calls, thereby making possible, among other things, the richness of the Internet and particularly the World Wide Web. Indeed, it's probably true to say that it was optical fibres that made the 'information superhighway' a reality.

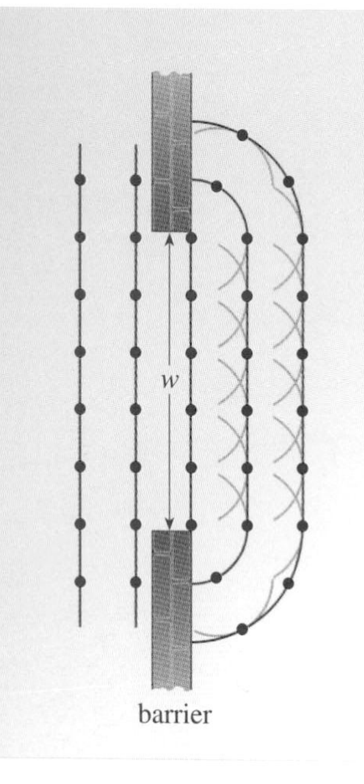

Figure 2.67 The Huygens construction for diffraction at an aperture of width w equal to five wavelengths.

6.6 Diffraction, superposition and interference

Wavefronts, in addition to being reflected by mirror-like surfaces and refracted when they meet an interface between two media of different refractive indices, are also modified whenever they interact with a finite sized 'object', whether that object is in the form of an obstacle, an aperture, or a collection of apertures. Once again the Huygens principle provides a way of picturing what is going on.

Imagine a plane wave approaching a barrier, like a breakwater in the sea. As you can see in Figure 2.67, the barrier obstructs that part of the wavefront that strikes it, letting through only that part of the wavefront that misses the barrier. On applying the Huygens principle to the wavefront at the barrier itself, you can see that the secondary point sources next to the barrier will give rise to wavefronts 'spreading' into the shadow area of the barrier. This spreading — which happens whenever a wave meets an aperture (or an obstacle) — is called **diffraction**.

All types of waves are diffracted: Figure 2.68 shows the diffraction of water waves in a device called a *ripple tank*. You are also probably familiar with the fact that sound waves are diffracted as they pass through a doorway; this is how you can hear what is said inside a room when you are standing outside the room and to the side of the doorway. Diffraction can also be demonstrated when a wavefront meets an *obstacle*, as shown in Figure 2.69.

Note for Open University students. For interest we have supplied you with a *virtual ripple tank* multimedia package to allow you to examine the diffraction of water waves.

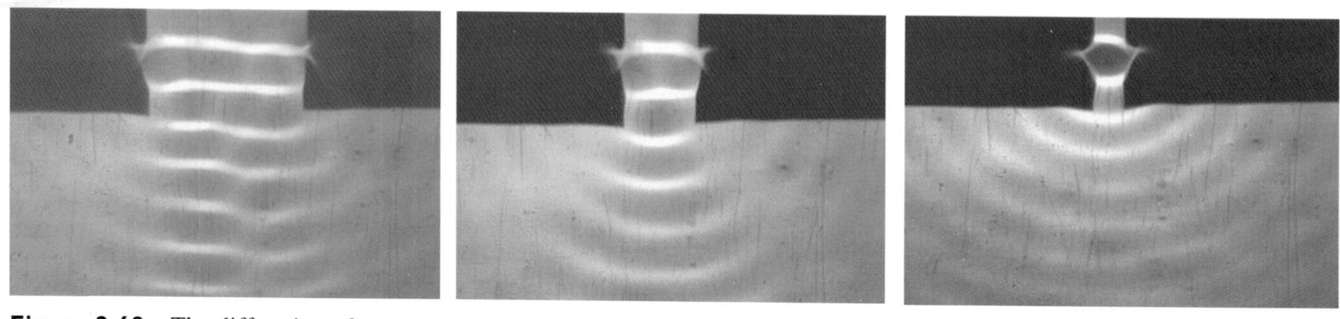

Figure 2.68 The diffraction of water waves. Note that as the slit width is reduced, the amount of diffraction increases.

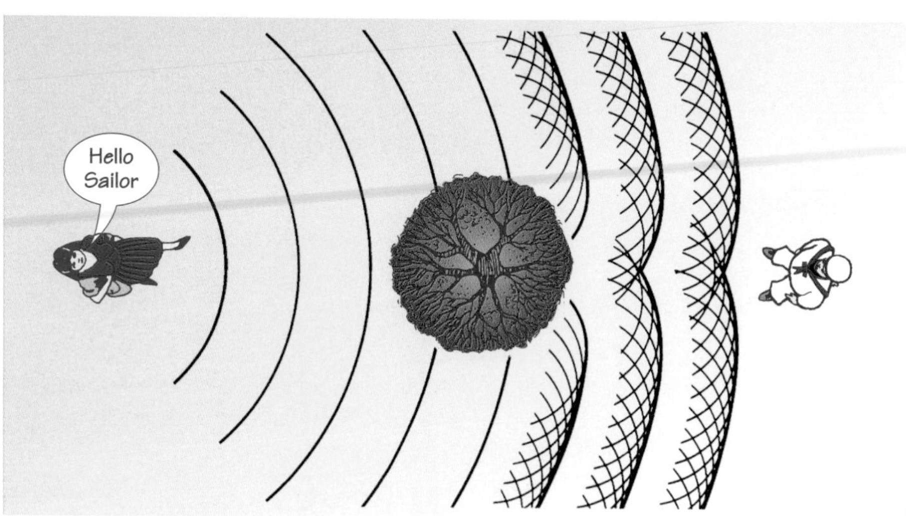

Figure 2.69 Waves are diffracted by obstacles as well as apertures. In this case, sound waves are diffracted by the tree.

Note that the extent to which waves are diffracted depends on the size of the aperture or obstacle relative to the wavelength of the radiation. If the aperture or obstacle is very large compared with the wavelength, then the diffraction effect is rather insignificant. Light waves are not appreciably diffracted by a doorway because the wavelength of visible light (about 400 to 700 nm) is very small in comparison with the width of the doorway. But light *is* diffracted, and provided the slit is narrow enough, the diffraction will become apparent. This provides striking evidence for the fact that light propagates like a wave.

The Huygens principle has given us a good way of predicting how a wavefront will be configured at some time after diffraction. However, it does not predict what will happen when waves from different sources or from different parts of the same source meet. For this, the principle of superposition must be used. As noted in Section 4.2, the principle of superposition is stated in terms of 'adding disturbances'. For electromagnetic radiation the disturbance in question can be thought of as variations in electric and magnetic fields. The effect of the superposition of two or more waves is called **interference**, and it is the final property of all wave motions that will be considered in this chapter.

Open University students should leave the text at this point, and use the multimedia package *Huygens' view of diffraction* which accompanies this book. The activity will occupy about two hours. You should return to the text when you have completed this activity.

6.7 Diffraction by two or more slits

Probably the most convincing evidence that light propagates like a wave was produced by the British physicist Thomas Young in 1801 (Figure 2.70). **Young's two-slit experiment** is important because it provides a simple and clear example of the effects of the superposition of waves.

Figure 2.70 Thomas Young (1773–1829) was a British physician and physicist who revived the wave theory of light, and in 1801 identified the phenomenon of interference. He was a child prodigy and his talents ranged far beyond medicine and physics. He had mastered most European and many oriental languages by the age of 20 and, after studying the texts of the Rosetta Stone, helped to decipher the ancient Egyptian language.

The two-slit experiment can be conducted using many different types of waves. The principle of the experiment, and the results obtained for light, are shown in Figure 2.71. Plane waves from a single, distant, source are diffracted at each of two slits, S_1 and S_2. Both the slits have a width that is small compared with their separation d, which is of similar magnitude to the wavelength. Because the waves are from a common source, they are **coherent** (i.e. the waves are in phase with each other at the slits), and for simplicity **monochromatic** waves (i.e. waves of a single frequency) will be considered. At any position in front of the slits, the waves diffracted by S_1 and S_2 can be combined using the principle of superposition. In the case of light waves, the resulting pattern of illumination on the screen takes the form of a series of light and dark regions called **interference fringes**. The overall pattern of fringes is often referred to as a **diffraction pattern**. (Note, however, that the same pattern is also sometimes referred to as an *interference pattern*. The reason for the dual nomenclature is that both diffraction *and* interference are necessary in order to generate the observed pattern, so either is an appropriate description. However, in this book we shall always refer to *diffraction patterns* made up of a series of *interference fringes*.)

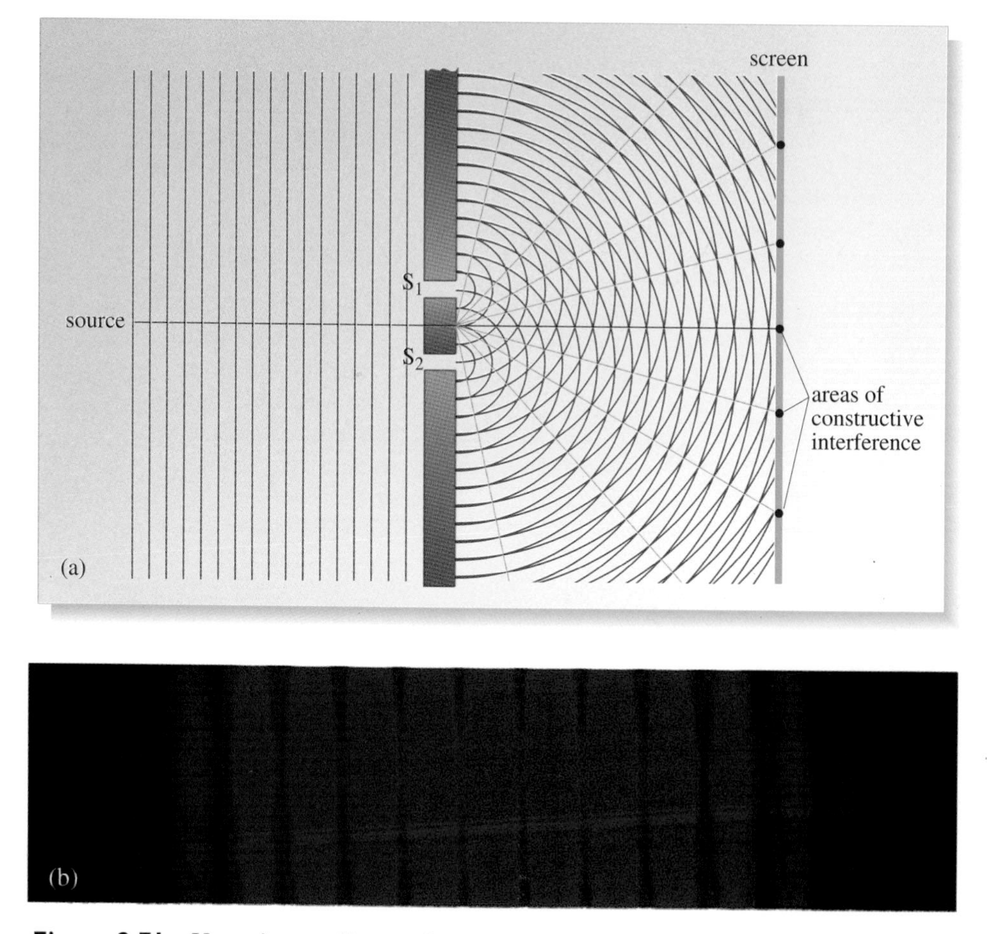

Figure 2.71 Young's two-slit experiment: (a) the principle of the experiment; (b) bright and dark fringes on the screen.

So how does the **double-slit diffraction pattern** arise? When the wave from slit S_1 arriving at a point on the screen is in phase with the wave arriving from S_2, the resultant disturbance will be the sum of the disturbances caused by the waves individually, and will therefore have a large amplitude (as shown in Figure 2.72a). This is known as **constructive interference**. When the waves are completely out of phase, the two disturbances will cancel. This is known as **destructive interference** (as shown in Figure 2.72b). On the screen, constructive interference will cause relatively high intensity, whereas destructive interference will lead to low intensity; this accounts for the observed pattern of fringes.

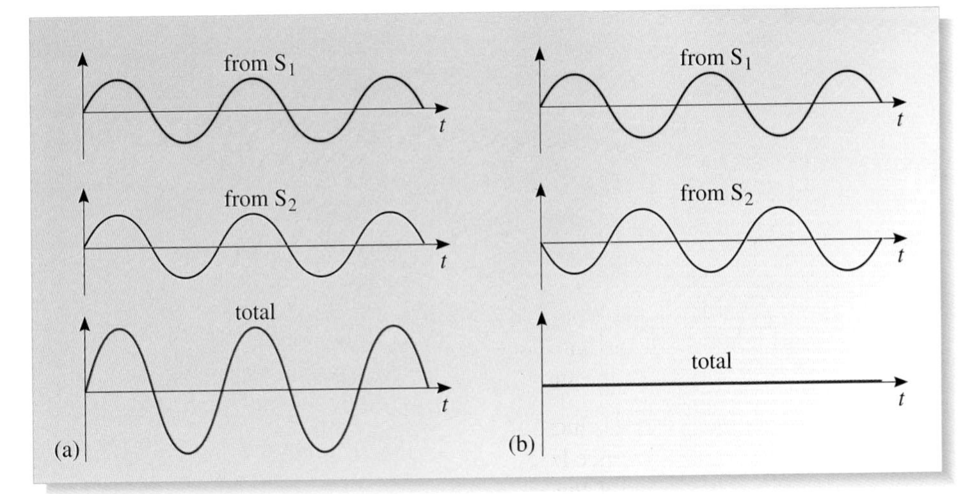

Figure 2.72 (a) Constructive, and (b) destructive interference.

In the context of two-slit diffraction, the general condition for constructive interference at any point is that the **path difference** between the two waves is a whole number of wavelengths; that is

$$\text{path difference} = n\lambda, \text{ where } n = 1, 2, 3, \text{ etc.}$$

The general condition for destructive interference is that the path difference is an odd number of half wavelengths; that is

$$\text{path difference} = (n + \tfrac{1}{2})\lambda, \text{ where } n = 1, 2, 3, \text{ etc.}$$

From Figure 2.73 you can see that, for two slits separated by a distance d, the path difference for waves from slits S_1 and S_2 arriving at the same point on the screen is given by

$$\text{path difference} = S_2N = d \sin \theta.$$

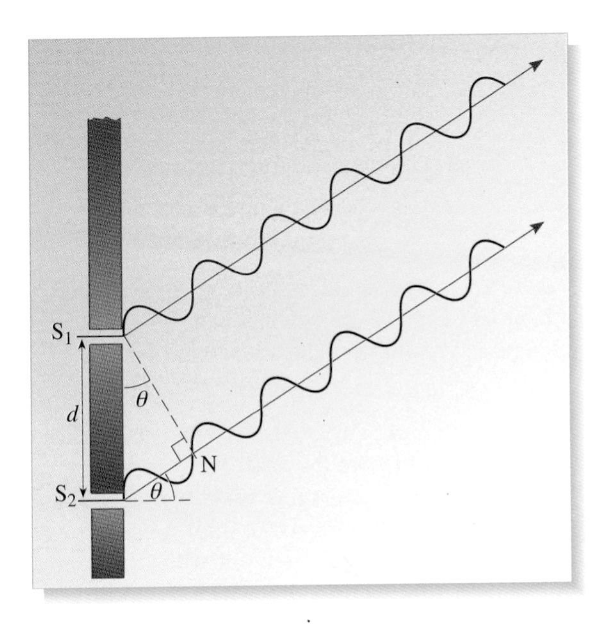

Figure 2.73 The path difference between the waves emerging from the two slits, and travelling towards the same point on the screen, is given by $S_2N = d \sin \theta$ (the thickness of the barrier is assumed to be negligible). The screen is so far from the slits that the two waves must follow (effectively) parallel paths if they are to meet at the screen.

So, for each occurrence of constructive interference

$$d \sin\theta_n = n\lambda. \qquad (2.31)$$

where n is an integer known as the **order of diffraction**, λ is the wavelength, d is the slit separation and θ_n is the angle at which constructive interference occurs relative to the direction of the incident waves. Equation 2.31 is an important equation, sometimes known as the **diffraction equation.**

- For a given value of n what would be the effect of: (a) increasing the slit separation while keeping the wavelength constant; (b) decreasing the wavelength of the light for the same slit separation?

○ (a) From Equation 2.31, if d is increased for constant n and λ, then θ_n must decrease; that is, the interference fringes will become closer together. (b) If λ is decreased for constant n and d, then θ_n will decrease; that is, the interference fringes will get closer together. These effects are illustrated in Figures 2.74a and b. ■

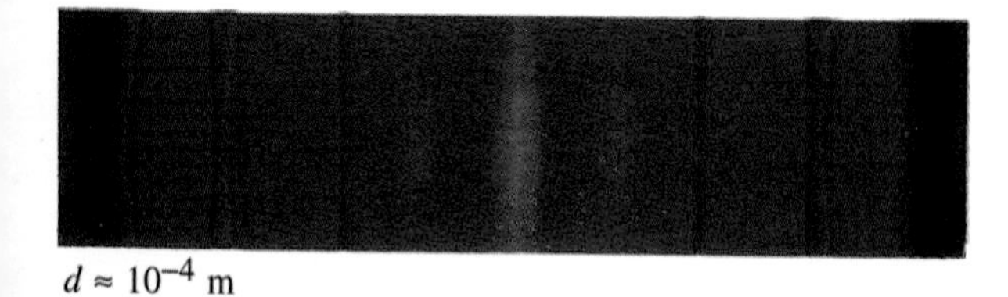

$d \approx 10^{-4}$ m

$d \approx 2 \times 10^{-4}$ m

$d \approx 4 \times 10^{-4}$ m

(a)

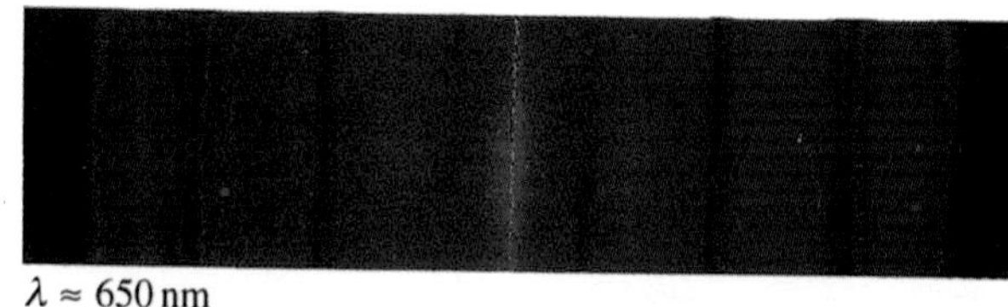

$\lambda \approx 650$ nm

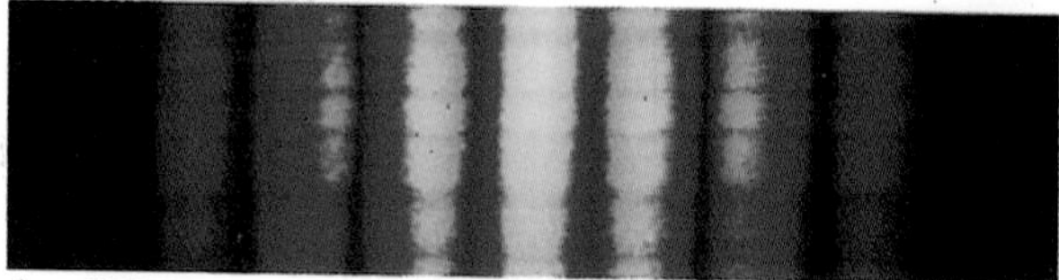

$\lambda \approx 550$ nm

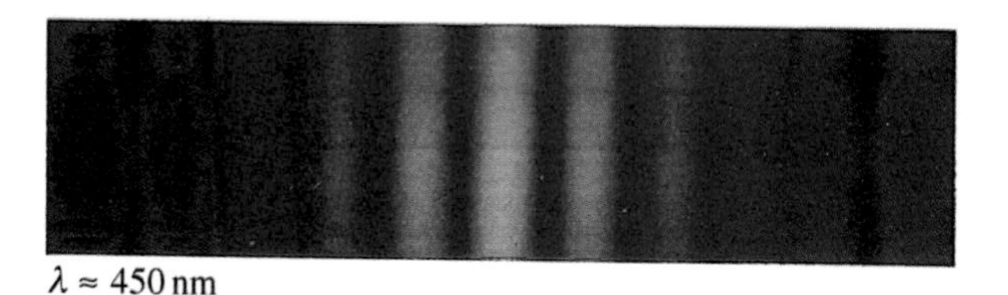

$\lambda \approx 450$ nm

(b)

Figure 2.74 (a) The diffraction patterns observed on a distant screen when double slits with various spacings are illuminated with red light. As the slit spacing increases, the angles at which maximum intensity occurs decrease. (b) The diffraction patterns observed when the double slit is illuminated with light of various colours. As the wavelength decreases, the angles at which maximum intensity occurs also decrease.

A further consequence of Equation 2.31 can be seen by rearranging it to give

$$\sin\theta_n = \frac{n\lambda}{d}.$$

Since $\sin\theta_n$ cannot be greater than 1, this means that an interference pattern is only observed for $n\lambda/d \leqslant 1$. By taking the case when $n = 1$, this means that $\lambda/d \leqslant 1$; that is, $\lambda \leqslant d$. In other words, if the wavelength is greater than the slit separation, no diffraction pattern will be obtained.

The fact that the two slits are illuminated by the *same* very distant light source in Young's experiment is highly significant. If this were not the case, the relative phases of the light at the two slits would be constantly changing in an unpredictable manner (the light at the two slits would be **incoherent**) and so the interference effect would be unobservable. However, the experiment can be carried out with a light source that is not monochromatic, for example with white light (which comprises all the colours in the visible spectrum). In this case, different coloured fringes are obtained, corresponding to interference maxima for each of the wavelengths involved. A double-slit diffraction pattern for white light is shown in Figure 2.75.

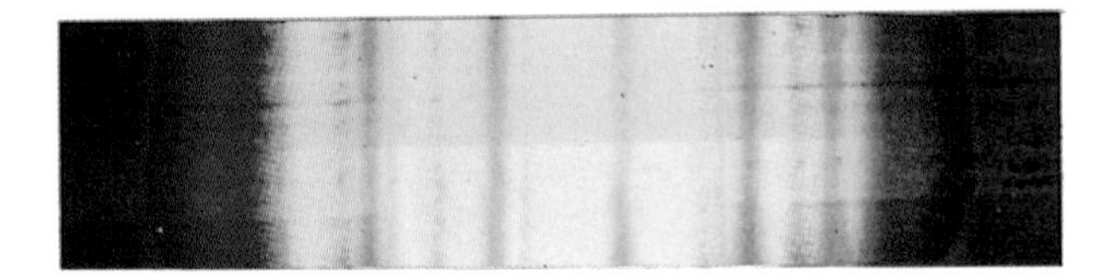

Figure 2.75 The double-slit diffraction pattern for white light.

Equation 2.31 also applies when not two but a large number of equally spaced slits are used. Such **diffraction gratings** typically have several hundred slits per millimetre and give much sharper diffraction patterns than double slits (Figure 2.76). The underlying principles are similar; in this case d in Equation 2.31 is the **grating spacing** — that is, the centre-to-centre separation of adjacent slits.

Figure 2.76 The diffraction patterns produced when a diffraction grating is illuminated with (a) monochromatic light, and (b) white light.

Question 2.18 A diffraction grating with 700 lines per millimetre is illuminated perpendicularly with red light from a helium–neon gas laser. If the second order of the diffraction pattern is seen at an angle of 62.4° with respect to the incident beam, what is the wavelength of the red laser light? ■

So far, a diffraction grating has been described as an array of slits which transmit light: this type is properly described as a *transmission* diffraction grating. Similar effects are obtained if a regular array of narrow lines is etched onto a mirror surface. This gives a *reflection* diffraction grating; Equation 2.31 applies whether the pattern is viewed by transmission or reflection. Note, however, that Equation 2.31 is valid only when the incident light is at a right-angle (i.e. normal) to the plane of the slits.

6.8 Diffraction by a single slit

It turns out that a series of bright and dark fringes is obtained even from diffraction of monochromatic light by a *single* aperture, as indicated in Figure 2.77. Here the waves from seven secondary sources within the slit have been drawn; their propagation is shown in two different directions. The slit width is w.

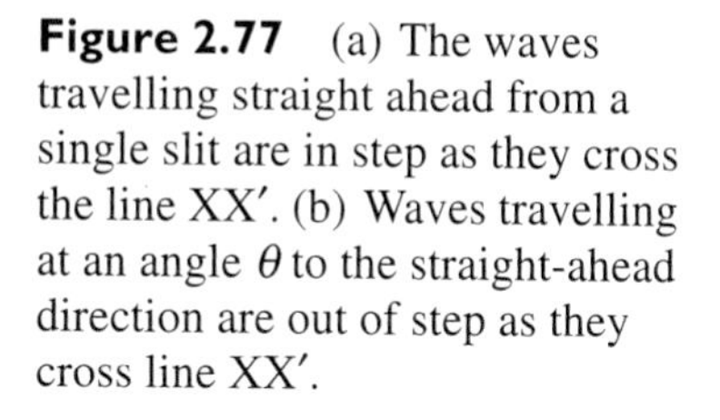

Figure 2.77 (a) The waves travelling straight ahead from a single slit are in step as they cross the line XX′. (b) Waves travelling at an angle θ to the straight-ahead direction are out of step as they cross line XX′.

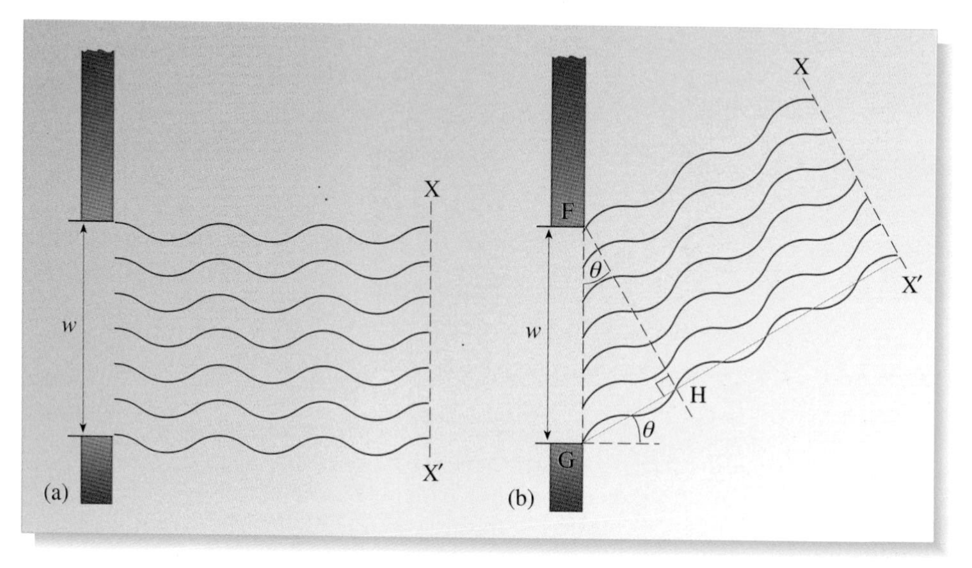

First, consider the secondary waves travelling straight ahead (Figure 2.77a). Since all seven secondary waves originate from the same wavefront, they are all in phase and will remain in phase as they travel away from the slit. Adding the disturbances produces constructive interference; that is, the resultant disturbance has a large amplitude and hence a large intensity. The result would have been equivalent no matter how many point sources had been drawn on the original wavefront.

Now consider the waves travelling at the angle θ shown in Figure 2.77b. This angle has been chosen very carefully such that the wave from the top of the slit is exactly out of phase with the wave from the midpoint, so these two disturbances cancel; that is, there is destructive interference. In the same way, the second wave from the top interferes destructively with the fifth wave from the top. At any point on the top half of the slit, there will be a wave from the bottom half that arrives out of step and interferes destructively. The result is that there is complete cancellation of all the waves travelling in the direction θ, and so the total light intensity in this direction is zero.

Applying simple trigonometry to triangle FGH shows that

$$\sin\theta = \frac{\text{GH}}{w}.$$

At this particular angle the distance GH is one wavelength, so the first *minimum* in the diffraction pattern occurs at an angle θ to the incident beam, where

$$\sin\theta = \frac{\lambda}{w}. \qquad (2.32)$$

For angles between zero and θ, there will be only partial cancellation of the waves, and the light intensity will have a value somewhere between zero and the intensity in the straight-ahead direction. For angles greater than θ, the waves will also partially cancel and the intensity will always be much smaller than in the forward direction.

The intensity can be calculated for *any* angle θ, although the mathematics required is beyond the scope of *The Physical World*. A typical plot of intensity versus diffraction angle is shown in Figure 2.78. Note that the central maximum is flanked by minima, outside which are much weaker subsidiary maxima. This is known as the **single-slit diffraction pattern**.

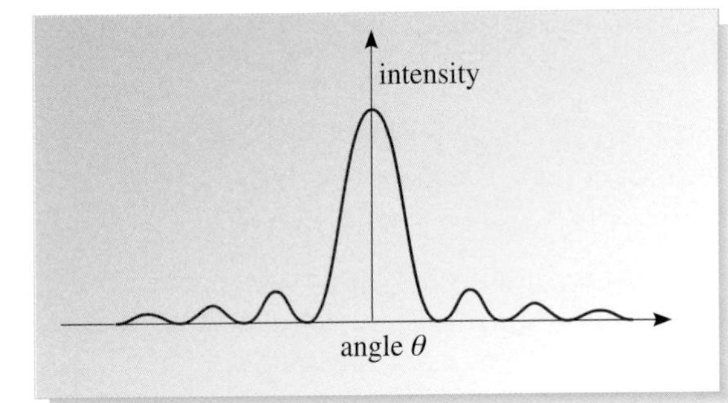

Figure 2.78 A single-slit diffraction pattern; the subsidiary maxima are all the same width, but of decreasing intensity moving away from the centre of the pattern.

Question 2.19 An aperture of variable width w is illuminated by plane waves of yellow light, wavelength 589.0 nm. (a) Find the three values of w such that the first diffraction minima appear, on each side of the straight-through direction, at 20.0°, 40.0° and 60.0°, respectively.

(b) At each of the three calculated values of w, find the angles of the first diffraction minimum for green light of wavelength 546.1 nm. ■

6.9 Applications of diffraction and interference

X-ray diffraction

Some diffraction gratings occur naturally in the world around us. For example, the regular spacing of the atoms in a crystal is about 0.1 nm, and the atomic layers in a crystal act as a very good diffraction grating for X-rays, which have wavelengths in this region (see Figure 2.79). The possibility of using X-rays in this way was proposed in 1912 by Max von Laue, and confirmed by Friedrich and Knipping. However, William and Lawrence Bragg later showed that crystals could be used as reflection gratings as well as transmission gratings, and successfully developed this technique to study the arrangement of atoms in crystals. X-ray diffraction was used by Rosalind Franklin, James Watson and Francis Crick to determine the structure of DNA, and by a number of other scientists to solve the structures of haemoglobin, insulin and many other biological molecules.

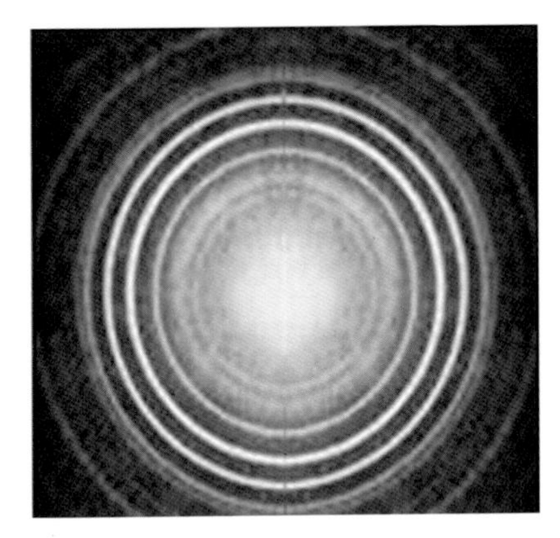

Figure 2.79 The diffraction pattern obtained when X-rays are diffracted from a target made of zirconium oxide. Because the zirconium and oxygen atoms in the crystal form a three-dimensional structure, the diffraction pattern is more complex than those produced using simple (one-dimensional) diffraction gratings.

Beetles and peacocks

The diffraction and interference of visible light is responsible for coloration in some insects and in substances such as opals and mother of pearl. Grooves on the back of the *Serica* beetle are responsible for the beautiful colours seen when the beetle is viewed from certain angles, as shown in Figure 2.80. When light is reflected from the grooves, they act as a diffraction grating. Constructive interference occurs at

Figure 2.80 Diffraction and interference of reflected light is responsible for the colours seen on the *Serica* beetle.

different positions for different wavelengths (i.e. colours), so, when viewed from a certain angle, some colours will appear bright but others will not.

- Grooves are visible in Figure 2.80 along the length of the *Serica* beetle's body. These are *not* the grooves responsible for the coloration, which can be ascertained just by looking at Figure 2.80. How?

❍ The orders in the diffraction pattern from a grating are always separated in the same direction as the lines on the grating: if the lines are separated horizontally, the diffraction orders are spaced horizontally too. In Figure 2.80, coloured bands occur *across* the beetle's body, so the grooves responsible for diffraction must be *across* the body too. ■

Question 2.20 The feathers of a peacock's tail (Figure 2.81) contain layers of cells in a regular pattern, repeating at intervals of about 500 nm. This forms a natural reflection diffraction grating for light. If white light is shining at normal incidence onto the tail, show that only the green to violet part of the spectrum can appear with first-order diffraction maxima. Explain briefly why the reflected colours change slightly as the viewing angle changes. ■

Figure 2.81 The layers of cells in the feathers of a peacock's tail form a natural reflection diffraction grating with a spacing of about 500 nm.

Butterflies and soap bubbles

Interference effects are not limited to those caused by the superposition of diffracted wavefronts. The coloration of the *Morpho rhetenor* butterfly (Figure 2.1), which was mentioned in Section 1, is caused by interference between waves reflected at different surfaces.

Similar effects are produced by **thin-film interference** in soap bubbles (see Figure 2.82a) and oil films. For simplicity, the latter will be considered first. Part of the light is reflected from the front of the film, part is reflected from the back surface, and the major part is transmitted into the air on the far side of the film. The colours arise from interference between the reflected waves. Constructive interference occurs when the two reflected waves are exactly in phase. Whether or not this is the case depends on the thickness of the film relative to the wavelength of the light, and on whether or not the wave displacement is inverted as it is reflected. It turns out that when waves of any sort come to a boundary at which the wave speed *decreases*, then reflection causes the waves to *invert*; at a boundary where the speed *increases*, the reflected wave is *not* inverted. So, in the case of reflection at the front and back of the soap bubble shown in Figure 2.82b, the wave reflected from the front surface is inverted with respect to the wave transmitted at the surface; in contrast, the wave reflected at the back surface is not inverted with respect to the transmitted wave (Figure 2.82c). In the case of both soap films and oil films, a flat film of uniform thickness would appear a uniform colour. As various colours are observed, this suggests that the film has a variable thickness.

Question 2.21 A horizontal soap film in air appears black at its thinnest point when viewed by reflected light. However, a thin film of oil floating on water appears bright at its thinnest portion when viewed in a similar way. What can you deduce from this about the relative speeds of light waves travelling in the materials involved? (You may assume that the path difference between the top and bottom of the film in each case is zero.) ■

The coloration of a *Morpho rhetenor* butterfly's wing is caused by interference between light reflected from several layers of transparent material in tiny fern-like structures above the wing, as shown in Figure 2.83a. The layers are about 60 nm thick and are separated by about 130 nm. Light is reflected at both surfaces of each layer, and the superposition of the reflected waves leads to interference effects, and hence to bright

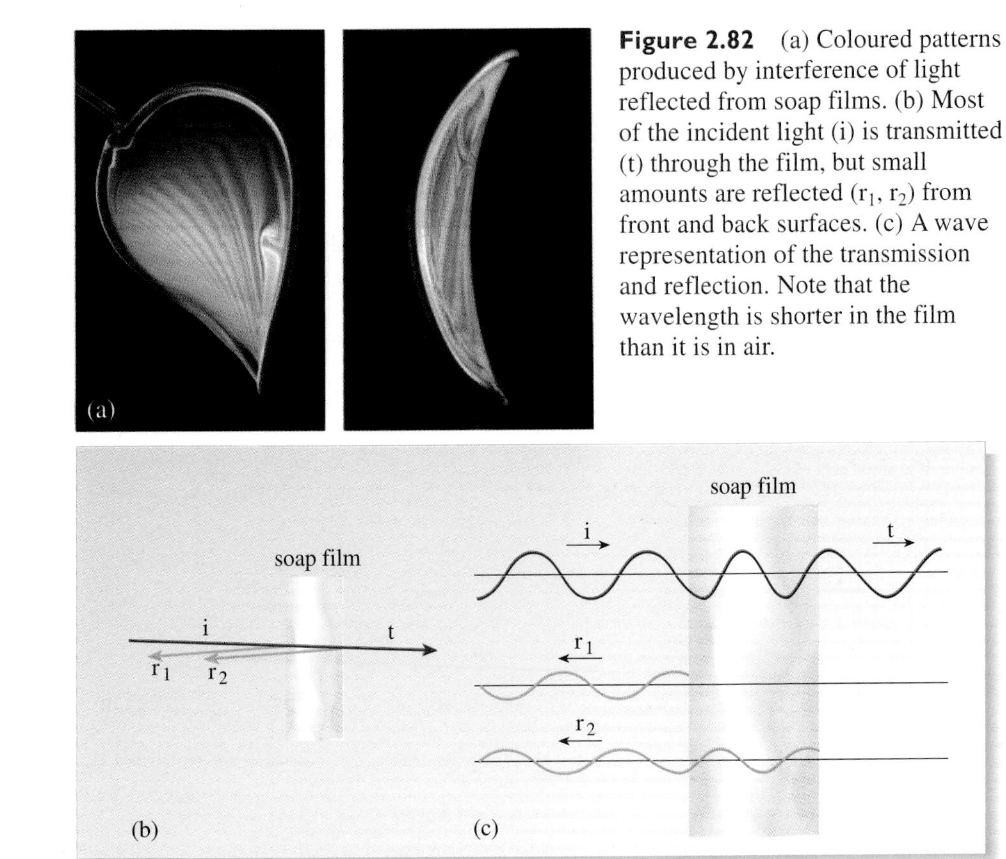

Figure 2.82 (a) Coloured patterns produced by interference of light reflected from soap films. (b) Most of the incident light (i) is transmitted (t) through the film, but small amounts are reflected (r_1, r_2) from front and back surfaces. (c) A wave representation of the transmission and reflection. Note that the wavelength is shorter in the film than it is in air.

iridescence at certain wavelengths. It turns out that the superposition of waves from the top and bottom of any one layer gives constructive interference for wavelengths in the ultraviolet rather than the visible part of the electromagnetic spectrum: reflections between waves from one layer and the next are the most likely candidates for the observed effects in the visible spectrum.

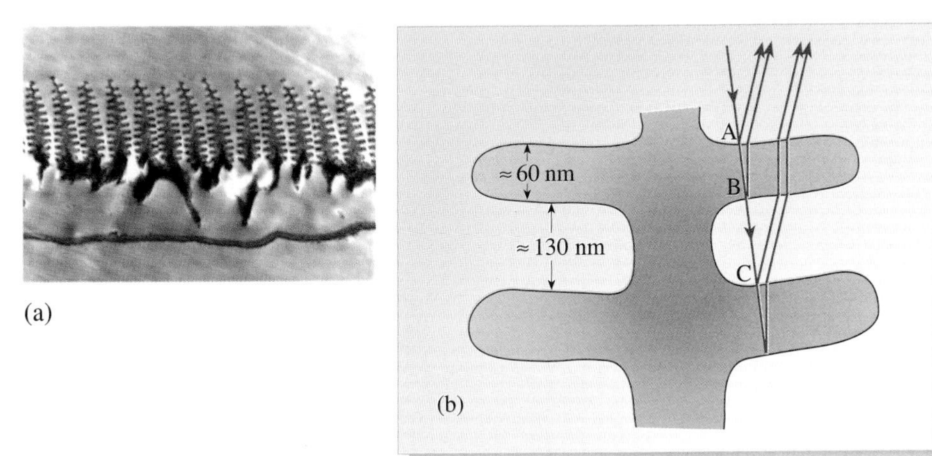

Figure 2.83 (a) A transmission electron micrograph of the detailed structure of the wing of the *Morpho rhetenor* butterfly. (b) Interference occurs between light reflected at different surfaces. Interference between light reflected at A and that reflected at C is more likely to cause effects in the visible part of the spectrum than interference between light reflected at A and that reflected at B.

Spectrometers

You have seen that a diffraction grating gives rise to various diffraction orders whose angular positions depend on the wavelength of the light. When light consisting of many different wavelengths passes through a diffraction grating, then each diffraction order in the pattern consists of a *spectrum* of the light source. This is of immense practical use in the analysis and investigation of unknown substances.

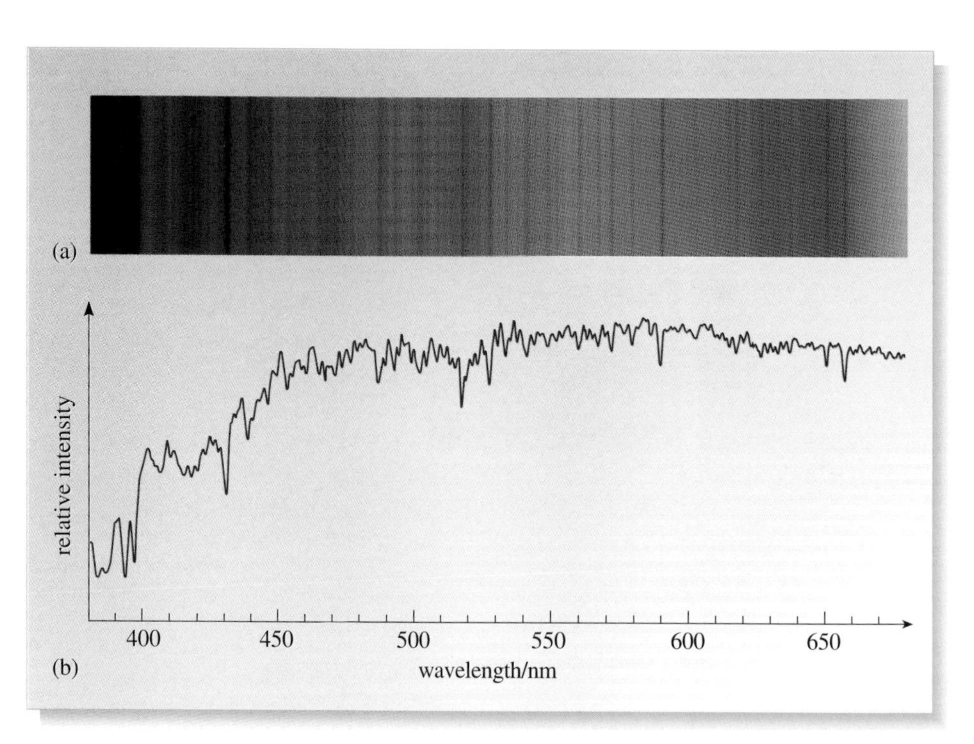

Figure 2.84 A spectrum of a star. (a) One order of the diffraction pattern of the light from the star. Dark bands indicate that light of these colours (wavelengths) is absorbed in the outer layers of the star by certain elements. Since the pattern of wavelength absorption appropriate to each element is known from Earth-based experiments, the elements in the star's atmosphere can be identified. (b) The diffraction pattern converted into a graph of intensity versus wavelength. 'Dips' are referred to as *absorption lines* and correspond to the dark bands in (a).

One of the consequences of quantum physics (see *Quantum physics: an introduction*) is that each type of atom or molecule emits, or absorbs, light of only certain specific colours. This constitutes a kind of 'spectral fingerprint' of the atom or molecule concerned. Diffraction gratings lie at the heart of devices known as **spectrometers**, which are used to analyse the spectra of unknown substances. By identifying just which colours (or wavelengths) are present in the spectrum of an unknown sample, the atomic or molecular composition of the sample can be unambiguously determined. This has been of particular use in the field of astronomy, where a large part of what astronomers have learned about stars and galaxies comes from observations of the light they emit (see Figure 2.84). Here on Earth the spectra of stars and galaxies are observed using spectrometers attached to telescopes, which are discussed further, along with other optical instruments, in Chapter 3.

Interferometers

An **interferometer** is a device that can be used to measure changes in length very accurately by means of interference effects (see Figure 2.85). Light from a monochromatic source S is split by a beam splitter (a part-silvered mirror) M. Half of the light proceeds by transmission to the mirror M_1, and the rest of the light is reflected to mirror M_2. After reflection at M_1 and M_2 the two waves meet again at M. The difference in distance travelled will be $2d_2 - 2d_1$, and the phase difference between the two waves will determine whether they interfere constructively, destructively, or with some intermediate cancellation, at O. An interferometer is usually set up such that the two waves initially interfere constructively, so giving *maximum* illumination at O. If one of the mirrors is moved by a small amount, then the phase difference between the two waves will alter, and the decrease in the amount of illumination at O enables the change in position of the mirror to be calculated.

Much of the early work using interferometers was done by A. A. Michelson, who received the Nobel Prize for Physics in 1907 for his precision optical instruments. In 1887 he collaborated with E. W. Morley, a chemist, in the famous **Michelson–Morley experiment**. A discussion of this experiment makes a suitable end to the

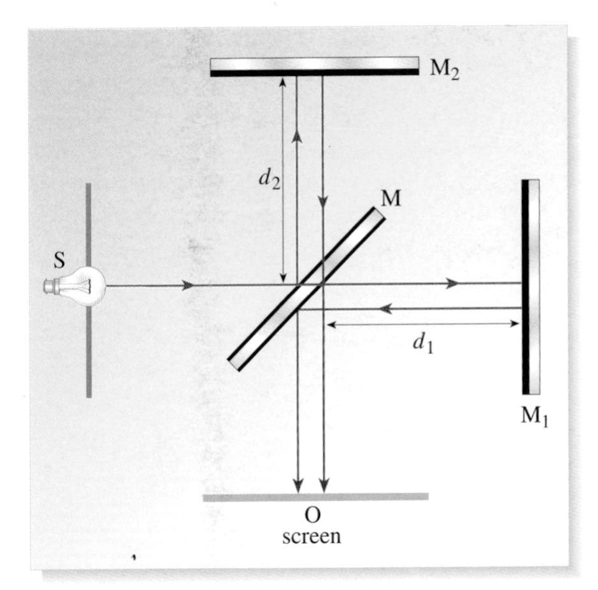

Figure 2.85 The principle of an interferometer.

chapter as it relies on the interference of light and it is also a significant experiment in terms of the history of scientific thought.

At that time it was widely believed that light waves travelled through a massless and incompressible medium called the 'ether'. The Michelson–Morley experiment was an attempt to measure the speed of the Earth relative to this ether. Michelson and Morley attempted to discover whether there is a difference in the speed of light measured in the direction of the Earth's rotation as compared with the speed at right-angles to this direction. If there were such a difference, the relative phase between the waves in the two arms of the interferometer would alter when the instrument was turned through 90°. This would show up as a *change* in the illumination at O between the two positions. The experimental result was somewhat controversial: *no change was detected*. This led to the downfall of the concept of the ether. The fact that the speed of light was found to be a universal constant paved the way for the special theory of relativity, which is discussed in Chapter 4.

7 Closing items

7.1 Chapter summary

1. Commonly, a wave is a periodic disturbance that transports energy from one point to another. If travelling through a medium, the wave does not generally cause a net movement of the particles of the medium in the direction of wave propagation.
2. In a transverse wave the vibrations take place in a direction perpendicular to the direction of propagation, whereas in a longitudinal wave the vibrations are in the direction of propagation.
3. The wavelength λ of a wave is the shortest distance between two points on the wave profile that are oscillating in phase at a given instant of time. The period T of a wave is the shortest interval in time between two instants when parts of the wave profile that are oscillating in phase pass a fixed point. The frequency f of a wave is the reciprocal of its period.
4. Sinusoidal, one-dimensional travelling waves can be described by the equation

   $$y = A\sin(kx \pm \omega t + \phi) \qquad (2.3)$$

   where the angular wavenumber $k = 2\pi/\lambda$ and the angular frequency $\omega = 2\pi/T$.

5 The speed, frequency and wavelength of a travelling wave are related by

$$v = f\lambda \cdot \tag{2.5}$$

6 The intensity of a wave is the amount of energy carried by the wave in unit time across unit area perpendicular to the direction of propagation. In many cases it is proportional to the square of the amplitude of the wave, and for a spherical wave is inversely proportional to the square of the distance from the source.

7 The Doppler effect is the change in observed frequency (and wavelength) due to the motion of the source of waves or the observer, or both:

$$\textit{moving source} \quad f_{\text{obs}} = f_{\text{em}}\left(\frac{v}{v \mp V}\right) \tag{2.12}$$

$$\textit{moving observer} \quad f_{\text{obs}} = f_{\text{em}}\left(\frac{v \pm V}{v}\right) \tag{2.15}$$

These formulae *do not* apply to light.

8 Gamma rays, X-rays, ultraviolet radiation, visible light, infrared radiation, microwaves and radio waves are all types of electromagnetic wave and are therefore transverse. They all travel at the speed of light in a vacuum ($3.00 \times 10^8\,\text{m s}^{-1}$), but they have differing wavelengths and frequencies.

9 The principle of superposition states that if two or more waves meet at a point in space, then at each instant of time the net disturbance at any point is given by the sum of the disturbances created by each of the waves individually.

10 A standing wave is one in which the positions of the nodes and antinodes do not move; it can be regarded as the superposition of two travelling waves.

11 Bounded systems have boundary conditions such that only certain modes of standing wave are allowed. Standing waves on strings, for example, have a node at each end.

12 Reflection, refraction and diffraction can be explained in terms of the Huygens principle, in which each point on a wavefront is regarded as a secondary source of waves. At a later time, a new wavefront can be constructed as the surface that touches all of the wavefronts emanating from the secondary sources.

13 The law of reflection states that when a wave strikes a reflecting surface, the incident and reflected rays and the normal to the surface all lie in the same plane, and the angle of incidence equals the angle of reflection.

14 Snell's law of refraction states that

$$\frac{\sin i}{\sin r} = \frac{v_1}{v_2} = \frac{n_2}{n_1} \tag{2.26 and 2.28}$$

so the direction of a wave changes as it crosses a boundary between media in which the speed of the wave is different. The speed of the wave, relative to the speed of light in a vacuum, can be characterized by the refractive index of the material.

15 When waves pass from a medium in which the wave speed is low to one in which it is high, there is a critical angle of incidence, beyond which total internal reflection occurs. This is the basis behind the operation of safety reflectors and optical fibres.

16 In any medium, different frequencies of light travel at slightly different speeds, so white light is dispersed (spread out) as it crosses a boundary where refraction occurs.

17 Waves are diffracted (spread out) when they encounter obstacles or apertures. In the case of light of wavelength λ passing through a single slit of width w, the position of the first minima in the diffraction pattern (on each side of the straight-through position) is given by

$$\sin\theta = \frac{\lambda}{w}. \qquad (2.32)$$

18 Situations in which interference occurs include those in which waves are diffracted by two slits, or by a grating, with slits separated by a distance d. The diffraction pattern consists of a number of light and dark interference fringes, and the positions of the maxima in the diffraction pattern are given by

$$d\sin\theta_n = n\lambda. \qquad (2.31)$$

19 Diffraction gratings occur in nature in various forms, and interference effects between reflected waves are responsible for the colours seen in thin films.

7.2 Achievements

Now that you have completed this chapter, you should be able to:

A1 Explain the meaning of all the newly defined (emboldened) terms introduced in this chapter.

A2 Write down a mathematical description of a travelling wave in terms of its amplitude, angular wavenumber (or wavelength) and angular frequency (or period).

A3 Describe the behaviour of travelling waves, and work out their velocity, frequency, period or wavelength from supplied data.

A4 Describe the nature of electromagnetic radiation.

A5 Work out the intensity or amplitude of a spherical wave at a certain distance from the source.

A6 Describe the Doppler effect and solve simple problems.

A7 Explain how the Huygens principle can be used to account for reflection, refraction and diffraction.

A8 State the law of reflection and use it to solve problems.

A9 Solve simple problems using Snell's law of refraction and the concept of refractive index.

A10 State the principle of superposition, and use it to explain interference effects and the production of standing waves on a string.

A11 Solve problems involving diffraction through single slits, double slits and diffraction gratings.

7.3 End-of-chapter questions

Question 2.22 A sinusoidal travelling wave has an amplitude of 3.0 cm, a period of 5.0 s and a speed of 15 cm s^{-1}. Give the mathematical function that describes this wave. Assume that the displacement is zero at $t = 0$ and $x = 0$.

Question 2.23 A sound wave with a frequency of 1.00 kHz in air strikes the surface of a lake and penetrates into the water. What are the frequency and wavelength of the wave in water? Assume that the speed of sound in water is 1480 m s^{-1} (for all frequencies).

Question 2.24 The sound intensity 0.25 m away from the speakers at an open-air concert is 10^{-3} W m^{-2}. How far away from the speakers should you stand in order that the music you hear has the same intensity as ordinary conversation with an intensity of approximately 3×10^{-6} W m^{-2}, assuming that no energy is absorbed by the air?

Question 2.25 A train travelling at 25 m s^{-1} is approaching a stationary observer standing at the side of the track. A whistle on the train is blowing and has a frequency of 1500 Hz, as heard in the cab of the train. What is the *change* in frequency that the observer hears as the train passes? Take the speed of sound in air to be 330 m s^{-1}.

Question 2.26 A very narrow beam of light containing waves of two wavelengths 486 nm (blue) and 656 nm (red) propagates through air and hits a glass block at an angle of incidence of 55°. The blue and red waves travel through glass at speeds of 1.80×10^8 m s^{-1} and 1.82×10^8 m s^{-1}, respectively. What is the angular separation between the directions of propagation of the two refracted waves?

Question 2.27 Refer back to Figure 2.64. Given that the angle of incidence i shown in Figure 2.64a is 27.0° when the angle of refraction r is 45.0°, calculate (a) the refractive index of the glass block and (b) the *critical angle*, as depicted in Figure 2.64b, for this sample of glass in air. Assume that the refractive index of the air is 1.00.

Question 2.28 Two sinusoidal waves with the same wavelength and amplitude are travelling in opposite directions on a string, with the same speed of 0.2 m s^{-1}. If the time interval between moments when the string is perfectly flat is 0.5 s, what is the wavelength? Describe the appearance of the string 0.25 s after it is flat.

Question 2.29 A beam of parallel light of wavelength 450 nm is incident normally on a diffraction grating with 600 lines per millimetre. What is the maximum number of orders that can be seen on each side of the central maximum on the far side of the grating? ■

Chapter 3 Optics and optical instruments

1 Seeing the very small and the very distant

As mentioned in the previous chapter, the intensity of the radiation from the Sun peaks in the visible part of the electromagnetic spectrum — the region of the spectrum to which our eyes are sensitive. As a consequence, it is probably fair to say that most of us receive the bulk of our information about the world around us through our sense of sight. It is true that, throughout the past century, our sensory range has been extended by the development of detectors sensitive to other regions of the electromagnetic spectrum (gamma-rays, X-rays, ultraviolet, infrared, microwaves and radio waves), or even to non-electromagnetic radiation, such as acoustic waves or electron waves. Yet the devices and instruments that have the longest history — and to which this chapter is devoted — are *optical* devices designed to directly aid or extend our visual senses in some way. Microscopes, for example, enable us to see objects and organisms — such as the cervical epithelial cells and the sea-urchin eggs shown in Figure 3.1 — that would otherwise be much too small to see with our naked eye. Telescopes, on the other hand, render visible numerous objects (see Figure 3.2) that would otherwise be too distant or faint for us to see directly. And the very fact that we can reproduce images in this book demonstrates the enormous value of yet another optical device — the photographic camera. More mundane, though undoubtedly of the utmost importance, was the discovery of how to construct optical spectacles (and subsequently contact lenses), without which our present way of life would, for many of us, be quite impossible.

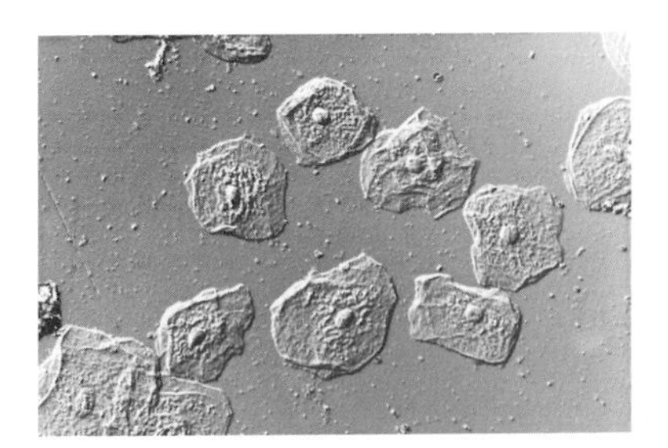

(a)

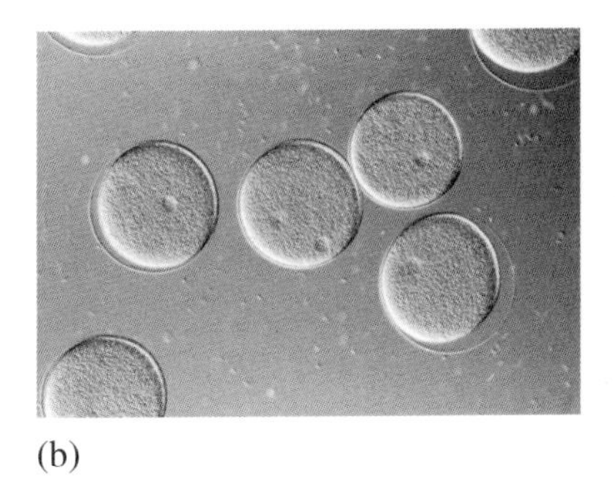

(b)

Figure 3.1 Optical microscope images of (a) epithelial cells from a cervical smear, and (b) sea-urchin eggs.

The ultimate aim of this chapter is to provide you with an understanding of the basic principles around which these optical instruments are designed and constructed. But to begin this discussion, we first need to examine how the laws of reflection and refraction that you met in the previous chapter can be used to explain the function and operation of the basic optical *components* (typically lenses and mirrors) from which such systems are built; this is the subject matter of Section 2. In Section 3 we

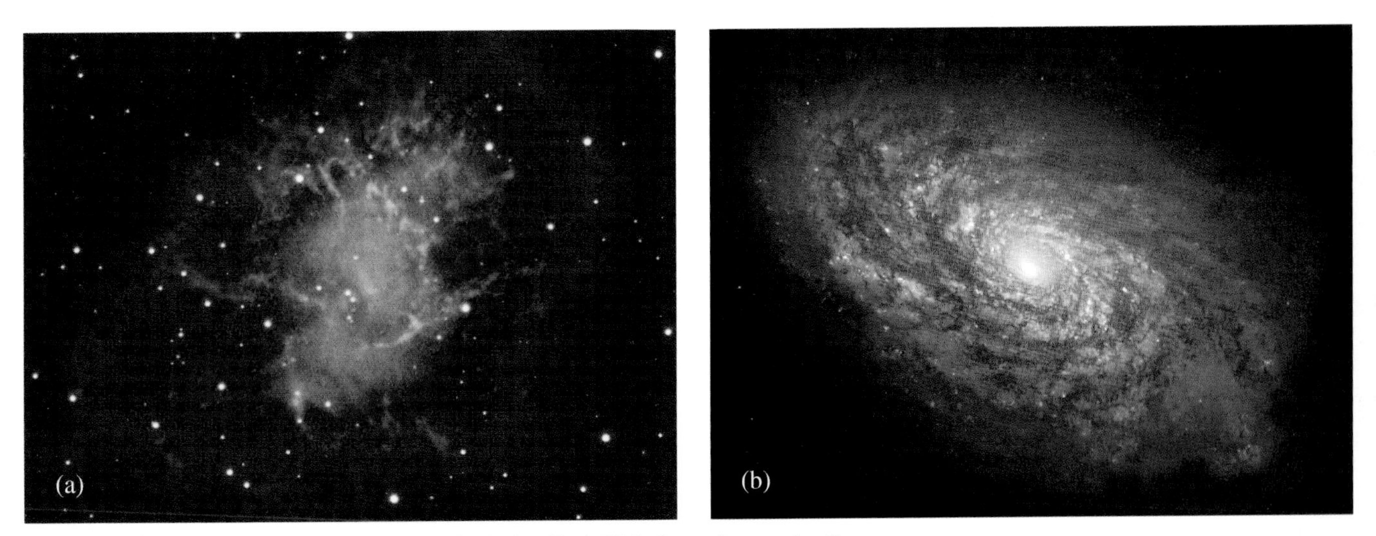

Figure 3.2 Optical telescope images of (a) the Crab Nebula — the result of an exploding star — and (b) a spiral galaxy similar to our own Milky Way galaxy.

then consider how the image formed by a lens (or mirror) is quantitatively related to the object, its size and its distance from the mirror or lens. The operation of various optical *systems* is then dealt with in Section 4.

2 Optical building blocks

In this section, we shall look first of all at a couple of basic *refraction* components (a parallel-sided glass block and a prism), so that the ideas developed here can be used to arrive at a functional description of perhaps the most important of all refraction components — the lens. At the end of the section, the behaviour of *reflection* components (that is, mirrors) is considered — mainly by analogy with lens properties.

2.1 The parallel-sided glass block

The parallel-sided block of glass hardly deserves the title 'optical component'; it's such a basic, simple, optical element. Yet it is an important starting point for studying the passage of light through rather more complex glass shapes. So let's begin by seeing how the ideas of refraction enable you to predict the passage of light through just such a block. Try the following question.

Question 3.1 By considering Snell's law, deduce how the three selected rays of light from a point source would be deviated as they pass through the parallel-sided block of glass shown in Figure 3.3. Answer the question by sketching-in the approximate paths of the three rays shown striking the first face of the block, (i) as they travel through the glass, and (ii) as they leave the glass block and re-enter the air. ■

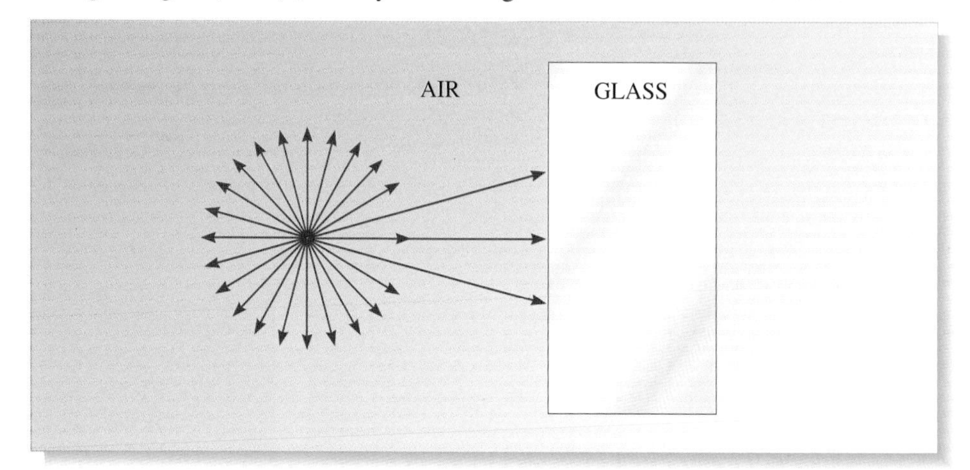

Figure 3.3 Three rays about to enter, pass through, and leave, a parallel-sided block of glass.

Now let's take a more careful look at this. Figure 3.4 shows the upwardly directed ray again, striking the air–glass interface obliquely, and being refracted *towards* the normal according to Snell's law

$$\frac{\sin i_{\text{air}}}{\sin r_{\text{glass}}} = \frac{n_{\text{glass}}}{n_{\text{air}}}.$$

where i_{air} and r_{glass} are as shown in the figure. The ray travels in a straight line through the glass block, and strikes the second (i.e. the glass–air) interface at an angle of incidence i_{glass} that is equal to the previous r_{glass}. At this second interface, the whole refraction procedure is reversed: the *relative* refractive index term in Snell's law now becomes n_{air}/n_{glass}, and the ray is bent *away* from the normal. But because the value of i_{glass} at this second interface is equal to r_{glass} at the first interface, it must follow that r_{air} at the second interface is equal to i_{air} at the first interface. In other words, the ray will emerge from the glass block at an angle identically equal to the angle of incidence with which it entered the block. As you can see in Figure 3.4, this means that the glass block has the effect of leaving the original (angular) *direction* of the ray unchanged, but *displacing the ray laterally.*

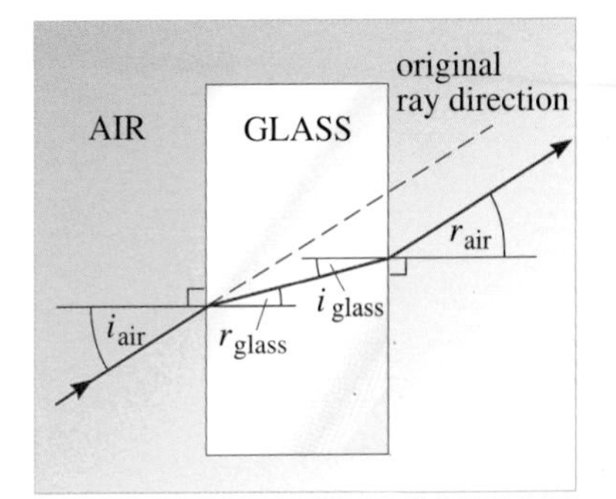

Figure 3.4 A parallel-sided glass block displaces a light ray *laterally*, but leaves the angular direction unchanged.

Question 3.2 Make sure you completely understand the preceding reasoning, by writing down the Snell's law equations for both the first and second interfaces in the glass block of Figure 3.4 and then showing *algebraically* that $r_{air} = i_{air}$. ■

2.2 Prisms

Now consider a triangular prism-shaped block of glass, as shown in Figure 3.5a. Since the two sides are no longer parallel, the refraction at the output side of the block will *not* now be equal and opposite to the refraction at the input side. Consequently, as you can clearly see in Figure 3.5a, there will now be both a lateral displacement of the ray, *and a resultant change in its direction*. As you might guess, the magnitude of this direction change depends on two properties of the prism: the size of the angle between the input and output faces (this is called the *prism angle*, labelled A in Figure 3.5), and the refractive index of the glass. You will also recall that this latter property is frequency-sensitive, so that the angle through which the prism bends blue light will be different from that through which it bends red light.

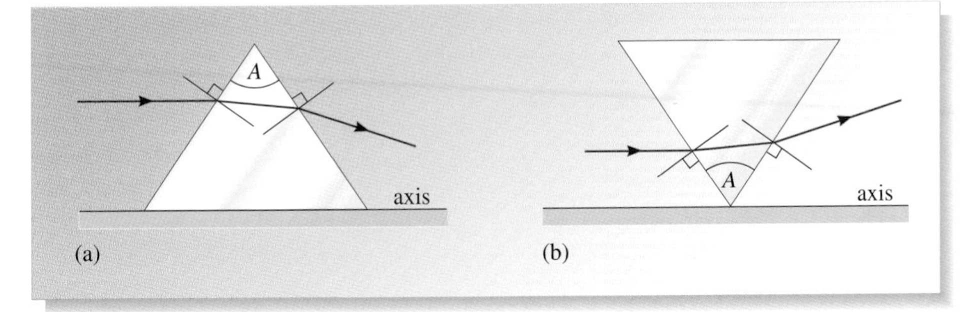

Figure 3.5 (a) A prism with its apex uppermost, deflects a ray towards the axis; (b) a prism with its apex at the bottom, deflects the ray away from the axis. In both cases, the ray is deflected away from the prism's apex.

Figure 3.5 also shows another very important property of the prism: it always bends light *away* from its apex. Consequently, the direction in which an incident ray is deviated depends on whether the prism has its apex (i.e. prism angle) above the base, or below it. If you think of the horizontal line drawn beneath the prisms in Figure 3.5 as an axis for the diagrams, then you can see that the apex-uppermost orientation shown in (a) deflects a ray that is initially parallel to the axis *towards* the axis, whereas the orientation shown in (b) deflects the initially parallel ray *away* from the axis. There is nothing startling about this: it's simply a consequence of the fact that the prism in Figure 3.5b has been turned upside-down relative to that in Figure 3.5a. But as you will now see, this simple idea provides a useful insight into the construction and function of glass lenses.

2.3 Lenses

Suppose a whole set of truncated prism pieces of glass are now stacked one on top of the other, as shown in Figure 3.6a. The top prism here is orientated apex uppermost (as in Figure 3.5a) and, moving down the stack, the prism angle gets progressively smaller. Exactly halfway down is a parallel-sided piece of glass, and below this the prism orientation switches over to that of Figure 3.5b, with the prism angle getting progressively larger to the bottom of the stack.

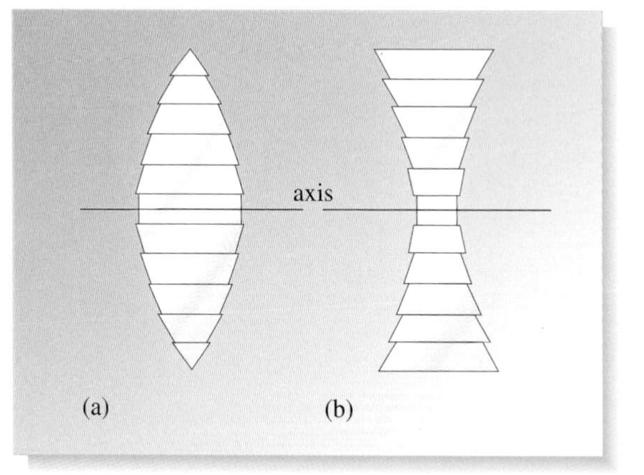

Figure 3.6 Prisms stacked in (a) a positive configuration, and (b) a negative configuration.

But an alternative stacking configuration is possible, as shown in Figure 3.6b. Here, the prisms in the top half have the upside-down orientation of Figure 3.5b, whereas the prisms in the bottom half have the orientation of Figure 3.5a. The two stacking configurations shown in Figure 3.6 will be called positive and negative configurations, respectively.

Suppose the prism angles are now carefully selected so that all incident rays parallel to the axis and in the plane of the page are bent in such a way as to make them, in the positive stacking case, *converge* to a single point on the axis beyond the stack, and in the negative stacking case, *diverge as if* from a single point on the axis on the input side of the stack. These two cases are shown in Figure 3.7.

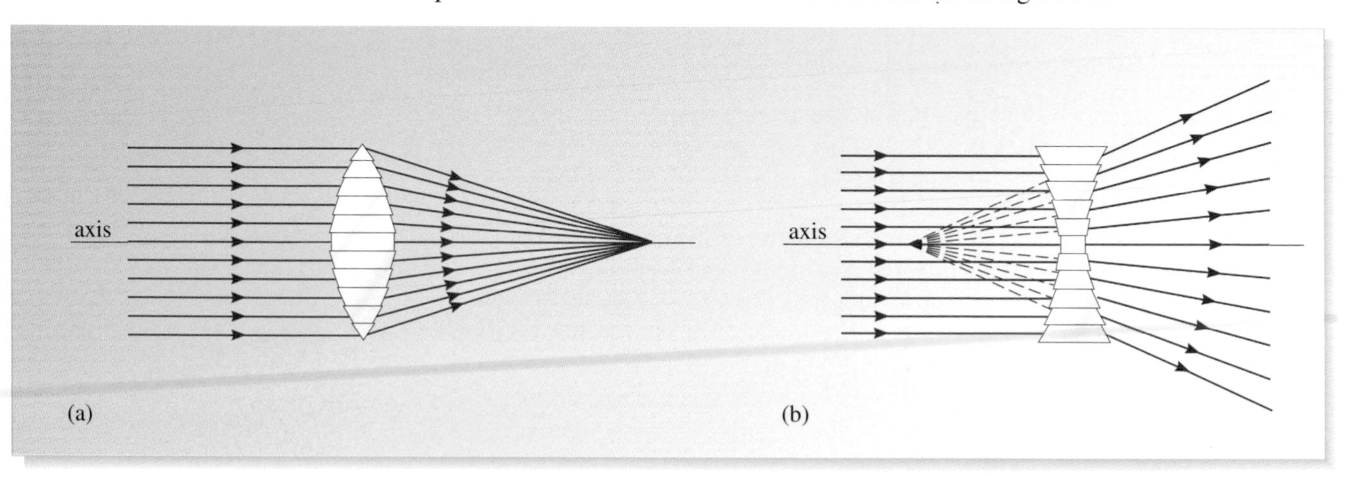

Figure 3.7 (a) A positive stacking configuration of prisms *converging* parallel rays to a point on the axis beyond the stack; (b) a negative stacking configuration *diverging* parallel rays so that they *appear* to originate from a point on the axis on the input side of the stack.

Now imagine increasing the total number of prisms in the stack, while at the same time making the change in prism angle between adjacent prisms smaller and smaller; eventually the surfaces of the stack will become smooth. The stack has become a single, specially shaped piece of glass called a **lens**, and, as Figure 3.8 shows, it can be either a 'positive' or 'negative' lens.

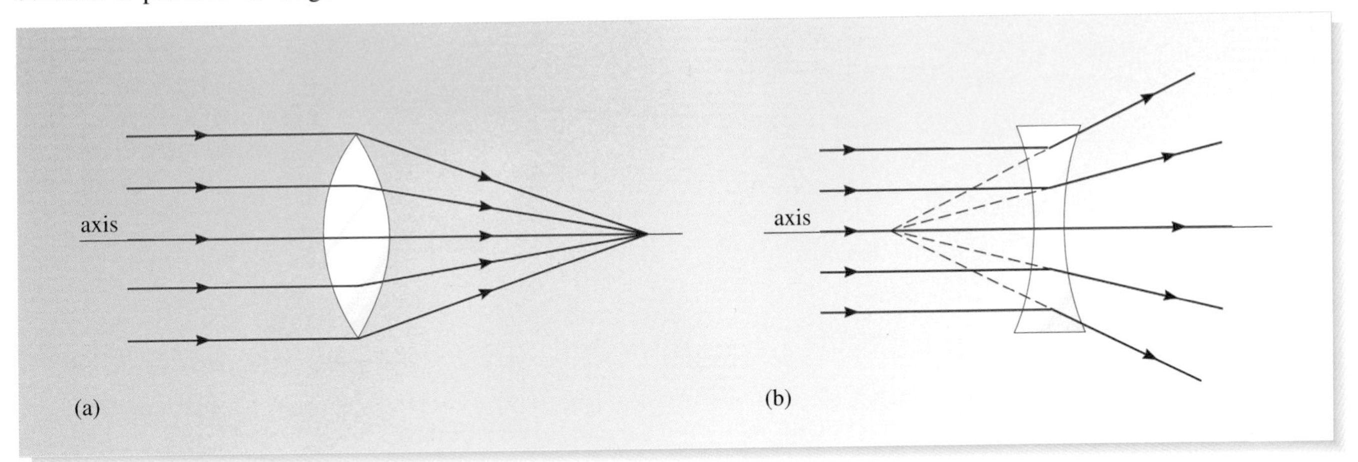

Figure 3.8 When the number of prisms in the stack becomes very large, the stack surfaces become smooth. The two prism-stacking configurations lead to either (a) a positive lens or (b) a negative lens.

There is an important point about the way in which the rays have been drawn in Figure 3.8 that you should notice. As you saw in Figure 3.5, when a prism bends a ray of light it does so in two stages: on entry to the prism, at the air–glass interface, and then again on leaving the prism, at the glass–air interface. The overall deviation is then the combination of these two changes. The same is true of the lenses shown in Figure 3.8: deviation of the rays occurs both on entry into the lens, and on exit from the lens. However, to show this accurately in all diagrams would tend to over-complicate the ray paths that need to be drawn. It has therefore become a standard convention in ray (geometrical) optics to treat the lens *as if* it deviated the rays all in one step at the plane perpendicular to the axis and passing through the centre of the lens. So Figure 3.8 does not really defy the law of refraction; it is merely drawn according to the usual convention — a convention that will be adopted throughout the rest of this book.

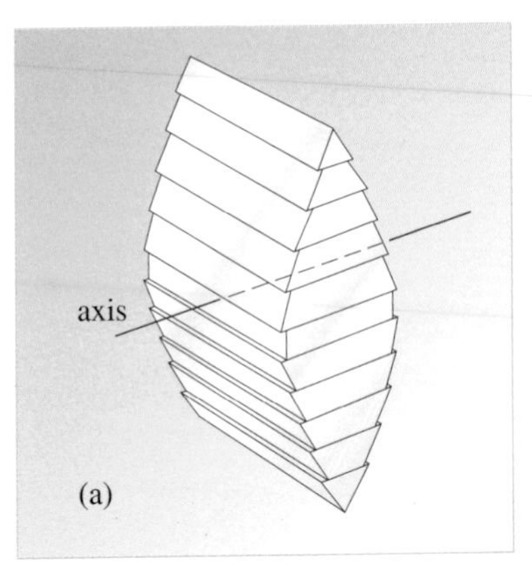

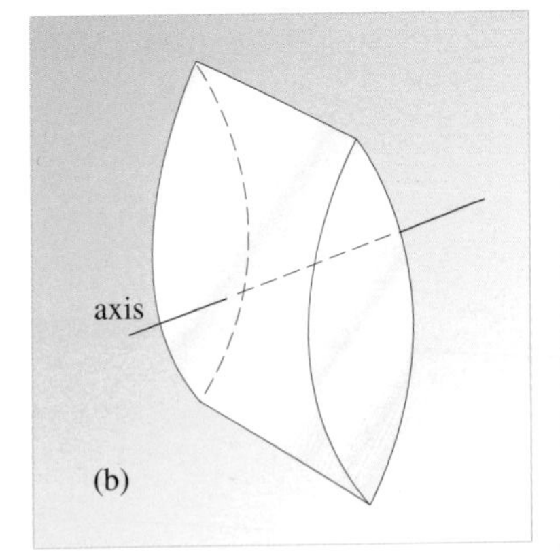

Figure 3.9 A three-dimensional perspective of (a) the positive prism-stacking configuration, and (b) the equivalent positive cylindrical lens.

Now strictly speaking, the two lenses shown in Figure 3.8 will only cause 'vertical' convergence or divergence of the incident rays of light. This, as the three-dimensional perspective of Figure 3.9a should make plain, is because of the way the lenses were built up by stacking prisms together. In fact, the lens shown in Figure 3.9b is called a **cylindrical lens**: it converges (or diverges, in the case of a negative lens) parallel rays to (or from) a horizontal line of points perpendicular to the axis. It has no 'horizontal' focusing effect.

Cylindrical lenses do have their uses in optics, but, for imaging purposes, lenses are much more useful if they can converge (or diverge) parallel rays to (or from) a single point. This requires the lens to have the same cross-sectional profile in *any* axial plane. The most practical way of achieving this is to make each lens surface approximate to a portion of the surface of a sphere. The lens will then be rotationally symmetrical about the axis, and is called a **spherical lens**. Lenses with two spherical surfaces come in a variety of forms: some typical lens cross-sections are shown in Figure 3.10. As can be seen from this figure, these lenses can still be divided into two basic classes or types — those that are *thicker* at the centre than at the perimeter (Figure 3.10a), and those that are *thinner* at the centre than at the perimeter (Figure 3.10b).

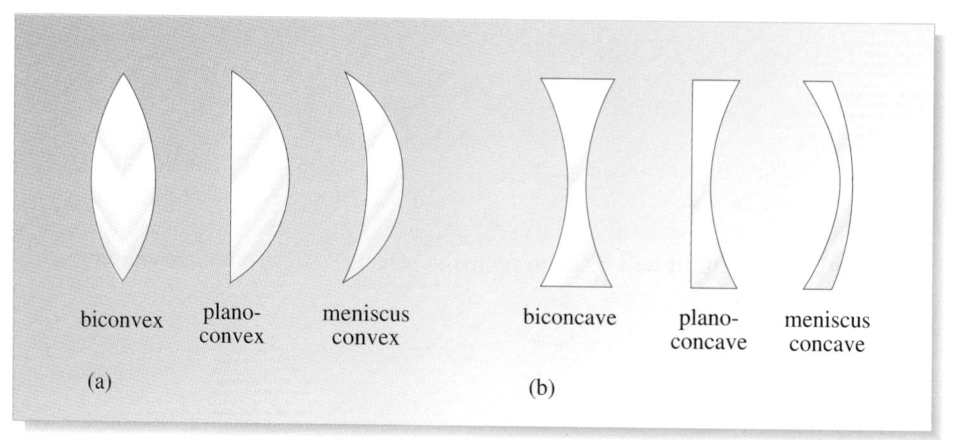

Figure 3.10 Various spherical lens shapes. The lenses in (a) all have profiles that converge parallel light rays; those in (b) diverge them.

All three lens profiles shown in Figure 3.10a are variations on the positive prism-stacking configuration shown in Figure 3.6a. They therefore all *converge* parallel rays to a point. For this reason, such a lens is known as a **converging lens**. This is the most unambiguous way to describe these lenses, since it describes them in terms of their function. However, the terms **convex lens** (because of the shape) and *positive lens* are also frequently used.

By contrast, the lens profiles shown in Figure 3.10b are all derived from the 'negative' prism-stacking configuration of Figure 3.6b; they therefore cause incident parallel rays to *diverge* as if from some point on the input side of the lens. Again, the most unambiguous name for such a lens is **diverging lens**, but you will also come across the terms **concave lens** and *negative lens*.

For both basic lens types discussed above, a special name is given to that point to which the parallel rays converge, or from which they appear to diverge. This point is known as the **focal point** or, in some texts, the *principal focus*. In addition, it is usual to define the axis about which the lenses are rotationally symmetrical as the **optical axis**. The *plane* that passes through the focal point and is also perpendicular to the optical axis is then known as the **focal plane**.

The precise 'bending power' of any lens (and hence the location of its focal point) will depend on the refractive index of the lens material at the optical frequency — colour of the light — of interest, and on the curvature of the lens surfaces (since this is a measure of the range of prism angles used in the implied underlying stacked prisms). These two variables are conveniently combined, for any given lens, into a single *operational* parameter: the distance from the optical centre of the lens to its focal point. This distance is defined to be the **focal length** f of the lens.

Question 3.3 Suggest a way in which the focal length of a converging lens might be measured. ■

Note that provided a lens is *thin* compared with its focal length (a good approximation for most general-purpose single lenses), the focal length can be thought of as a general property of the lens itself. Hence, even when the two surfaces of a particular lens have *unequal* curvature, the focal length of the lens will be the same irrespective of which surface the rays meet first. This is useful to know: it means that it doesn't matter which way round you position the lenses in your optical systems!

2.4 Curved mirrors

Although refraction of a beam of parallel rays by an object with plane surfaces (for example, a single prism) may be able to change the direction of travel of that beam, it cannot by itself focus the beam down to a point. To achieve this, it is necessary to use refraction objects that have *curved* surfaces (i.e. lenses). A similar distinction applies for *reflection* at **mirror** surfaces. A **plane mirror** will change the direction of the parallel rays, but it will have no focusing effect. Reflection at *curved* mirror surfaces, on the other hand, *can* be used to focus the rays.

For example, when parallel rays fall on a *concave* **spherical mirror** then, as shown in Figure 3.11a, they will be reflected to pass through a single point on the optical axis of the mirror. This point is known as the *focal point* of the mirror, and the distance along the axis from this point to the reflecting surface is called the *focal length f* of the mirror. The mirror behaves like a converging lens. The concave mirror is also known as a **converging mirror** or *positive mirror*. Note that a *concave* mirror has the same function as a *convex* lens; to avoid confusion, you're probably safer thinking of lenses and mirrors in terms of their function (converging or diverging) rather than in terms of their shape.

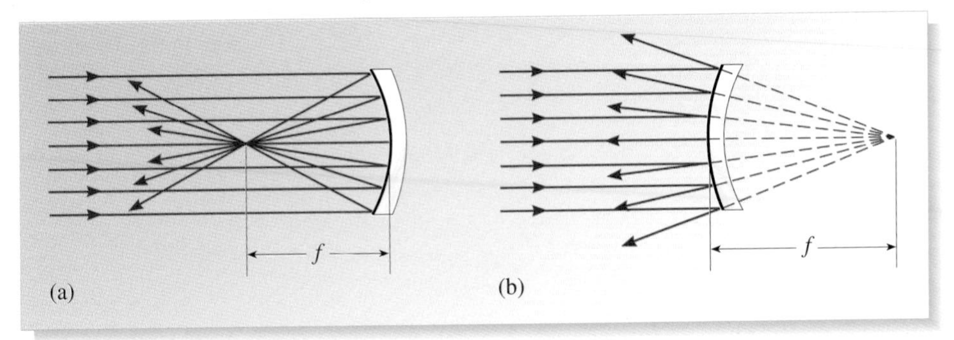

Figure 3.11 Curved mirrors focus parallel rays of light. Note the convention introduced in Chapter 2 of using a thick line to represent the front (i.e. reflecting) surface of a mirror. The concave-shaped mirror in (a) *converges* the rays to pass through the focal point in front of the mirror; the convex-shaped mirror in (b) *diverges* the rays so that they appear to have originated from the focal point behind the mirror.

By contrast, the parallel rays falling on the convex spherical mirror shown in Figure 3.11b are reflected in such a way that they *appear* to have originated from a single point on the optical axis behind the mirror. Again, this point is called the focal point of the mirror, and the distance along the axis from this point to the reflecting surface is the mirror's focal length. This (convex) mirror functions in the same way as a diverging (i.e. concave) lens. It is known as a **diverging mirror** or *negative mirror*.

2.5 Aberrations

As noted above, the 'bending power' of a lens is in part dependent on the refractive index of the material *at the optical frequency of interest*; the reason for this qualification is that, as you know from the previous chapter, the refractive index is different at different frequencies. Consequently, the focal length of a lens will be different for light of different colours. This is illustrated in Figure 3.12: the focal point for red light is further away from the lens than is the focal point for blue light. (Check that this is consistent with your understanding of dispersion; look again at Section 6.4 of Chapter 2 if necessary.) Hence, if the incident rays are of white light, the focal length, and thus the focal point, is not uniquely defined: there will be a rainbow-like blurring of the focal spot. This is the basis of a lens imperfection — an *aberration* — known as **chromatic aberration**. One advantage that mirrors have over lenses is that, since the reflection process is *not* frequency dependent (unlike the refraction process), they do not suffer from chromatic aberration.

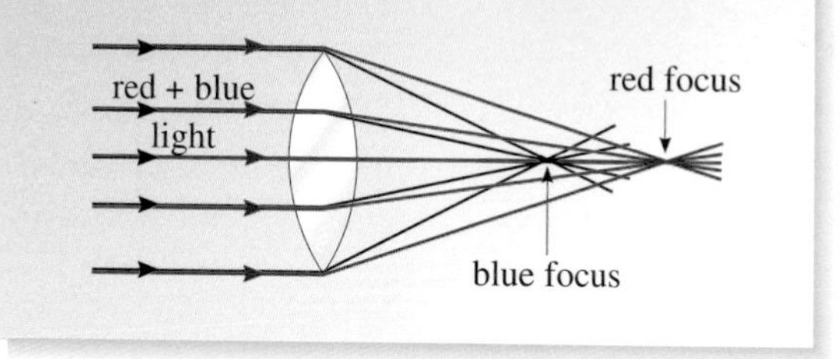

Figure 3.12 Chromatic aberration in a lens: blue light is bent more strongly than red light. Mirrors do not suffer from chromatic aberration.

There is one other important type of aberration that will be introduced here. When the concept of prism stacking was explained as a way of understanding lenses, the point was made that the prism angle of adjacent prisms had to be adjusted so that all the rays were deviated through angles suitable for converging them to a common point. Then, as the number of prisms in the stack becomes very large, the surfaces (in three dimensions) become spherical. Unfortunately, these two statements turn out to be mutually contradictory: spherical surfaces do *not* bring all parallel rays to the same focal point. In fact, rays near the optical axis are brought to a focus further away from the lens than are rays more distant from the axis. This imperfection is illustrated in Figure 3.13a; it is known as **spherical aberration**. The more the rays are restricted to the central, axial, part of the lens, the less significant is this effect.

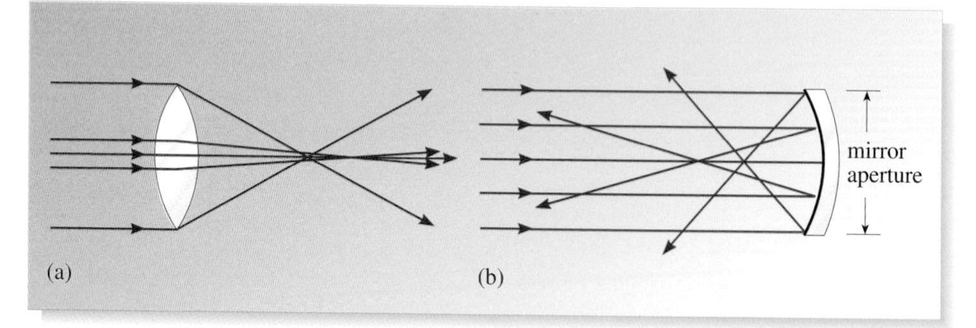

Figure 3.13 Spherical aberration: (a) with a converging lens, rays close to the optical axis are brought to a focus further away from the lens than are rays more distant from the optical axis; (b) with a converging spherical mirror, a similar effect occurs (both effects are exaggerated here for clarity).

Spherical mirror surfaces are not ideal either, and they also give rise to spherical aberration. As with lenses, incident parallel rays close to the optical axis will be focused further away from the mirror than will rays more distant from the optical axis (see Figure 3.13b). The problem can be neglected if the mirror **aperture** constitutes only a small part of the total sphere surface or, equivalently, if only those rays near to the optical axis are considered.

Although these aberrations are important in real imaging systems, the remainder of this chapter will assume that their effect is small and can be neglected. This will enable us to concentrate on the image-forming capabilities of an optical system, without being distracted by its possible imperfections.

3 Objects and images

In the previous section, you saw how rays parallel to the optical axis of a lens or curved mirror converge on — or appear to diverge from — a point on the optical axis known as the focal point. But in imaging systems we are often not so much interested in what the lens or mirror does to *parallel* rays, but in what it does to rays *diverging* from an object. A first step in investigating this question further is to initially restrict the argument to point-source objects on the axis, and then extend the discussion to include off-axis point objects, and extended objects.

3.1 Locating the image

You know that rays parallel to the optical axis of a lens are bent to pass through the lens's focal point. You also know that these parallel rays can be thought of as originating from an *infinitely distant point source* located on the axis of the lens. These two ideas are the key to understanding what happens when a point source is a *finite* distance away from the lens. By specifying a lens's focal length, we have, in effect, specified the lens's 'bending power' with respect to the rays striking it. So, with that idea in mind, see if you can answer the following question.

● Figure 3.14 shows a point-source object O a distance u from a converging lens, and on the optical axis of that lens. The lens's focal point is marked by the letter F on the diagram. In this situation the rays from the source are also brought to a focus on the axis, but *not* at the focal point. Will the focus be to the left or the right of F?

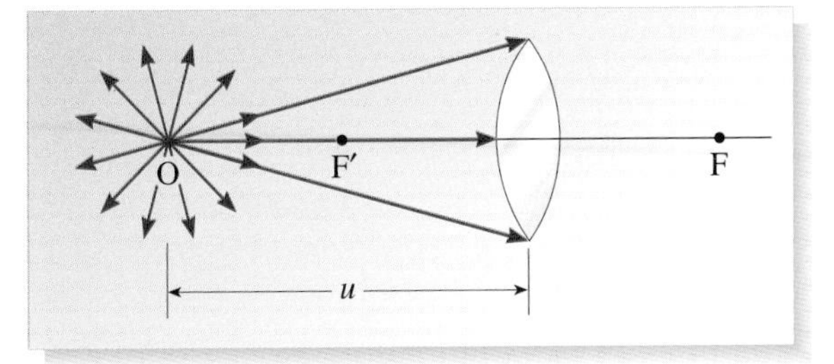

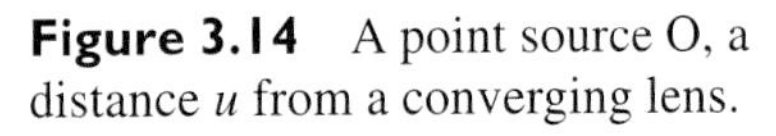

Figure 3.14 A point source O, a distance u from a converging lens.

❍ Your reasoning should have been as follows. The 'bending power' of the lens is fixed. Rays parallel to the optical axis are bent to pass through F. Rays from O, however, are not parallel but diverging as they strike the lens. Hence (putting it somewhat pictorially), some of the 'bending power' will have to be employed in first rendering the rays parallel, so only the 'bending power' left over can be used to converge the rays. Thus, the rays will converge on a point *to the right* of F. ■

The point on which the rays will converge is said to be the **image** of the point-source object O. This reasoning can be further extended. If the **point object** is moved even closer to the lens so that it is at the point a distance f *in front* of the lens (such a point is often called the *conjugate* focal point, and labelled F′ in Figure 3.14), then the rays will emerge from the lens parallel to the optical axis and the image will be formed to the right of the lens, but *at infinity*.

Question 3.4 How would the rays emerge from the lens in Figure 3.14, if the object were moved yet closer to the lens, so that it was *within* its focal length (i.e. to the left of the lens, but to the right of F′)? Where would the image be now? ■

Exactly the same kind of reasoning can be used to work out what would happen with a *diverging* lens. In some ways this is more straightforward: the rays from a point source must diverge, so a diverging lens can only make them diverge even more, irrespective of the position of the object. You know that, for such a lens, parallel rays (equivalent to a point object at infinity) are diverged so that they appear to originate from the focal point F. An object closer to the lens will produce rays that are already diverging, so the lens will increase this divergence: the rays will appear to come from a point *between* F and the lens. So with a diverging lens, the rays never actually pass through the image point, but instead diverge *as if* they had originated from the image point. This type of image and that discussed in relation to Figure 3.14 are differentiated as follows: when the rays *converge* to the image point, the image is said to be a **real image**; when the rays appear to *diverge* from the image point, the image is said to be a **virtual image**. Figure 3.15a shows a converging lens producing a real image of an object, and Figure 3.15b shows a diverging lens producing a virtual image of the object. (In both cases, the object's distance from the lens is greater than the lens's focal length.)

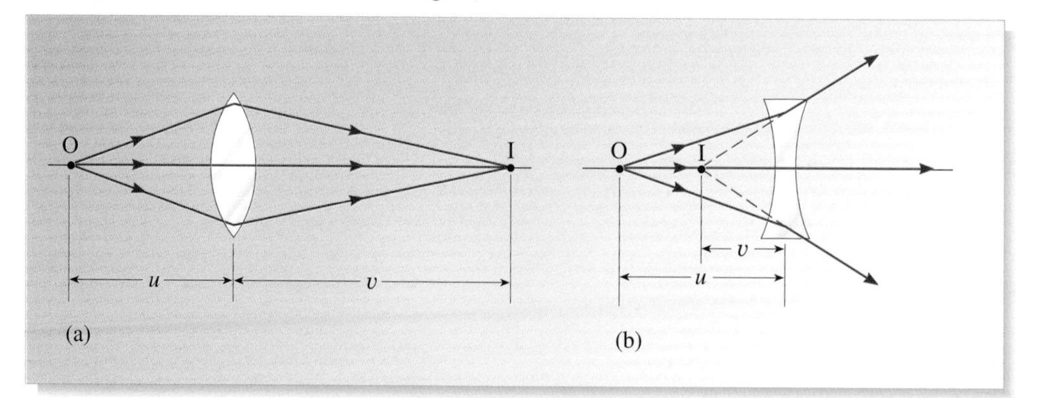

Figure 3.15 (a) A converging lens forming a *real* image of a point-source object; (b) a diverging lens always forms a virtual image of a point-source object.

It is conventional to denote (as in Figure 3.15) the distance of the *object* from the lens (as measured along the optical axis) by the symbol u, and the distance of the *image* from the lens by the symbol v. It turns out that there is a general relationship between u, v and the focal length f. You will often hear this relationship referred to as the **lens equation** (despite the fact that it works equally well for curved mirrors). It is enormously useful in optical imaging; you should learn it by heart! It is

$$\frac{1}{v}+\frac{1}{u}=\frac{1}{f}. \tag{3.1}$$

However, in using this equation, we need a way of recognizing the difference between real and virtual images (and real and virtual objects, in systems using more than one lens), and between converging and diverging lenses. The convention used throughout this book is known as the **real is positive** convention (and by implication, the 'virtual is negative' convention). The basic rules of this convention can be listed as follows:

(i) a converging lens or mirror has a positive focal length;

(ii) a diverging lens or mirror has a negative focal length;

(iii) **real objects** (objects from which rays *diverge*) are assigned positive u-values;

(iv) real images (formed by *converging* rays) have positive v-values;

(v) virtual images (images from which rays appear to have *diverged*) have negative v-values;

(vi) when a lens or mirror converges rays as if to a real image, but the rays are intercepted by a second lens or mirror before the image is formed, then what would have been the real image behaves as a **virtual object** for the second lens or mirror and must be assigned a negative u-value (see Figure 3.16 for an example);

(vii) the equation $1/v + 1/u = 1/f$ automatically gives the correct sign to the value being calculated (provided, of course, that the other two values have been substituted with the correct sign).

The real-is-positive convention is probably the most common convention, but other conventions are sometimes used in other texts. If you have already met, and are familiar with, a different convention, you are probably better off sticking to it. Provided you use it consistently, you should always (eventually) get the same answers as we do.

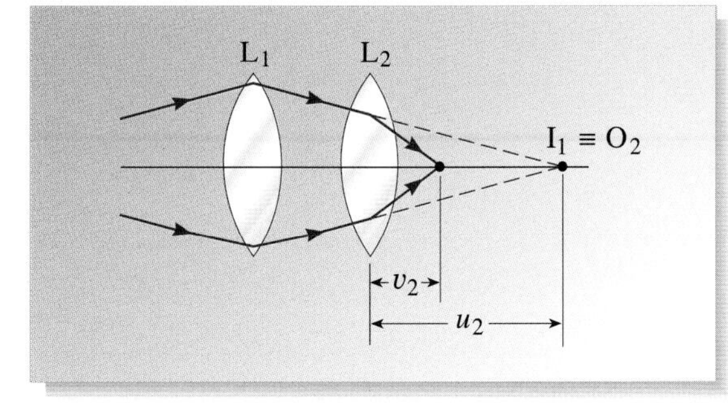

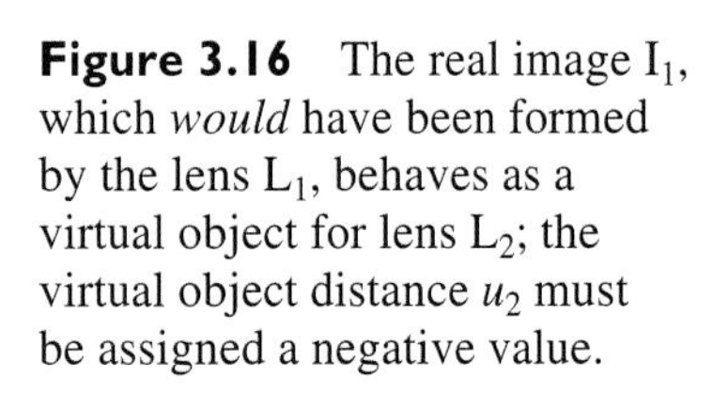

Figure 3.16 The real image I_1, which *would* have been formed by the lens L_1, behaves as a virtual object for lens L_2; the virtual object distance u_2 must be assigned a negative value.

There's one final point of nomenclature to mention at this stage. It's frequently necessary to specify the location of an image (or object, or optical component) relative to a particular given optical component — a lens, say. It's common practice to do this by describing things on the 'input' side of the lens (the side from which the rays are approaching) as being *in front of* that lens, and things on the other side of the lens as being *behind* or *beyond* the lens. Such a description is unambiguous in most cases.

Question 3.5 Use the lens equation, together with your understanding of image formation, to confirm that:

(a) a converging lens produces a parallel beam of light from a point source placed a distance f in front of the lens;

(b) a converging lens focuses parallel light to a *real* image a distance f beyond the lens;

(c) a diverging lens acts on parallel light to form a *virtual* image a distance f in front of the lens. ■

3.2 The wavefront approach

We've just been considering how optical waves are modified by optical components, mainly lenses, and to do this, *ray* construction arguments were used. But, as you know, rays are simply lines drawn perpendicular to wavefronts: so it should be possible to analyse the imaging properties of lenses just as well using a wavefront approach, and indeed it is. In fact, following the same story through again, but this time using wavefront arguments, will probably help you to consolidate your understanding of lens behaviour.

Consider first a point-source object infinitely distant from a converging lens. You know that such an object gives rise to parallel rays at the lens. But equally such an object gives rise to plane wavefronts at the lens's surface. (Have another look at Sections 6.1 to 6.3 in the previous chapter if you need to refresh your memory on this point.) But the converging lens has a convex shape — fatter at the centre than at the perimeter. So, that part of the plane wavefront falling on the centre of the lens has to pass through more glass than the part of the wavefront falling on the lens's perimeter (see Figure 3.17). Consequently, the wavefront will be *delayed* more at the lens's centre than at its edge (since the speed of the wave in glass is slower than in air. The result of this is that the plane wavefronts are transformed into spherical wavefronts that converge to a point on the optical axis; this, of course, is the focal point.

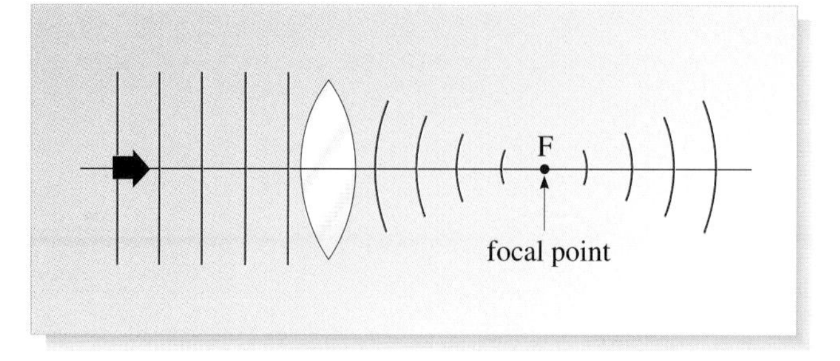

Figure 3.17 A converging lens transforms plane wavefronts into spherical wavefronts.

Question 3.6 Sketch the diagram (equivalent to Figure 3.17) for the case of plane waves whose wavefronts are perpendicular to the axis of a *diverging* lens. ■

We can now move on to consider what happens when the point object is not at infinity, but is instead a finite distance u in front of the lens. Now, the incident wavefronts will be spherical, as shown in Figure 3.18. Nevertheless, the lens will still delay some parts of this wavefront more than others. With a converging lens (Figure 3.18a) the central part of the wavefront is delayed more than the outer parts. This will have the effect of reducing the curvature of the wavefronts; if the bending power of the lens is great enough relative to the position of the source (it's the

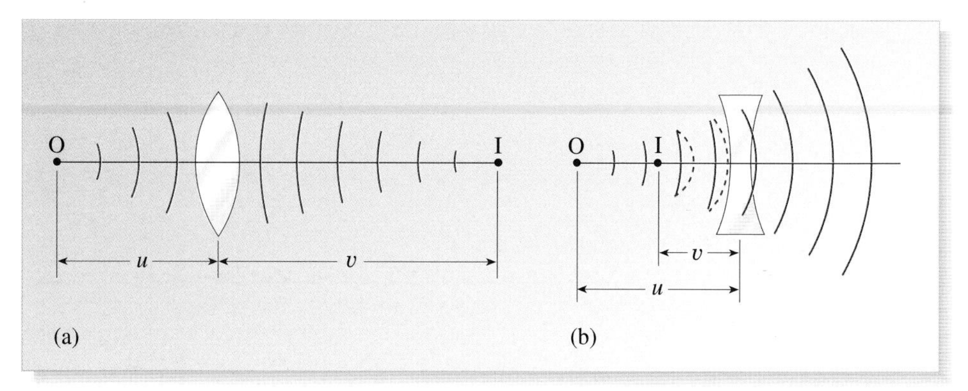

Figure 3.18 The modification of spherical wavefronts by (a) a converging lens, and (b) a diverging lens.

source's distance from the lens that determines the curvature of the incident wave-fronts at the lens surface), the curvature may be completely neutralized (so that plane waves emerge from the lens), or even reversed, as shown in Figure 3.18a. With a diverging lens (Figure 3.18b) the central part of the wavefront is delayed *less* than the outer parts; the curvature of the wavefronts will always be *increased*.

You should now compare Figure 3.18a and b with Figure 3.15a and b. Can you see that the one is simply the wavefront version of the other? There is just one small proviso that should be made with regard to the above discussion: components with spherical surfaces do *not* (perhaps surprisingly) convert plane waves into *truly* spherical wavefronts, and therefore do not generate a well-defined, perfectly focused, point image. However, distortions are negligible if the diameter and thickness of the lens are small compared with its focal length. This is simply spherical aberration again, but expressed in terms of wavefront ideas.

3.3 Extended objects

When dealing with practical imaging problems, it is often helpful to be able to call on both the wavefront *and* the ray constructs. In many situations though, the ray construction has a slight advantage over the wavefront approach, in that it usually allows us to follow much more clearly the passage of light through the various optical components. (You probably felt, for instance, that it was easier to see what was going on in Figure 3.15b than in Figure 3.18b.)

The advantages of using rays become even more apparent when the object — and hence the corresponding image — is not a point source on the optical axis, but is instead a *collection* of point sources all joined together to form an **extended object**. How, for instance, would you go about trying to find the location of the image of the extended object OB shown in Figure 3.19? The wavefront approach here would be totally unwieldy. But, provided the rays are chosen carefully, the ray construction can produce the required answer quickly and easily. So which rays should we choose?

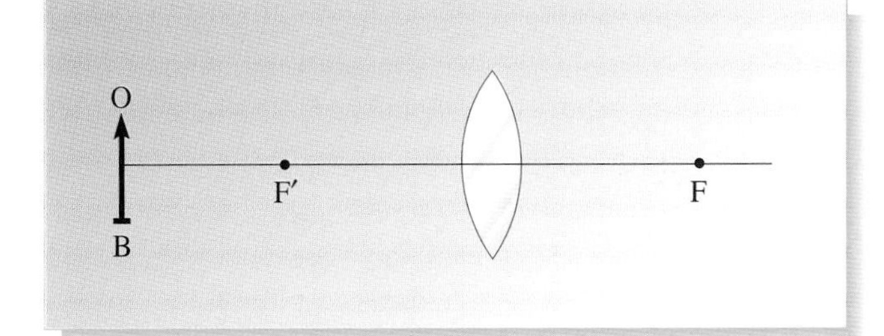

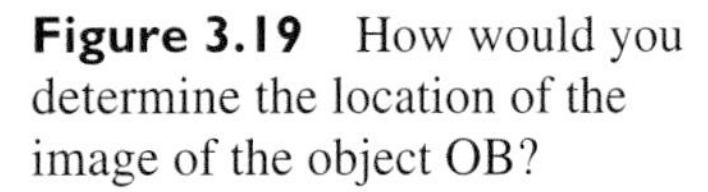

Figure 3.19 How would you determine the location of the image of the object OB?

Tracing key rays through lenses

It turns out that there are three sets of rays whose paths are easy and useful to trace through simple lenses. They are as follows.

1 All rays entering a lens parallel to its optical axis will, on emerging, pass through the focal point (in the case of a converging lens), or will appear to have originated from the focal point (in the case of diverging lens). This is shown in Figure 3.20a. It follows directly from the way the focal length is defined.

2 Because of the reversibility-of-light-paths argument (see Chapter 2, Section 6.1), any ray that passes through the focal point before striking a converging lens, or that would have passed through the focal point had it not been intercepted by a diverging lens, will emerge parallel to the optical axis. This is the reverse of the situation in point 1 above, and is illustrated in Figure 3.20b.

3 Rays travelling along the optical axis of a lens pass through the lens undeviated. But in addition, *all rays* that pass through the optical centre of the lens can also be assumed, to a good approximation, to be undeviated by the lens (see Figure 3.20c). The approximation is sound if the thickness of the lens is small compared with its focal length (this is sometimes called the **thin-lens approximation**), and provided that the angle the ray makes with the optical axis is not too large (this is known as the *paraxial* approximation). These conditions are satisfied in a large majority of practical situations.

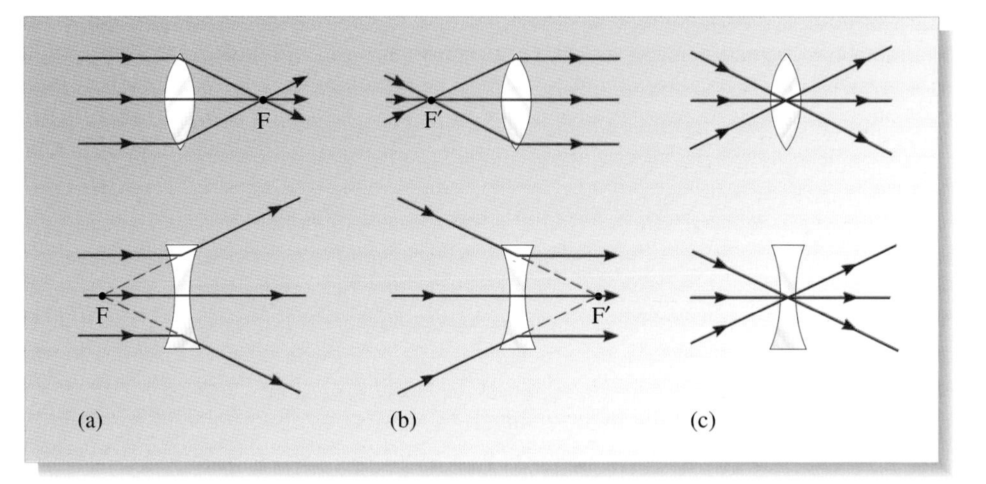

Figure 3.20 Some rays are particularly easy to trace through lenses: (a) rays parallel to the axis converge to (or diverge from) the focal point; (b) rays passing through the focal point in front of a converging lens (or converging towards the focal point beyond a diverging lens) emerge parallel to the axis; (c) all rays that pass through the optical centre of a (thin) lens do so without deviation.

Locating the extended image

You are now in a position to work out where the extended image of the object OB in Figure 3.19 is located. Remember, *all* the points on OB will behave like point sources and will radiate energy (i.e. send out rays) in all directions. But for our purposes, only the top point O and the bottom point B need to be considered, and only those rays travelling in the key directions identified above. This is shown in Figure 3.21.

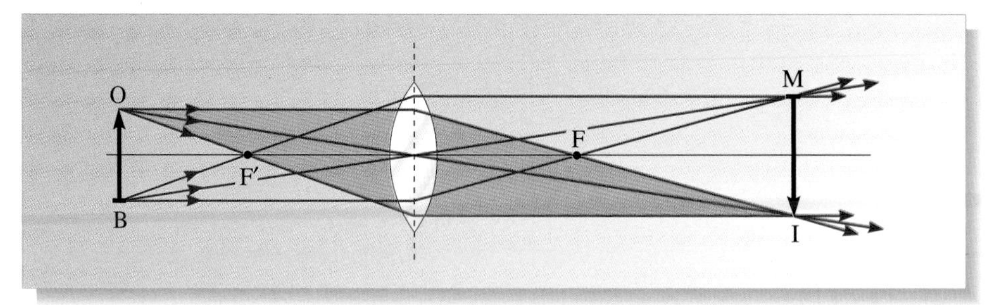

Figure 3.21 The location of the image IM is found by determining the point of intersection of key rays from each of the points O and B. (The rays from O have shading superimposed on them to try to help you distinguish between the two sets of rays.)

The intersection of the three key rays originating from the point O gives the image I of this point. Similarly, the image point M is found by locating the intersection of the three rays emanating from B. In your mind's eye you can now complete the rest of the image by envisaging the location of image points corresponding to object points intermediate between O and B. It should be clear to you from this construction that *the image is inverted.*

There is another point worth noting about this construction. Drawing three rays from each of the points O and B was unnecessary: any two rays would have been quite sufficient to find the intersection point at the image. This is generally true. In practice, you should select whichever two rays are easiest or most convenient to work with.

Note that we now have two ways of determining an image position. Application of the lens equation in the case of extended objects still gives the correct relationship between the focal length and the distances of the object and image from the lens *as measured along the optical axis*; that is, in the context of Figure 3.21 u and v, relate to the distance of the midpoint of OB, and the midpoint of IM, respectively, from the lens centre. Of course, an *accurate* geometric construction, along the lines of Figure 3.21 would enable the value of v to be measured directly from the diagram. But this is often time consuming. So a more useful way to proceed in most cases is to use the lens formula to evaluate the relevant distance(s), and then to use a *sketch* diagram to determine whether or not the image is inverted relative to the object, and possibly also whether or not it is enlarged relative to the object.

Extended images and curved mirrors

The set of three key rays can also be used to locate the image of an extended object when a curved mirror is employed as the optical element. There is just one difference. Rays do not normally pass *through* mirrors (as they do lenses); so the concept of an optical centre through which rays can pass without deviation has to be relinquished, in the case of mirrors, in favour of a new, more appropriate property. This more appropriate property is the **centre of curvature** of the mirror. It is defined as the geometrical centre point of that sphere whose surface, in part, constitutes the spherical mirror. A ray from this point to the mirror would then be travelling along a radius of the sphere. As the radius of a sphere is always perpendicular to its surface, such a ray would be reflected straight back along its own path. Figure 3.22 shows how the three key rays are used to locate the inverted, reduced image produced by a converging mirror.

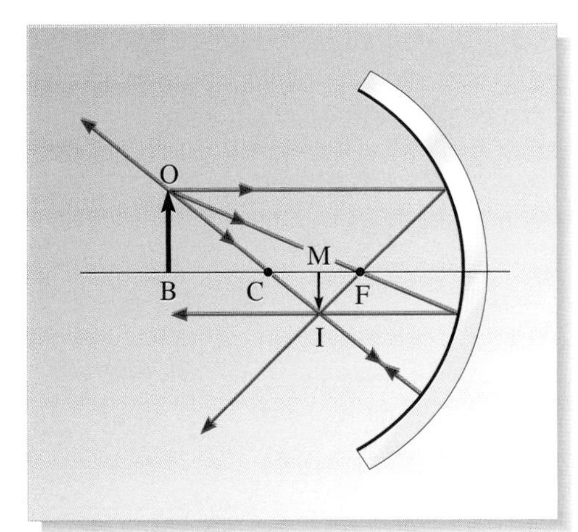

Figure 3.22 A converging mirror produces a reduced, inverted image of OB, located between the mirror's focal point F and its centre of curvature C. A spherical mirror's focal length is half its radius of curvature.

An interesting, and often very useful, point to note about a mirror's centre of curvature is that its distance from the mirror is equal to *twice* the focal length of the mirror. This is easily verifiable by examining a point object located at the centre of curvature. A ray from this object always strikes the mirror normally and is therefore returned back along its path, as just discussed above. Consequently, the point *image* must also be located at the centre of curvature. If the mirror's **radius of curvature** (the distance from the centre of curvature to the mirror's surface) is represented by ρ, then here $u = v = \rho$. Putting this common value for u and v into the lens equation,

$$\frac{1}{\rho} + \frac{1}{\rho} = \frac{1}{f}$$

so $$f = \rho/2. \qquad (3.2)$$

Question 3.7 What is the nature of the extended image produced by a converging mirror, when an extended object is perpendicular to the optical axis and located in the plane that passes through the mirror's centre of curvature? ■

Magnification

As you have just seen, when the object and image have a finite extent, then it is possible to deduce from a ray construction whether the image is reduced or enlarged in size relative to the object. But it is also possible to derive a simple mathematical expression that allows this size change to be calculated without having recourse to a ray diagram. Look at Figure 3.23.

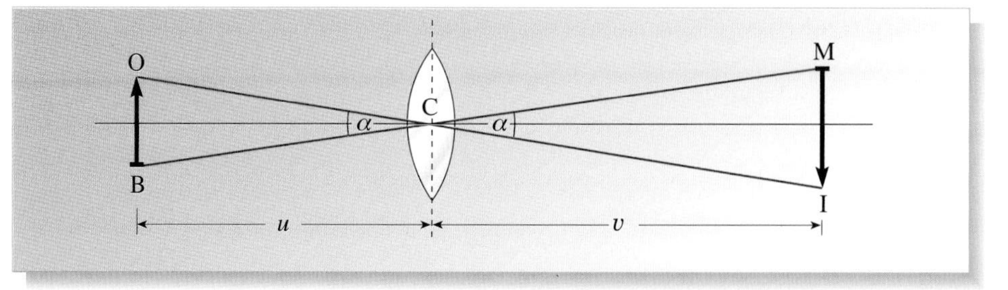

Figure 3.23 The linear magnification m is defined as IM/OB, which is also equal to v/u.

This diagram is a repetition of Figure 3.21, but with only one of the key rays drawn from each of the points O and B. These two rays, which pass undeviated through the centre of the lens, define the size of both the object and image. But notice that the angle α at the apex of the triangle OCB must be equal to the angle at the apex of the triangle ICM. Also, the base angles of the triangle OCB are equal to each other, and are also equal to the base angles of the triangle ICM. Clearly, the two triangles are *similar*. (Similar is used here in its technical sense to mean that the triangles have identical *shape*, and differ only in *scale*.) Because of this, it follows that

$$\frac{u}{\text{OB}} = \frac{v}{\text{IM}}$$

so $$\frac{\text{IM}}{\text{OB}} = \frac{v}{u} \qquad (3.3)$$

where any sign conventions attached to v and u are here ignored.

But IM/OB is simply the ratio of the size of the image to the size of the object; this is clearly a measure of the enlargement or reduction in the size of the image relative to the object: it is called **linear magnification** or more commonly just **magnification**, and represented by the symbol m. But from Equation 3.3, it is also apparent that the magnification can be written in terms of the object and image *distances* as

$$m = \frac{v}{u} \tag{3.4}$$

where again, only the *magnitudes* of v and u are taken into account. This important result is quite general, and can be applied to any lens or mirror.

Question 3.8 Refer again to Figure 3.22, and try to prove that the magnification IM/OB = v/u for this mirror. (*Hint* Draw in another ray from O that strikes the mirror at the place where the optical axis intercepts the mirror.) ■

If you refer back to Question 3.7, you can now see why the inverted image was the same size as the object: because both the object and the image are located in the plane passing through the mirror's centre of curvature, v equals u, and so the magnification $v/u = 1$. Incidentally, magnifications between 0 and 1 are perfectly reasonable: they indicate simply that the size of the image is *less* than the size of the object.

3.4 Focusing inclined plane waves

A problem encountered frequently in optical imaging is that of a plane wavefront arriving at an optical component (say a converging lens) at a small angle of inclination θ to the optical axis. Such an arrangement is shown in Figure 3.24.

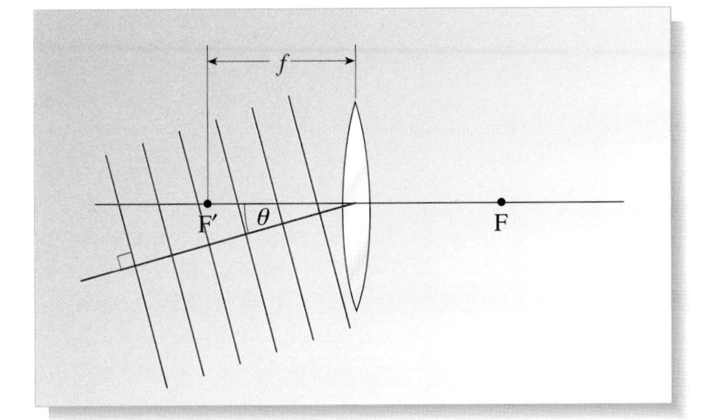

Figure 3.24 An inclined plane wavefront arriving at a converging lens.

- What kind of point source gives rise to plane wavefronts inclined at an angle to the optical axis?

- ○ Infinitely distant point sources give plane wavefronts. But for the plane wave to be inclined at an angle to the optical axis, the point source itself must be positioned *off the axis*. In other words, an off-axis, infinitely distant, point source gives rise to inclined plane wavefronts. ■

You now have all the tools necessary to work out how this lens focuses the inclined plane wave. Try answering the following question.

Question 3.9 Using both the ray *and* wavefront constructions, sketch on Figure 3.24 the propagation pattern of the radiation after it has passed through the lens. Derive an algebraic expression for the distance from the optical axis of the point at which the plane wave is brought to a focus. ■

So, summing up, a plane wavefront inclined at an angle θ to the optical axis of a converging lens is imaged to a point in the focal plane of the lens, at a distance $f\tan\theta$ from the optical axis. Exactly equivalent results are obtained for diverging lenses (except that the image point is virtual), and for converging and diverging mirrors.

4 Optical systems

Although it is possible to obtain useful images with individual optical elements, things get much more interesting when several components are combined to make an optical *system*. This section presents a quick survey of some of the more important of these systems, showing you how the rays of light pass through them, and indicating the type and location of the various images produced by them.

Many optical systems are, of course, designed to be used in conjunction with the human eye — that is, with the eye as an integral part of the system, looking *through* the microscope or telescope, and not just observing an image thrown onto a screen. Indeed, you were warned earlier (in the answer to Question 3.3) of the invalidity of attempting to measure the focal length of a lens while looking through it. Such a procedure not only introduces a second lens into the system (your eye-lens), but furthermore introduces a lens with rather peculiar properties. So it would seem prudent to begin our survey of optical systems by examining the basic optics of the eye itself.

4.1 The human eye

The eye's optical structure

The human eye is roughly spherical in shape, consisting essentially of a gel-like mass protected by a relatively hard shell (called the *sclera*). A schematic plan-view of a horizontal section through a typical eye is shown in Figure 3.25.

Although the detailed biophysics of the eye is quite complex, the basic optical structure is relatively straightforward. A positive (converging) lens system produces focused images of the outside world on a photosensitive surface, called the *retina*. The retina, which covers most of the inside wall of the eyeball, comprises an array of photochemical detectors that are sensitive to different colours of light, and which, when activated, send signals through a mass of connected nerve fibres, via the *optic nerve*, to the brain. There is a *blind spot* on the retina at the point where the optic nerve enters the eye. There is also a tiny area of the retina, called the *fovea*, where the visual detectors are packed much more closely together than elsewhere; this enables much higher quality (more acute) vision at this point, and under normal circumstances the eyeball rotates to concentrate on any object of interest, and so project its image onto this area of the retina. This operation defines the *visual axis* of the eye. Since the visual axis is aligned quite closely with the optical axis of the eye-lens, this means that the eye-lens system is mainly operating on rays entering the eye at only small angles of inclination to its optical axis. This, of course, is one of the requirements necessary in order to minimize the effects of lens aberrations.

The eye-lens system itself is really composed of three separate components: the *cornea*, which is the transparent front part of the eye casing; the *crystalline lens*,

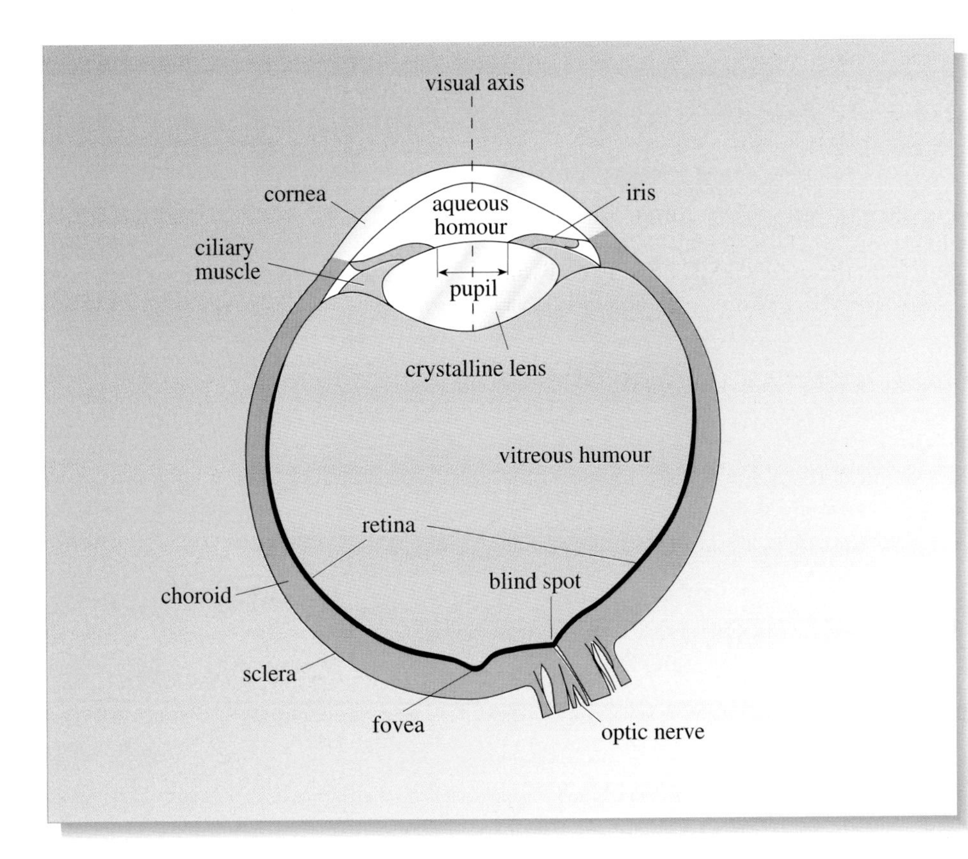

Figure 3.25 The (left) human eye viewed from above.

which is a complicated structure made up of well over 20 000 layers of transparent cellular material (and is not at all like a crystal!); and the chamber between the cornea and the crystalline lens, which is filled with a watery fluid known as the *aqueous humour*. The cornea has a refractive index of $n_c \approx 1.38$, only a few per cent more than the refractive index of the aqueous humour ($n_{ah} \approx 1.34$). The material of the crystalline lens is not homogeneous, but near its centre has a refractive index of $n_{lens} \approx 1.41$. The gelatinous material that fills the volume between the lens and the retina is called the *vitreous humour*; it has a refractive index $n_{vh} \approx 1.34$ — that is, approximately the same as that of the aqueous humour.

It often comes as a surprise to people to learn that the major part of the eye's focusing power is contributed not by the crystalline lens, but by the cornea + aqueous humour combination, acting as a *fixed focal-length* converging lens. Yet what makes the eye an exceptional optical system is the ability of the crystalline lens to *change* its focal length (on instructions from the brain) by changing the curvature of its surfaces. The ciliary muscle can squash or stretch the thousands of layers that make up the lens, thereby changing its shape. So although the crystalline lens may make only a minority contribution to the total focusing power of the eye-lens system, it is this contribution that lends the system its distinctive capability — that of being able to vary its focal length.

When people talk colloquially about another person's beautiful, or piercing, or limpid eyes, they are really referring, of course, to the *iris*. This is the coloured diaphragm just in front of the crystalline lens that gives the eye its characteristic hue — blue, green, brown, hazel or grey, for instance. The hole at the centre of this diaphragm, which is the hole through which light enters the eye, is called the *pupil*. In humans, it is roughly circular; it looks black because the inside of the eye is normally dark. The amount of light entering the eye can be altered (again on instructions from the brain) by retracting or advancing the iris across the lens. The typical eye can change its pupil size from a diameter of about 2 mm in bright light, to about 8 mm in very dim light.

The eye's optical-power range

When two lenses are placed next to each other, it is often more convenient to treat them as a single, **equivalent lens**. But how can the focal length of this equivalent lens be determined? Try the following question.

Question 3.10 Imagine two thin converging lenses, of focal lengths f_1 and f_2, placed next to each other. Suppose a parallel beam of light is incident normally on the first lens. By treating the image produced by this first lens as the object for the second lens, calculate the position of the final image, and hence deduce the **equivalent focal length** of the two-lens combination. (Assume that the separation of the lenses is very much less than either of the focal lengths.) ■

Because of this result, it is common within certain optical professions (especially ophthalmology) to define the bending power or, more concisely, the **power of a lens** P, as the *reciprocal* of the focal length; that is

$$P = 1/f. \qquad (3.5)$$

P is then frequently expressed in units of **dioptres** (abbreviated to D), where 1 dioptre ≡ 1 metre^{-1} (1 D ≡ 1 m^{-1}). The advantage of this is that, following directly from the result of Question 3.10, the combined total power of a system of lenses can be expressed simply as the *sum* of the individual powers; that is

$$P_{total} = P_1 + P_2 + P_3 + \cdots \qquad (3.6)$$

Note that the 'real is positive' convention must still be applied: for example, a diverging lens has a negative focal length and therefore also has a negative power.

Question 3.11 A converging lens has a focal length of 25 mm. What is its power in dioptres? ■

Equation 3.6 is useful when discussing the eye. The power of the cornea + aqueous humour is fixed for any given eye, with a typical value of around 40 D. The power of the crystalline lens, however, can be varied (in a normal eye) from about 20 D to about 24 D.

● So what is the typical *range* of power of the combined eye-lens system in a normal eye?

❍ The total power range must be (40 + 20) D to (40 + 24) D — that is, from +60 D to +64 D. ■

Using this result, it is possible to calculate the range over which the overall focal length of the eye-lens can be varied. A power of +60 D is equivalent to a positive focal length of (1000/60) mm, or approximately 16.7 mm; similarly, a power of +64 D gives a focal length of approximately 15.6 mm (remember, the higher the power, the shorter the focal length). Now it turns out that the distance from the optical centre of the eye-lens system to the fovea in a typical human eye is approximately 17 mm. Hence, since the *maximum* focal length of the eye is also approximately 17 mm, it follows that parallel light (from an infinitely distant object) will be sharply focused onto the fovea when the crystalline lens is fully stretched (since the smaller the curvature of a lens, the longer its focal length). Note that the shape (and hence focal length) of the crystalline lens is governed by the tension of the ciliary muscles.

As the object moves from infinity towards the eye, the lens will have to decrease its focal length to keep an image of the object in focus on the fovea. This is necessary because the image distance — from the eye-lens to the fovea — is fixed. This ability of the eye to change its focal length and so 'accommodate' a whole range of different object distances is called, perhaps not surprisingly, **accommodation**. The most distant point on which the eye can clearly focus is called the **far point**; for the normal eye it is at infinity. By contrast, the *closest* point on which the eye can clearly focus is called the **near point**.

● Assuming that the distance between the fovea and the centre of the eye-lens system is 16.7 mm, and that the shortest focal length the eye-lens system can adopt is 15.6 mm, calculate how close an object can be brought to the eye if the image is still to remain in focus on the retina.

○ Using $1/v + 1/u = 1/f$, with $v = 16.7$ mm and $f = 15.6$ mm, gives $u = 237$ mm. ■

In fact, this closest distance of distinct vision, even in the normal eye, varies dramatically with age, ranging from perhaps 70 mm as a teenager, through around 250 mm in middle age, to perhaps 1 m or more by age 60. It is conventional therefore, when discussing the optics of the eye, to assume a *standardized* near-point distance for the normal eye of 250 mm. This distance is commonly called the **normal near-point** distance. It is also assumed that the normal eye's far point is at infinity. It is this range that corresponds to the power variation of 4 D (from typically 60 D to 64 D). These limits of accommodation are summarized for the normal (standard) eye in Figure 3.26.

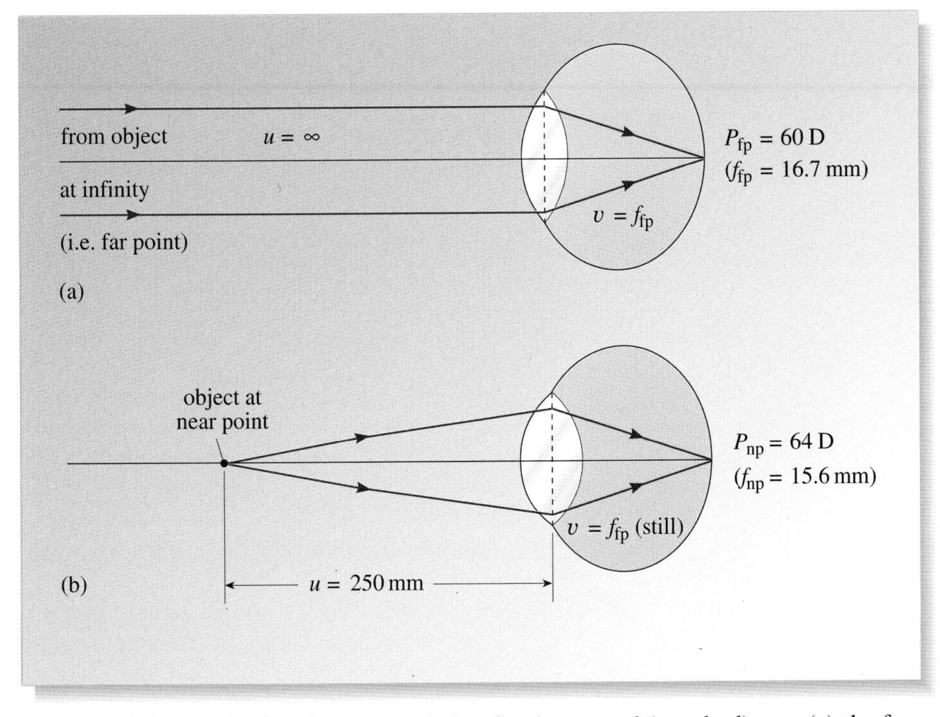

Figure 3.26 The limits of accommodation for the normal (standard) eye: (a) the far point is assumed to be at infinity; (b) the near point is assumed to be a distance $u = 250$ mm from the lens; the lens has to be 'fatter' here than in case (a) if the image is still to be in focus on the retina.

Correcting common eye defects

Unfortunately, not everyone's near point and far point accord with those assumed for the normal eye, even when their eyes are otherwise perfectly healthy. The two most common deficiencies are *myopia* or **short-sightedness**, in which the far point is located much closer than infinity, with the consequence that all objects beyond even quite modest distances are out of focus, and *hyperopia* (sometimes *hypermetropia*) or **long-sightedness**, in which the near point is a good deal further away than 250 mm, thus making it impossible to focus on close objects. Both these conditions are summarized in Figure 3.27. As you can see, the short-sighted eye has too great a power (too short a focal length) for the size of the eyeball; even when the muscles

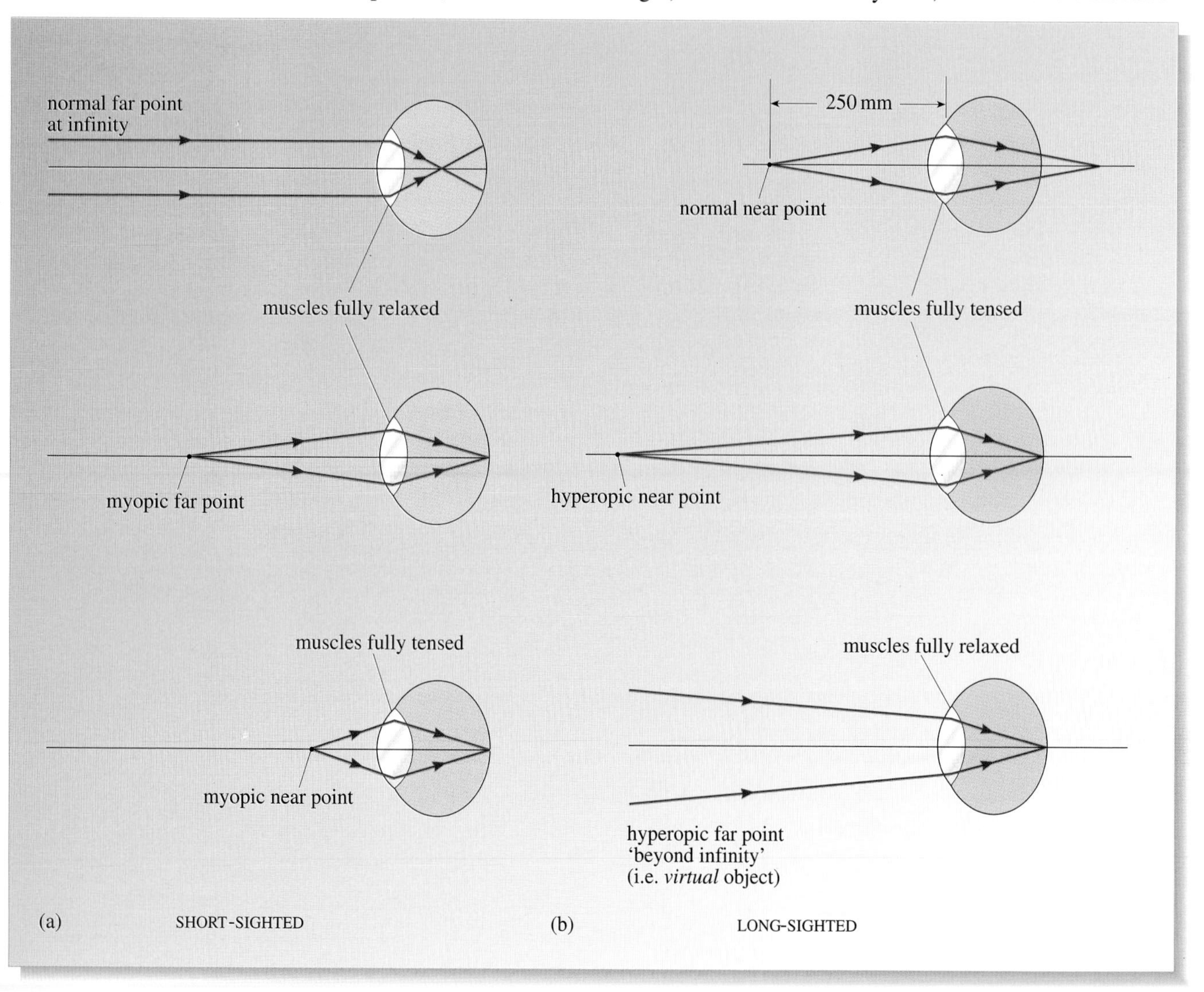

Figure 3.27 (a) The short-sighted eye is too powerful: light from the normal far point (at infinity) is always focused in front of the retina. In fact, it is not uncommon for an eye with this defect to have a far point that is closer to the lens than the normal eye's near point (at 250 mm). The short-sighted near point can then be extremely close, perhaps less than 100 mm. In each of the above cases, the defect can be remedied by inserting an appropriate lens in front of the eye — that is, by wearing spectacles or contact lenses. (b) The long-sighted eye is too weak: light from the normal near point (at 250 mm) is always focused beyond the retina. The actual near point for such an eye is frequently well over 1 metre from the eye-lens, and the far point can be 'beyond infinity' in the sense that, when fully relaxed, this eye can focus already-converging rays (which can, of course, only, have come from a virtual object).

are fully relaxed, the eye brings parallel light to a focus *in front* of the retina; tensing the muscles would only make matters worse. Conversely, the long-sighted eye has too weak a power (too long a focal length) for the size of the eyeball; even when the muscles are as tensed as possible, the eye still focuses an object located at the normal near point (250 mm) to a 'would-be image' *beyond* the retina; relaxing the muscles would only make matters worse.

- What type of lens would be necessary to correct short-sightedness, and what type would correct long-sightedness?

- ❍ The short-sighted eye is too powerful, too strongly converging: a *diverging* lens is necessary to weaken the overall power. The long-sighted eye is too weak, not strongly converging enough: an additional *converging* lens must be used to increase the overall power. ■

Figure 3.28a shows how a concave (diverging) lens is used to enable parallel light to be focused onto the retina of a short-sighted eye. The lens has to make the light *appear* to have diverged from the *far* point; that is, the focal length of the lens must be given by $f = -d_{\text{fp}}$ (where d_{fp} is the distance from the lens to the myopic far point). With a long-sighted eye, it is necessary to use a convex (converging) lens to make the light from an object placed at the *normal* (250 mm) near-point distance *appear* to have originated from the eye's actual near point. This is illustrated in Figure 3.28b. Note that the correcting lens is always assumed to be positioned as close as possible to the eye.

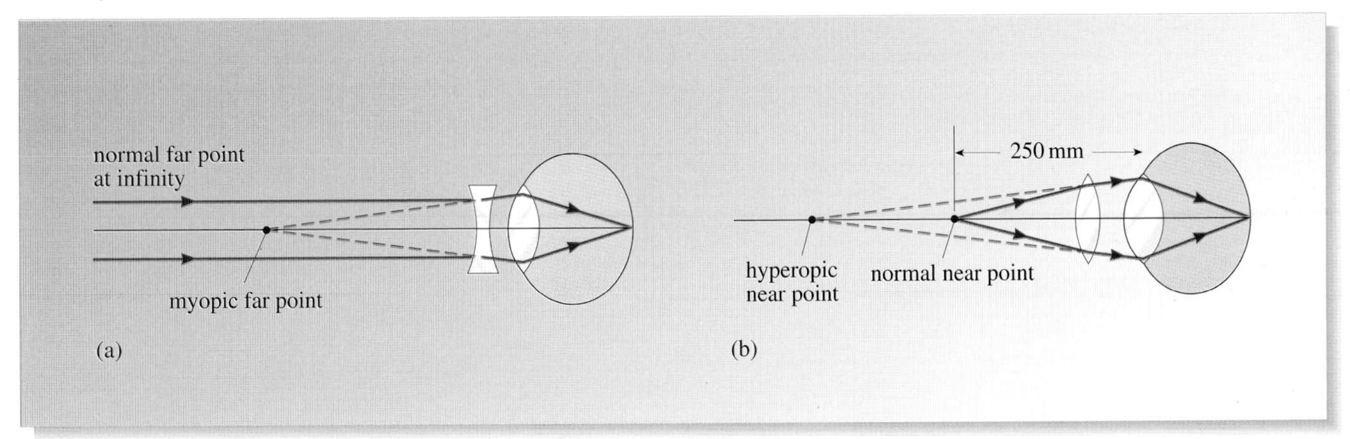

Figure 3.28 (a) Short-sightedness is corrected with a diverging lens; (b) long-sightedness is corrected with a converging lens.

Question 3.12 (a) If a particular long-sighted eye's near-point distance is represented by d_{np}, derive an expression for the focal length of the converging lens required to correct the long-sightedness.

(b) Calculate the power of the correcting lens needed for a long-sighted person whose near point is 1.50 m from the eye-lens. ■

As we get older, our ciliary muscles weaken and our eye-lens tissues stiffen, with the result that we lose the ability to accommodate our vision to objects close to us. This condition is known as *presbyopia*. It is not quite the same as being long-sighted, because the far point is not affected: it remains at infinity (or slightly closer). Consequently, although a prescription for converging-lens spectacles would solve the problem for *close* vision (by, in effect, bringing the near point in to a distance of about 250 mm, exactly as in the long-sighted case), these lenses would also have the unfortunate effect of converging the rays from *distant* objects to a focus *in front* of the retina. Such spectacles would have to be used as 'reading-only' spectacles. Alternatively, *bifocal spectacles* might be prescribed. In these, two different focal-length lenses are combined into a single 'lens' element: the lower part of each 'lens' has a higher power (typically around 3 D higher) than the upper part. The upper part is then fine for long-distance vision, and the lower part is suitable for reading or other close work.

The retinal image

Before concluding this section on the eye, there are just two further qualities pertaining to the retinal image that are worthy of mention. The first relates to the image's orientation. As is clear from Figure 3.29, when the eye produces a focused image of the object OB, the point O is imaged to the point I, and the point B to the point M. The reason for spelling out the obvious here is to emphasize the fact that *the image on the retina is inverted relative to the object.* So why don't we see the outside world upside-down? Well, the truthful answer is that we do — optically speaking, that is. But the act of *seeing* is rather more than just pure optics; it also involves the brain *making sense* of the optical pattern projected onto the retina. And because we've all been making sense of such images since birth, we've learnt to interpret inverted retinal images as corresponding to objects the right way up. (It's also worth noting in passing, that the brain has additionally learnt to compensate for the curvature-distortion in the image caused as a result of the curved surface of the retinal wall; notwithstanding such curvature, we *do* see straight objects as straight.)

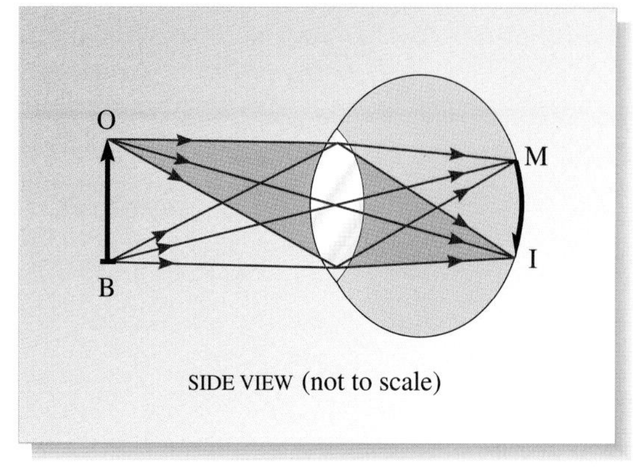

Figure 3.29 The upright object OB produces an *inverted* image IM on the retina. The brain has learnt to interpret such upside-down images as corresponding to upright objects. The brain has also learnt to compensate for the curvature of the image arising from the curvature of the retinal wall of the eyeball.

The second point of interest relates to the actual physical *size* of the retinal image. If you look again at Figure 3.29, you should be able to convince yourself that only two key rays are required to determine this size, namely the one ray from O and the one ray from B that each pass through the optical centre of the eye-lens system. This is illustrated in Figure 3.30a, where the object is a £1 coin (diameter 22 mm) placed at the eye's 'normal' near point (i.e. 250 mm from the lens). It's an amazing fact that such a close object produces a retinal image that is a mere 1.5 mm in diameter!

Question 3.13 Justify this last statement. Assume the eye-lens-to-fovea distance is 17 mm. ■

Even more amazing is that the lettering on the coin, which is quite clearly visible at this distance despite the fact that it is less than 1.5 mm high on the coin itself, produces a retinal image only 0.1 mm high. Check this for yourself.

However, when it comes to judging the size of an object, it is not sufficient for the brain to make use of a simple built-in scaling multiplier: a retinal image 1.5 mm high corresponds to a 22 mm-high object, so an image 3 mm high must therefore correspond to a 44 mm-high object, and so on. The problem with such an approach is that the eye can focus on the same object *over a range of different distances*. So, in Figure 3.30b, the same 22 mm diameter £1 coin, located 2.5 m from the eye, produces a retinal image 0.15 mm in diameter. Its *subjective* size is smaller, despite the fact that the object's real size is unchanged. In a situation like this, it is more convenient to express an object's subjective size in terms of the angle α that it subtends at the eye-lens.

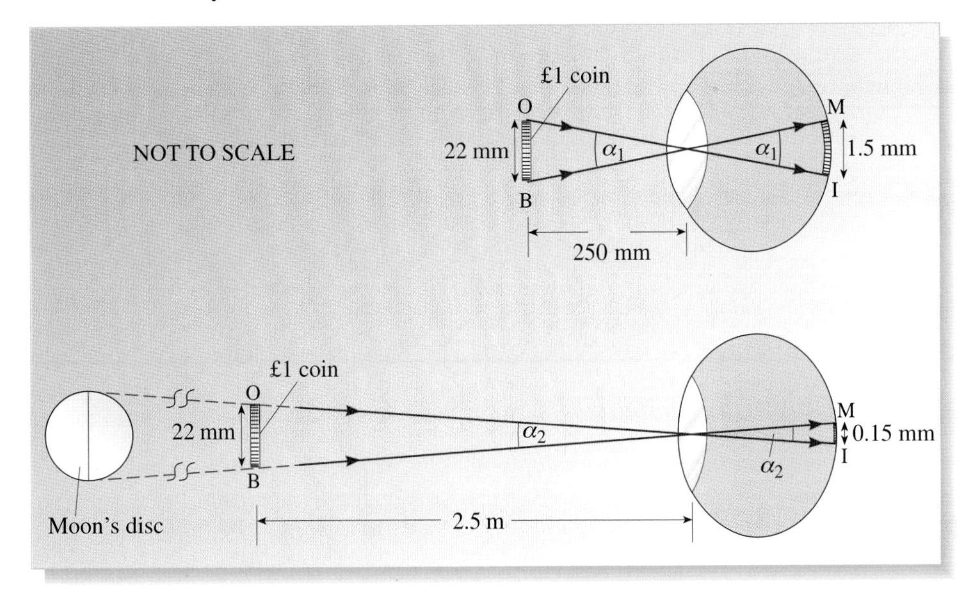

Figure 3.30 (a) A £1 coin placed at the normal near point produces a retinal image 1.5 mm in diameter. (b) The same coin placed 2.5 m from the eye is subjectively smaller, producing an image only 0.15 mm in diameter. In fact, its subjective size would be the same as that of the Moon's disc. It is common to express an object's apparent size in terms of its visual angle α— that is, the angle it subtends at the eye.

● What angle α_1 does the £1 coin subtend in Figure 3.30a, and what is the new angle α_2 subtended by the coin when it is moved to the position shown in Figure 3.30b?

❍ For small angles — which is a condition normally assumed to be true for optical systems — the angle subtended by an object OB at a distance d is given by OB/d radians. In Figure 3.30a the angle α_1 must be 22/250 radians = 0.088 rad = 5.0°. In Figure 3.30b, α_2 is 22/2500 radians = 0.0088 rad = 0.5°. ■

The angle α is sometimes called the **visual angle** of the object. Figure 3.30b shows that the visual angle of the £1 coin at 2.5 m is the same as the visual angle of the Moon's disc, about 0.5°. This is also the same as the angle subtended by the Sun's disc (but **never** attempt to look directly at the Sun). The subjective sizes of the Sun, the Moon, and a £1 coin at 2.5 m, are all the same. The methods used by the eye + brain system to determine whether an object is small and close, or bigger and further away, or huge and very distant, constitute a fascinating area of study, which has proved fertile ground for optical illusionist and visual psychologist alike.

4.2 The magnifying glass

The visual angle subtended at the eye by a given object becomes larger (and so the retinal image becomes larger) as the object is brought closer to the eye. So, in order to see *very* small objects, all you have to do is bring them *very* close to the eye... isn't it? Well no. As you've already seen, once an object is brought closer than your near point, your eye is unable to bend the rays sufficiently to produce a focused image on the retina. And there's no point in having a large retinal image if it's also out of focus.

What you need is a more powerful eye-lens so that you can bend the rays from a very near object strongly enough to produce a focused image. So the remedy is clear: simply insert an additional lens between the object and your eye, such that it will supplement the bending power of your eye-lens system.

Figure 3.31 The magnifying glass has been the favourite tool of the fictional detective since the time of Sherlock Holmes

- Should this be a converging or diverging lens?
- ❍ It must *increase* the total bending power, so it must be a converging (positive) lens. ■

Such a lens is perhaps the simplest of optical instruments, a **magnifying glass** (Figure 3.31).

Basic principles

With the help of a magnifying glass, your eye should now be able to focus on objects located *within* your near-point distance. But what value of focal length should be chosen for this lens, and where should the object be positioned relative to the lens? In order to answer these questions, the problem can be approached from a slightly different perspective.

The object is located at less than your near-point distance. But you want to use the extra lens to make the object *appear* to your eye *as if* it were located between your near point and infinity; that is, you want the extra lens to produce a *virtual* image of the object, located somewhere between infinity and your near point.

- Where must the object be located, relative to the converging lens's focal point, if the lens is to produce a virtual image?
- ❍ Converging lenses are normally thought of as producing real images. To produce a *virtual* image, the object must be located at a distance *less than the focal length* from the lens. Have another look at the answer to Question 3.4 if you are unsure of this point. ■

When the lens is next to the eye, the whole position is summed up in Figure 3.32. Part (a) of this figure shows the best that is possible *without* the extra lens: the object is positioned at the near point (a distance d_{np} from your eye), and it subtends an angle α_{OB} at the lens. Parts (b) and (c) of Figure 3.32 show the two extremes for the location of the virtual image. In (b), the image is located at the near point, and the object located at the appropriate place between the lens and its focal point F to achieve this. The magnifying lens is said to be in **near-point adjustment**. In (c), the virtual image is located at infinity (the normal eye's far point) and hence the object has to be located in the focal plane of the lens. This is called **far-point adjustment**. You can see that in both these cases the angle subtended at the lens by the image (α_{IM}) is larger than α_{OB}. (Note that throughout this analysis, the simplifying assumption is made that the correcting lens is positioned as close to the eye-lens as possible, so that the angle subtended at the correcting lens is, to all intents and purposes, the same as the angle subtended at the eye. This is a fairly reasonable

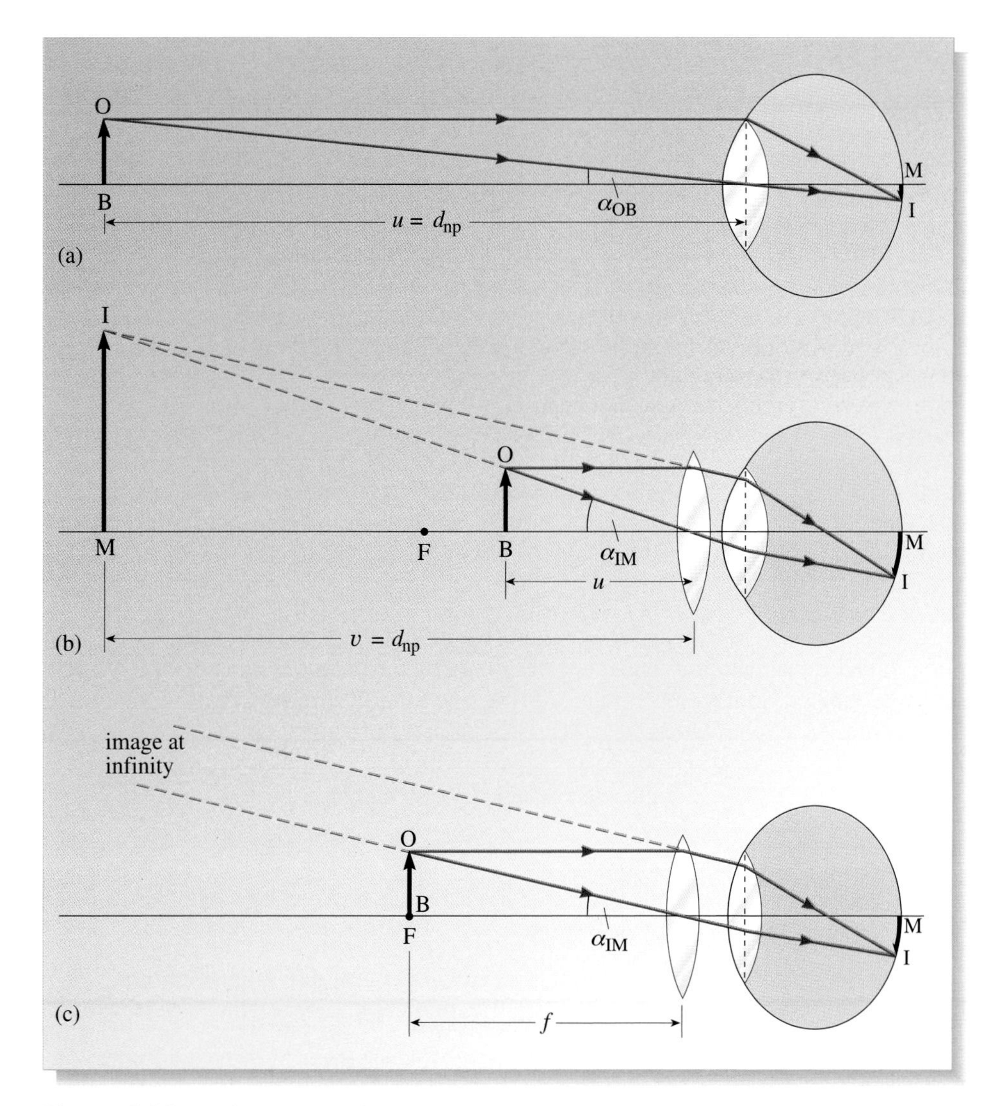

Figure 3.32 Using a magnifying glass. (a) Without a magnifying glass, the closest to the eye that an object can be positioned and still seen in focus is the near point, a distance d_{np} from the lens. The object has a visual angle of α_{OB}. (b) Using a lens to produce a virtual image at the near point. The object is at the appropriate place between F and the lens. (c) When the object is located exactly at F, the virtual image is located at infinity.

assumption, since this is how most optical instruments are normally used: you usually put your eye as close as possible to the eyepiece of a microscope or telescope, for example. However, it is not the way that Sherlock Holmes used his magnifying glass!) Notice that in Figure 3.32b and c the object is closer than the near point, so that a focused image cannot be formed on the retina without the use of a magnifying glass.

One of the important points you should pick up from Figure 3.32 is that, for all optical instruments used in conjunction with the eye, there is not just one satisfactory focus adjustment: any adjustment that positions the image somewhere between the near point and infinity is acceptable. However, it is fairly common practice to use microscopes in near-point adjustment (because this gives the maximum magnification, as you will see in the next section), and telescopes in far-point adjustment (so as to keep the eye as relaxed as possible).

Magnifying power (angular magnification)

So the question about *where* the object should be placed is answered. In near-point adjustment it should be at an object distance u that satisfies the lens equation $1/u = 1/f - 1/v$, where v must be set equal to $-d_{np}$ (remember the image is virtual and at the near point); that is, for near-point adjustment,

$$\frac{1}{u} = \frac{1}{f} + \frac{1}{d_{np}}. \tag{3.7}$$

In far-point adjustment, on the other hand, it should be in the lens's focal plane (i.e at $u = f$). And of course, any u value between f and the value given by Equation 3.7 would also be acceptable.

But this hasn't yet answered the question as to what the focal length of the magnifying glass should be. To do that requires a relationship between the focal length of the lens and its magnifying power, and 'magnifying power' has not yet been formally defined. We all feel we know what it means — it's a measure of how much bigger the optical system makes the object seem — but could you write down a mathematical expression for it?

Well you've already seen that the apparent size of an object is determined by the size of the retinal image, which in turn is determined by the *visual angle* subtended at the eye by the object. So it's convenient to define the **magnifying power** M of an optical instrument as

$$M = \frac{\text{angle subtended at eye by image}}{\text{largest angle that can be subtended at eye by object}} = \frac{\alpha_{IM}}{\alpha_{OB}}. \tag{3.8}$$

In the example shown in Figure 3.32a the largest angle that can be subtended at the eye is when the object is at the near point.

The magnifying power M is also sometimes called the **angular magnification**. You must not confuse this with the *linear* magnification m of a lens, given by $m = v/u$.

This now gives us a way of calculating the magnifying power of the magnifying glass. The largest angle that can be subtended at the eye by the object (and still have the retinal image in focus) is achieved when the object is at the near point (see Figure 3.32a again). Hence

$$\alpha_{OB} = \frac{\text{OB}}{d_{np}} \tag{3.9}$$

where OB is the height of the object, and α is assumed to be small enough for the small-angle approximation to be valid (in other words, $\tan\alpha \approx \alpha$ when $\alpha < 0.2$ radian $\approx 12°$). Now look at Figure 3.32b. The angle α_{IM} subtended by the image will always be the same as the angle subtended by the object *in its new position much closer to the eye*; that is

$$\alpha_{IM} = \frac{\text{OB}}{u}. \tag{3.10}$$

Combining Equations 3.9 and 3.10 to find M gives

$$M = \frac{\alpha_{IM}}{\alpha_{OB}} = \frac{\text{OB}/u}{\text{OB}/d_{np}} = \frac{d_{np}}{u}. \tag{3.11}$$

This will be true for all possible values of u between the near-point and far-point adjustments. In far-point adjustment (Figure 3.32c), $u = f$ and

$$M(\text{far-point adjustment}) = \frac{d_{\text{np}}}{f}. \tag{3.12}$$

In near-point adjustment, u is given by the previously derived Equation 3.7. This gives

$$M(\text{near-point adjustment}) = \frac{d_{\text{np}}}{u} = d_{\text{np}}\left(\frac{1}{u}\right) = d_{\text{np}}\left(\frac{1}{f} + \frac{1}{d_{\text{np}}}\right)$$

that is
$$M(\text{near-point adjustment}) = \frac{d_{\text{np}}}{f} + 1. \tag{3.13}$$

Comparison of Equations 3.12 and 3.13 shows that the angular magnification in near-point adjustment is just 1 greater than that in far-point adjustment. For magnifying powers greater than about 5 or 6, this difference is neither here nor there. From either equation, you can see that the magnifying power is greater the *smaller* the focal length of the magnifying lens.

Question 3.14 A magnifying glass has a focal length of 50 mm. Calculate its magnifying power, in both near-point and far-point adjustment, for a normal (standard) eye. ■

In practice, because of aberrations in the lens, the maximum reasonable magnification that can be achieved with a single lens is around ×5 or ×6. For higher magnifications, a *compound-lens* system, or *microscope*, must be used. The word 'compound' is frequently included so as to differentiate this microscope from the magnifying glass, which is sometimes called a **simple microscope**.

4.3 The compound microscope

The first **compound microscope** may have been invented as early as 1590 by a Dutch spectacle maker named Zacharias Janssen, although Galileo Galilei followed closely behind, announcing his invention of a similar device in 1610. However, the first practical compound microscope is generally taken to be the one developed around 1673 by Anton van Leeuwenhoek, who used single lenses in the form of small spherical glass beads. In 1674 he used his instrument to discover red blood cells, and others soon used similar instruments to discover many new microscopic creatures. Today the microscope is an important tool in many branches of science (Figure 3.33).

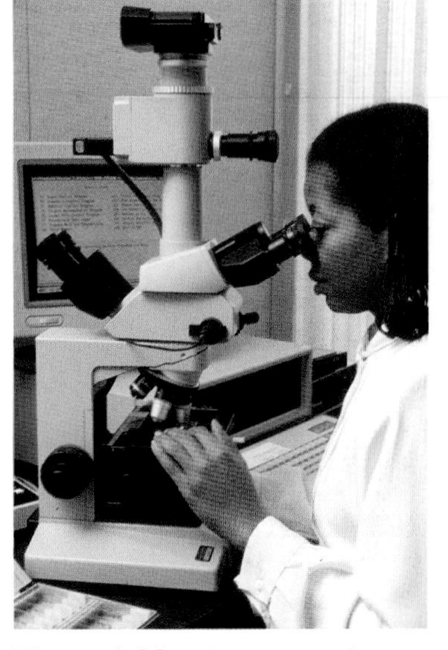

Figure 3.33 A compound microscope being used in a scientific laboratory. In practice, the lenses used in microscopes and telescopes are combinations of lenses, designed to minimize chromatic and spherical aberration.

The basic idea behind a compound microscope is to distribute the total magnification process over two stages. The first stage requires the production of a *real*, enlarged image of the object.

- ● What kind of lens can produce a real, enlarged image?
- ❍ Only a converging lens can produce a real image, and then only when the object is placed a distance *greater* than the focal length away from the lens. (An object *within* the focal length produces a *virtual* image, as with the magnifying glass.) Furthermore, the amount by which the image is enlarged relative to the object — the *linear* magnification — is given by v/u; so for high magnification, v should be as large as is reasonable. ■

The second stage then treats this real image as the object for a second lens, used as a magnifying glass.

- Where must the second lens be located relative to this real image if it is to be used as a magnifying-glass lens?

- ❍ The distance between the magnifying-glass lens and the object (which in this case will actually be the real image produced by the first lens) must be *less* than, or equal to, the focal length of the magnifying-glass lens (as stated in the previous answer above). ■

Whenever an optical device is discussed in this chapter, the term *final image* will generally mean the last image produced by the device itself. Strictly speaking, of course, the truly final image is the one produced on your retina!

If the real image is located in the focal plane of the magnifying-glass lens, the final image will be at infinity: the microscope will be in far-point adjustment. As the distance between the real image and this second lens is reduced, the final image will 'move in' from infinity until it is eventually located at the eye's near point; that is, the microscope is now in near-point adjustment. Any position between these two extremes will give a satisfactory retinal image. Figure 3.34 shows these two limits of adjustment for a typical compound microscope. Note that it is conventional to call the lens nearer the object the **objective lens**, or more simply the *objective*, and the lens nearer the eye the **eyepiece lens**, or just the *eyepiece*. On the diagram, the lower-case subscripts o and e denote, respectively, objective and eyepiece.

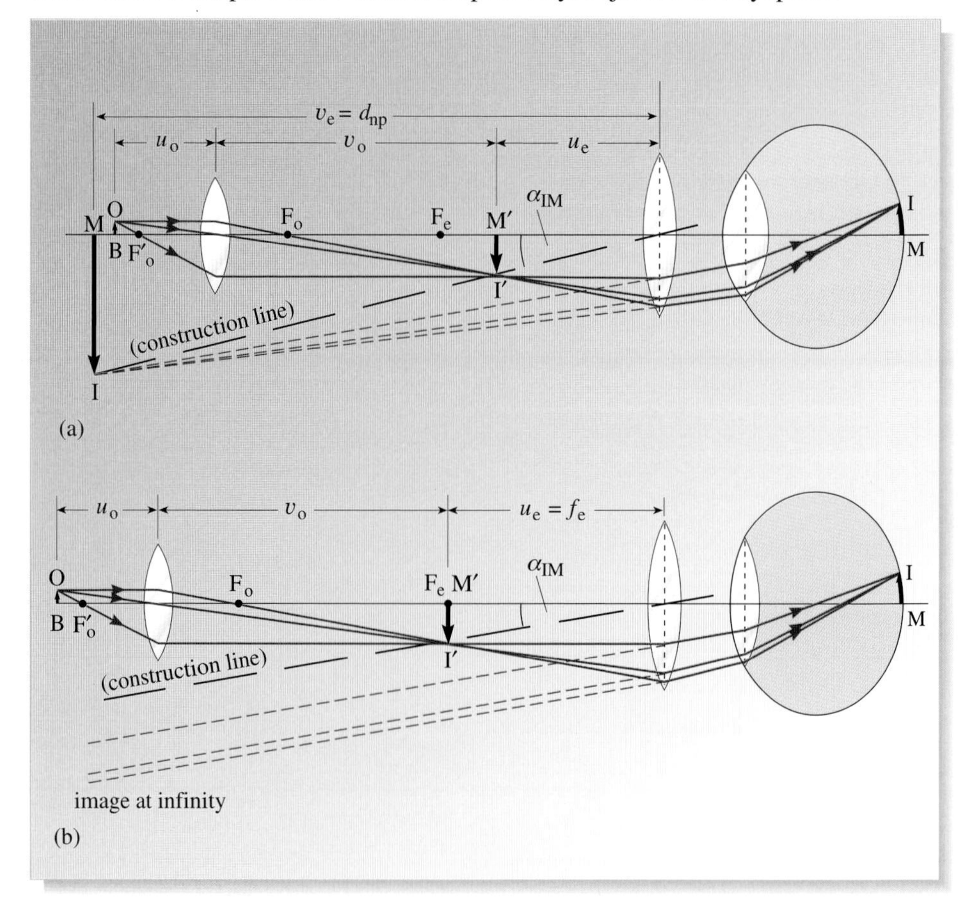

Figure 3.34 A compound (two-lens) microscope set up in (a) near-point adjustment, and (b) far-point adjustment; the separation between the lenses is greater in (b) than in (a). In both cases, the construction line drawn through the centre of the eyepiece lens, and points I′ and I, defines the angle subtended at the eye by the image.

It's now a relatively straightforward task to work out the overall magnifying power (*angular* magnification) of the microscope. This can be done either intuitively, or formally using algebra, and both approaches are given below.

The intuitive argument goes as follows. The magnifying power of the eyepiece lens by itself will be given by the expression worked out in the previous section for the magnifying glass (Equation 3.11); the only difference here is that the subscript e (for eyepiece) must be added to all the variables; that is, $M_e = d_{np}/u_e$. But this value of M_e represents the ratio of the angle subtended at the eye by the final image to the angle that would be subtended by the microscope's *intermediate image* I′M′ if it were placed at the eye's near point. However, what is required is the *overall* magnifying power — that is, the angular magnification of the final image relative to the *object* placed at the near point. But the intermediate image I′M′ is m_o times bigger than the size of the object OB, where m_o is the *linear magnification* of the objective lens, and is given by v_o/u_o. In other words

overall angular magnification =
(linear magnification of objective) × (angular magnification of eyepiece)

i.e. $$M = m_o \times M_e. \tag{3.14}$$

Now for the formal approach. M is defined as α_{IM}/α_{OB}, where α_{OB} is the angle subtended at the eye by the object when it is placed at the near point. So

$$\alpha_{OB} = \frac{OB}{d_{np}} \tag{3.15}$$

where OB is the 'height' of the object. From Figure 3.34a, you should be able to see that the angle subtended by the final image IM is the same as the angle subtended by the intermediate image I′M′, and can be expressed as

$$\alpha_{IM} = \frac{I'M'}{u_e} \tag{3.16}$$

where I′M′ is the 'height' of the intermediate image. But I′M′/OB is just the linear magnification of the objective lens, and is equal to v_o/u_o; that is

$$I'M' = OB \times \frac{v_o}{u_o}. \tag{3.17}$$

Substituting Equation 3.17 into 3.16 gives

$$\alpha_{IM} = \frac{OB \times v_o}{u_e \times u_o}. \tag{3.18}$$

Finally, dividing this expression for α_{IM} by the expression for α_{OB} in Equation 3.15 gives the overall magnifying power M. As you can see, the height OB cancels leaving

$$M = \frac{v_o}{u_o} \times \frac{d_{np}}{u_e}. \tag{3.19}$$

The first term here (v_o/u_o) is just the linear magnification m_o of the objective, whereas the second term (d_{np}/u_e) is simply M_e, the angular magnification of the eyepiece (see Equation 3.11). Equation 3.19 is the algebraic expression (and justification) of the intuitively deduced Equation 3.14.

It's a simple matter to re-express Equation 3.19 for the far-point and near-point adjustment conditions. In far-point adjustment, u_e is set equal to f_e, so

$$M(\text{far-point adjustment}) = \frac{v_o}{u_o} \times \frac{d_{np}}{f_e}. \tag{3.20}$$

In near-point adjustment, $1/u_e = 1/f_e - 1/v_e = 1/f_e + 1/d_{np}$ (remember that v_e is negative because the final image is virtual). Thus

$$M(\text{near-point adjustment}) = \frac{v_o}{u_o}\left(\frac{d_{np}}{f_e} + 1\right). \quad (3.21)$$

Question 3.15 Refer back to Figure 3.34. Is the image seen by the user of the microscope an upright or inverted image? ■

Since the magnifying power M_e of the eyepiece lens is limited to about ×5 or ×6 (recall the discussion relating to the magnifying glass), the microscope's overall magnifying power can be increased beyond this only by making v_o/u_o greater than unity, preferably by a factor of several times. But don't forget that u_o has to be greater than f_o for a real intermediate image to be formed. Consequently, v_o should be several times greater than f_o. So, if the length of the microscope is not to become unwieldy, the focal length of the objective lens should be kept reasonably short. Unfortunately, the shorter the focal length of this lens, the more likely it is to suffer from aberrations. In practice, if the microscope length is to be kept to less than 300 mm say, then an *overall* magnifying power of ×50 is about the best that can be achieved with two single lenses.

Question 3.16 An experimenter constructs a microscope from an objective lens of focal length 15 mm, and an eyepiece lens of focal length 50 mm. Calculate the overall magnifying power and the distance between the two lenses, for both far-point *and* near-point adjustment, when the object is located (a) 20 mm in front of (below) the objective lens, and (b) 18 mm in front of the objective lens. Assume a normal near-point distance of 250 mm. ■

'Bench-standing' microscopes are very often used in *near-point* adjustment, especially in those cases where the user wishes to make a drawing of the object, or objects, being observed; the reason for this is that, in this adjustment, the plane of the final image and the plane of the drawing paper will be roughly the same distance away from the eye, making the copying process easier on the eye.

Furthermore, this adjustment gives the maximum possible magnification for any given object distance. However, if the instrument is to be used for very long periods, it is less tiring on the eyes to use far-point adjustment, though this will incur a small loss of magnifying power.

4.4 The refracting telescope

As discussed earlier, there are two possible reasons why the retinal image of an object might be small: either the object itself is intrinsically very small (in which case a microscope is used to enlarge the angle subtended at the eye), or the object is *very distant*. In this latter case, a two-lens system can again be used to enlarge the visual angle. The first person to construct such an instrument may have been the same Zacharias Janssen who invented the microscope. However, the earliest indisputable evidence is from 1608, when Hans Lippershey sought a patent on a device for seeing at a distance. Galileo Galilei soon built a version for himself (using two lenses and an organ pipe as a tube), and with it discovered the four largest satellites of Jupiter and the phases of the planet Venus (Figure 3.35). Such a system is called a **refracting telescope**; as you will see in the following section, telescopes that use an objective *mirror* rather than an objective *lens* are referred to as *reflecting* telescopes.

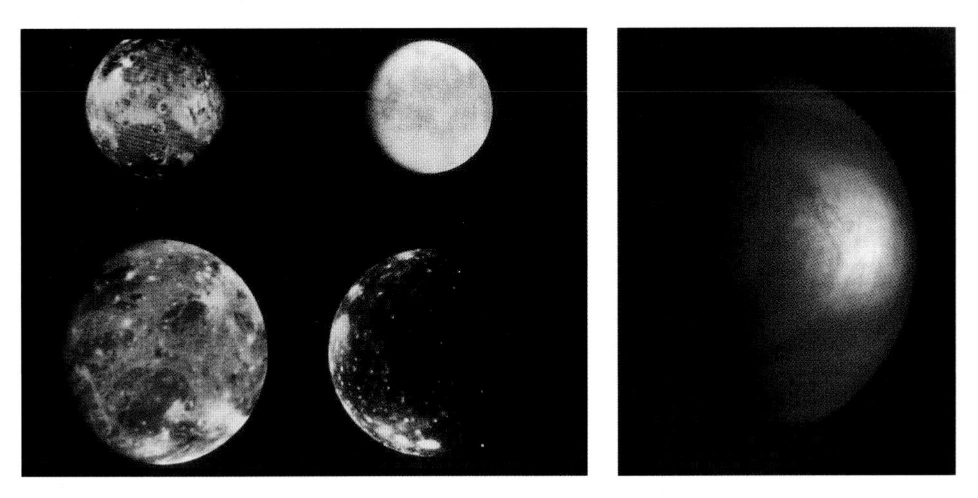

Figure 3.35 (a) The four Galilean satellites of Jupiter (Io, Europa, Ganymede and Callisto) first seen by Galileo Galilei in 1610; (b) the phases of the planet Venus — a phenomenon also discovered by Galileo, which demonstrated that Venus orbits the Sun at a closer distance than the Earth.

Distant extended sources

Suppose you want to view a very distant object, yet one that has an appreciable spatial extent — say a crater on the Moon, for example. Because the object is very distant, the rays from any individual point on the object — say the 'top' of the crater — will all be parallel by the time they reach your eye. Furthermore, these rays will be inclined at an angle α_{OB}, say, to the rays from the 'bottom' of the object; that is, the object subtends the angle α_{OB} at your eye. For the unaided eye, α_{OB} is likely to be very small (as noted earlier, the Moon's complete disc subtends an angle of only about 0.5°). As with the microscope, the purpose of the telescope is to increase this angular size. Figure 3.36 shows how it is done in the case of a refracting telescope. In fact, the microscope and refracting telescope systems are almost identical; the

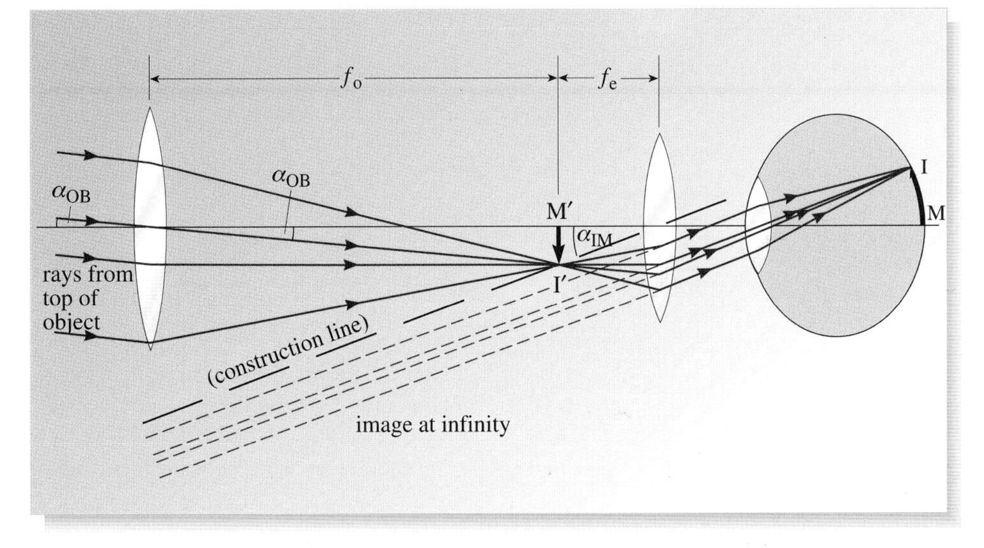

Figure 3.36 A two-lens telescope in far-point adjustment. In fact, the version constructed by Galileo was slightly different from the one shown here in that it used a *diverging* eyepiece lens. However, the principle was broadly similar to this one, the design of which is due to Johannes Kepler around 1630, and is now used in most refracting telescopes. For simplicity, the rays from the bottom of the object under observation are here assumed to be parallel to the optical axis. In practice, of course, the axis is more likely to be *centred* on the object; you could then think of the illustration here as showing 'half an object'.

objective lens forms a real, inverted intermediate image I′M′ of the object, and the eyepiece lens again behaves like a magnifying glass, producing an enlarged, virtual image of this intermediate image. The systems differ only in the positioning of this intermediate image. In the telescope, because the object is effectively at infinity, the image I′M′ will be formed in the focal plane of the objective lens. Strictly speaking, this image is not (linearly) magnified: I′M′ will generally be much *smaller* than the actual object. But what matters, of course, is that the *visual angle* of this image is much greater than it was for the object. The visual angle of the final image will be the same as the visual angle of the intermediate image I′M′. The angle is labelled α_{IM} in Figure 3.36.

Now, just as with the microscope, the eyepiece lens can be adjusted so as to position the final image anywhere between the eye's near point and infinity (its far point). However, since telescopes are normally used for long periods of observing, it is more relaxing for the eye if the telescope is set up in far-point adjustment. This is how it is shown in Figure 3.36, with the distance between the intermediate image I′M′ and the eyepiece equal to the focal length f_e of the eyepiece.

● What adjustment would you make to put the telescope into near-point adjustment?

❍ You would push the eyepiece in closer to the objective lens, so that the distance between I′M′ and the eyepiece was *less* than f_e. In fact, you'd make this distance as small as you could without the image going out of the focus range of your eye. ■

From Figure 3.36 you should be able to see that, when the refracting telescope is in far-point adjustment, the intermediate image I′M′ is in the focal plane of both the objective and eyepiece lenses. Consequently, the length of the telescope is $(f_o + f_e)$. Calculating the magnifying power is very easy in this adjustment. The angle subtended by the object at the unaided eye will, to all intents and purposes, be the same as the angle subtended at the objective lens (since the length of the telescope is negligible compared with the distance to the object). Hence, from the diagram

$$\alpha_{\mathrm{OB}} = \frac{\mathrm{I'M'}}{f_o}. \tag{3.22}$$

But also from the diagram,

$$\alpha_{\mathrm{IM}} = \frac{\mathrm{I'M'}}{f_e}. \tag{3.23}$$

Now since the angular magnification is $\alpha_{\mathrm{IM}}/\alpha_{\mathrm{OB}}$,

$$M = \frac{\mathrm{I'M'}/f_e}{\mathrm{I'M'}/f_o}$$

so $$M = \frac{f_o}{f_e}. \tag{3.24}$$

This is a very important result. Clearly, to achieve the highest possible magnifying power, the ratio f_o/f_e should be made as large as possible.

Distant point sources

When a telescope is used to look at the stars, there is a particular challenge: the stars are not only very distant but they are also very *faint* objects. So the telescope must be used to increase the apparent brightness of these objects. Actually, as far as stars are concerned, that is *all* the telescope can do. The visual angle of a star is so small that even after maximum magnification, the angle subtended at the eye is smaller than the blurring produced by the imperfections in the lenses. As far as the telescope is concerned, stars really are point sources of light.

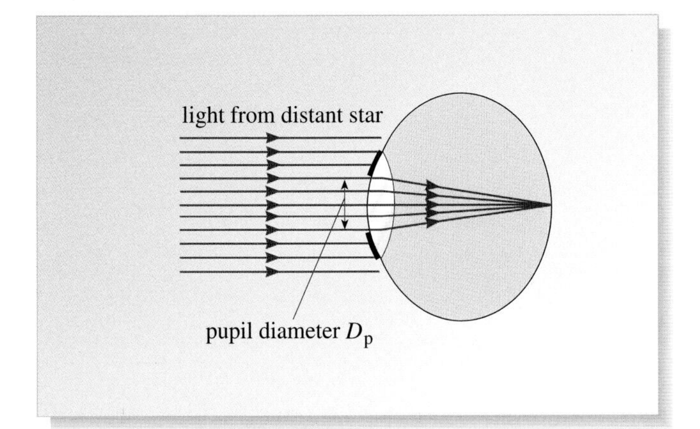

Figure 3.37 The amount of light gathered by the unaided eye is proportional to the area of the pupil.

So how does the telescope increase the star's apparent brightness? Well, imagine looking directly at a star with your unaided eye. The rays of light from this point source will enter your eye in a direction parallel to the visual axis. Now the rays from this star arriving anywhere at the Earth's surface will all be parallel to each other (since the diameter of the Earth is negligible compared with the distance to the star). But your eye can only sample a very small fraction of this total light; the amount of light gathered will be determined by the area of the pupil (Figure 3.37). If D_p represents the *diameter* of the pupil, this area will be given by $\pi(D_p/2)^2$.

Now look at Figure 3.38, which shows the typical telescope arrangement of two lenses separated by the sum of their focal lengths. The total light gathered by this telescope is determined by the area $\pi(D_o/2)^2$ of the objective lens (where D_o is the diameter of this lens). Furthermore, this light is then concentrated down into the smaller area $\pi(D_e/2)^2$, where D_e is the diameter of the beam *at the eyepiece* (not

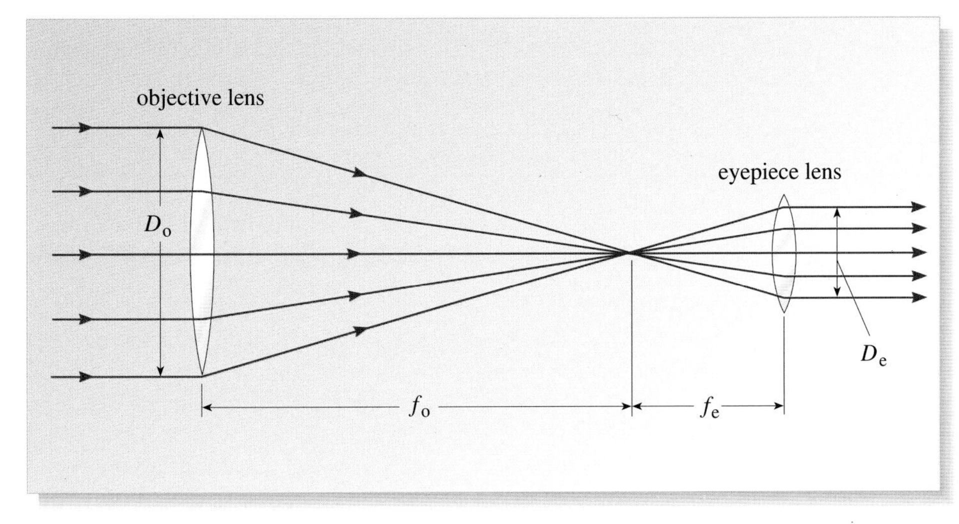

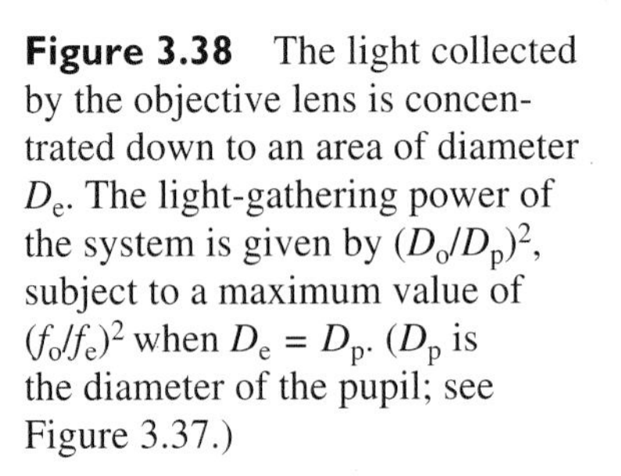
Figure 3.38 The light collected by the objective lens is concentrated down to an area of diameter D_e. The light-gathering power of the system is given by $(D_o/D_p)^2$, subject to a maximum value of $(f_o/f_e)^2$ when $D_e = D_p$. (D_p is the diameter of the pupil; see Figure 3.37.)

necessarily the diameter *of* the eyepiece). You can see from the similar triangles in the diagram, that $D_o/D_e = f_o/f_e$. If the focal lengths and diameters of the lenses are adjusted so that D_e is less than or equal to the pupil diameter D_p, then it is possible to look through the telescope and 'see' *all* the light gathered by the objective lens. In effect, this system of lenses has enabled us to increase the apparent brightness of the star by increasing the amount of light gathered onto the retina. The factor by which the light gathered into the eye is increased by the telescope is known as the telescope's **light-gathering power**. Provided that $D_e \leqslant D_p$, so that all the light passing through the objective lens enters the eye, the light-gathering power is given by the ratio of the objective lens's area to the area of the pupil; that is

$$\frac{\pi(D_o/2)^2}{\pi(D_p/2)^2}.$$

Hence, for $D_e \leqslant D_p$

$$\text{light-gathering power} = \left(\frac{D_o}{D_p}\right)^2. \tag{3.25}$$

For any given pair of lenses, the light-gathering power has a maximum value when the beam diameter at the eyepiece D_e is equal to the pupil diameter D_p. If D_e is any greater than this, then some of the light gathered by the objective lens will spill outside the diameter of the pupil and will effectively be lost. Put another way, the *effective* aperture of the objective will be less than its actual aperture. So, from the similar triangles argument mentioned before (i.e. $D_o/D_e = f_o/f_e$), it must follow that the *maximum* light-gathering power of this system of lenses, adjusted for parallel light in and parallel light out, must be given by $(f_o/f_e)^2$. In other words

$$\text{maximum light-gathering power} = \left(\frac{D_o}{D_e}\right)^2 = \left(\frac{f_o}{f_e}\right)^2 \tag{3.26}$$

when $D_e = D_p$.

Question 3.17 One of the most powerful refracting telescopes in the world has an objective lens 1 m in diameter with a focal length of 20 m. The eyepiece lens has a focal length of 8 mm and a diameter of 2 mm.

(a) What is the length of this telescope when it is set up in far-point adjustment to look at the night sky, and what is its corresponding magnifying power?

(b) Calculate the telescope's light-gathering power (in this adjustment) relative to a dark-adapted eye with a pupil diameter of 8 mm. ■

Problems with refracting telescopes

Equation 3.26 indicates that a refracting telescope with an objective lens only ten times the diameter of the pupil will increase the apparent brightness of a star one hundredfold. To be able to see the faintest objects, clearly D_o should be made as large as is reasonably possible. Unfortunately, this is easier said than done. To begin with, there are serious technological problems in producing very large lenses. In order to ensure that the initial block of glass, from which the lens is to be made, is perfectly transparent and optically homogeneous throughout, the molten glass may need several years (!) of gradual and controlled cooling. Next comes the problem of grinding and polishing: it is not easy to sustain a perfect spherical curvature for a very large focal length lens over the whole of its surface area. And when you have a large lens, it is inevitably a thick lens, which absorbs strongly in the blue and violet region of the spectrum. It is also a very heavy lens, which means that it would have a tendency to sag under its own weight. In practice, usable objective lenses with a diameter D_0 much larger than 1 metre cannot be made. Figure 3.39 shows a photograph of one of the largest refracting telescopes in the world, the 0.9 m refractor at the Lick Observatory, California. Note the extremely long body of the telescope in relation to its ≈1 m diameter.

Figure 3.39 The 0.9 m refractor at the Lick Observatory, California.

As for the maximum possible value of f_0, the limits are set not just by the problems of grinding the lens to the right curvature, but also by the need to make the whole instrument movable. It is clear from Figure 3.36 that the physical length of a refracting telescope cannot be less than f_0. Hence, it would hardly be realistic to plan a telescope with an objective focal length of 100 m!

Finally, there is the problem of optical aberrations. As you saw earlier, light passing through different parts of a lens is not focused to the same point by spherical surfaces (because of spherical aberration). Even from the same part of the lens, waves of different frequency are focused to different points (due to chromatic aberration). By combining several lenses of different optical strengths and different refractive indexes such aberrations can be reduced. However, the problems are still formidable, and increase with the increasing size of the lenses and with the angle of the rays with respect to the optical axis. Thus, in practice, refracting telescopes have only a relatively narrow field of view within which resolution is good.

The net result of all these problems is that refracting telescopes are no longer built for serious astronomical work. You may come across small hobby telescopes built to this design, but in general these are less satisfactory than a pair of binoculars of comparable cost.

4.5 The reflecting telescope

A lens is not the only object that can collect and focus light and thus produce visual images. People have known about and used looking glasses (mirrors) for much of recorded history, but it took no less a genius than Isaac Newton to realize how a curved mirror could be used to construct a **reflecting telescope**, and that this would overcome some of the most important shortcomings of refracting telescopes.

A concave spherical mirror will reflect parallel rays approaching along its axis of symmetry so that they come together at one point (the focus) lying between the reflecting surface and its centre of curvature. The main advantage of focusing by reflection is that the angle of reflection is the same for all frequencies in the incident radiation. There is no analogue to the chromatic aberration that takes place in lenses. Hence, if the objective lens of a telescope is replaced by a reflecting spherical **objective mirror**, the chromatic aberration on the input side of the telescope is automatically and completely eliminated. As noted earlier, though, spherical aberration is still present because rays reflected from the points further away from the axis of symmetry will be focused nearer to the reflecting surface. Even so, the spherical aberration of a converging mirror is always less than the spherical aberration of a converging lens of the same focal length, and it can usually be neglected for converging mirrors that are only small parts of the hemisphere. Unfortunately, by reducing the size of the mirror to reduce spherical aberration, some of the potential light-gathering power is lost, and the useful field of view is also limited.

There are two ways of dealing with the problem of spherical aberration. Either a parabolic shape for the mirror can be chosen (as in Figure 3.40a), or the focusing of a spherical mirror can be corrected by introducing a suitable predistortion into the incoming wavefront. This is done by placing in front of the mirror a transparent plate of such a shape that it refracts the initial parallel rays near the optical axis differently from those further away from it (as shown in Figure 3.40b). This correcting plate is known as a **Schmidt plate**, and the reflecting telescopes in which a Schmidt plate is used are called **Schmidt telescopes** (or **Schmidt cameras** when used in conjunction with photographic detectors).

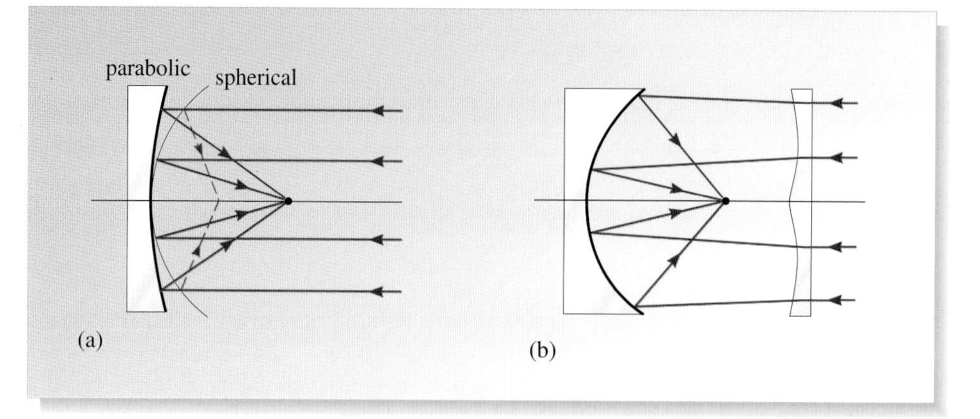

Figure 3.40 (a) The reduction of spherical aberration using a mirror of parabolic shape. Parallel rays of light are all brought to the same focus, irrespective of their distance from the optical axis. (b) The Schmidt correcting plate compensates for spherical aberration. Parallel rays of light further from the optical axis are bent, with respect to rays closer to the optical axis, by the Schmidt plate. The net result is that all rays are brought to a common focus.

In case you are wondering how you could actually see the image of a star produced by a spherical converging mirror without being in the way of the oncoming light, this problem was solved simply and neatly by Newton as shown in Figure 3.41. He put a small flat (plane) mirror (the **secondary mirror**) just before the focus of the main reflector (the **primary mirror**) and at an angle of 45° to the optical axis. He thus moved the image towards the side wall of the telescope tube, where he then fixed an eyepiece for direct observations. A telescope using this arrangement is known as a **Newtonian telescope**.

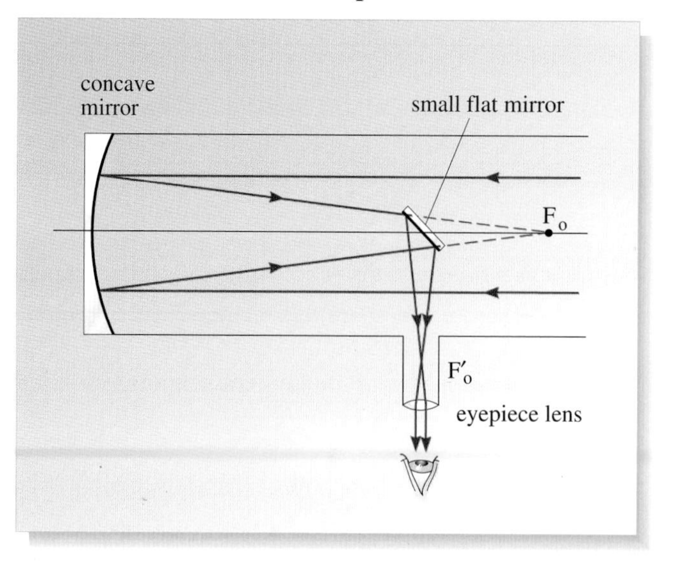

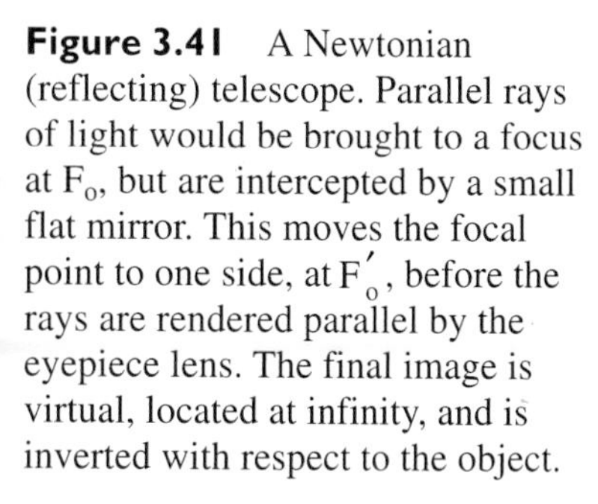

Figure 3.41 A Newtonian (reflecting) telescope. Parallel rays of light would be brought to a focus at F_o, but are intercepted by a small flat mirror. This moves the focal point to one side, at F'_o, before the rays are rendered parallel by the eyepiece lens. The final image is virtual, located at infinity, and is inverted with respect to the object.

A further improvement was introduced by the French astronomer Guillaume Cassegrain, one of Newton's contemporaries. His idea is illustrated in Figure 3.42, and is now used in most large modern telescopes. In place of Newton's flat and tilted secondary mirror, Cassegrain used a slightly *diverging* secondary mirror placed on the optical axis of the main one. The light is therefore reflected back towards the centre of the primary mirror, where it passes through a hole on the optical axis and then onto an eyepiece. This has the effect of extending the path of the reflected light before it is brought to a focus at the extended focal point F_{ext}. Hence the **effective focal length** of the objective mirror (which determines the size of the image) in a **Cassegrainian telescope** can be approximately double that for a Newtonian telescope of the same length. Both Newtonian and Cassegrainian telescopes may be constructed using either parabolic objective mirrors or using spherical objective mirrors with Schmidt correcting plates.

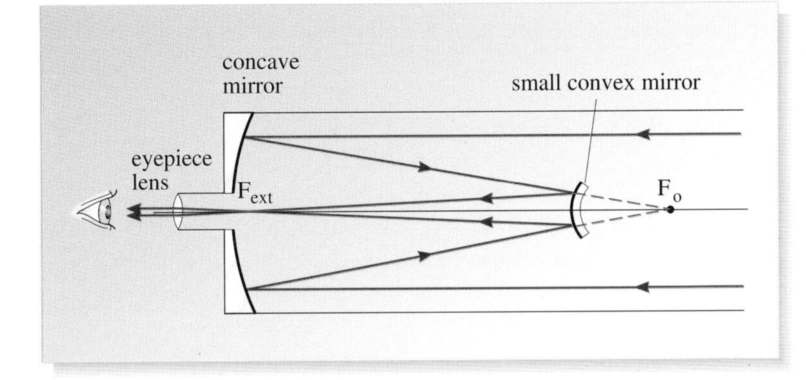

Figure 3.42 A Cassegrainian (reflecting) telescope. Parallel rays of light would be brought to a focus at F_o, but are intercepted by a small convex mirror. This diverges the rays of light somewhat so that they are not brought to a focus until the point F_{ext}. The rays then diverge before entering the eyepiece lens and emerge from it parallel. The final image is once again at infinity, virtual and inverted. The effective focal length of this telescope is equal to the distance from the objective mirror to the small convex mirror *plus* the distance from the small convex mirror to the point F_{ext}. This can be roughly twice the length of the telescope body.

The angular magnification and light-gathering power of a reflecting telescope are also given by Equations 3.24 and 3.25, the only difference being that f_o and D_o are now the effective focal length and diameter of the objective *mirror* rather than the objective *lens*.

If a telescope is to be used with a photographic or electronic detector, instead of the eye, then a *real* image must be allowed to fall onto the light-sensitive surface of the detector. In this case there is no point in using the telescope in far-point adjustment with an eyepiece lens, since that produces a virtual image located at infinity. The simplest solution is to remove the eyepiece lens entirely, and place the detector in the focal plane of the objective mirror (i.e. at F_o' in Figure 3.41 or at F_{ext} in Figure 3.42).

This also has the advantage of removing any aberrations introduced by the eyepiece lens. Alternatively, the secondary mirror may also be removed, and the detector may be placed directly at the **prime focus** of the main mirror (i.e. at F_o in Figure 3.41 or Figure 3.42). This has the additional advantage of removing one more optical component, and with it the inherent aberrations and absorption losses that it contributes. Figure 3.43 shows photographs of two famous optical (reflecting) telescopes.

Figure 3.43 (a) The 5 m diameter reflector (the Hale telescope) at Palomar Mountain, California; (b) the 1.2 m diameter Schmidt telescope, also at Palomar Mountain.

Figure 3.44 The Gemini North telescope, at an observatory on Mauna Kea, Hawaii, with its 8 m diameter primary mirror.

In comparison with refracting telescopes, the reflectors start with the important advantage of zero chromatic aberration. But they also score heavily on some aspects of practical construction and technology. It is much easier to produce mirrors (rather than lenses) of very large diameter because the glass does not have to be perfectly transparent or optically homogeneous. (Large mirrors are also much lighter than large lenses.) The grinding and polishing is carried out on only one surface, which is finally covered by a thin reflecting layer of aluminium. In the late 1990s several telescopes with mirrors between 8 and 10 m in diameter were constructed (Figure 3.44), and at the time of writing (2000) were the largest in the world.

On the debit side for reflecting telescopes, there is greater loss of optical intensity than in the refractors, because the reflecting surfaces are never quite perfect and may have appreciable absorption (10 per cent or more). Aluminium surfaces also deteriorate rather quickly and have to be renewed relatively frequently. On the other hand, a perfectly polished lens remains serviceable for many years.

4.6 The camera

This brief survey of optical systems concludes with a discussion of perhaps the most familiar of all optical devices — the photographic **camera**. The first permanent photograph was taken in 1826 by Joseph Nicéphore Niépce using a box camera with a small converging lens. Modern cameras (Figure 3.45) have come a long way since the designs of the early nineteenth century, but the principles remain broadly the same.

Figure 3.45 A modern camera.

In many ways, the camera is simply a 'mechanical eye', being composed of elements that mimic the most important components in the human eye. It has a converging lens, which focuses a real, inverted image onto a photosensitive surface called *film* (see Figure 3.46); it has an iris diaphragm — often called a *stop* — which, like the iris in the eye, can be adjusted to change the size of the central *aperture-stop* (equivalent to the eye's pupil), and so control the amount of light entering the camera; and, though this is pushing the analogy rather hard, the camera has a *shutter*, which operates a bit like the eyelid and can be opened and closed for a predetermined length of time to allow light to enter the otherwise light-tight box and expose the film. But of course, the factor that probably gives the camera its overriding importance is its ability to render the image permanent.

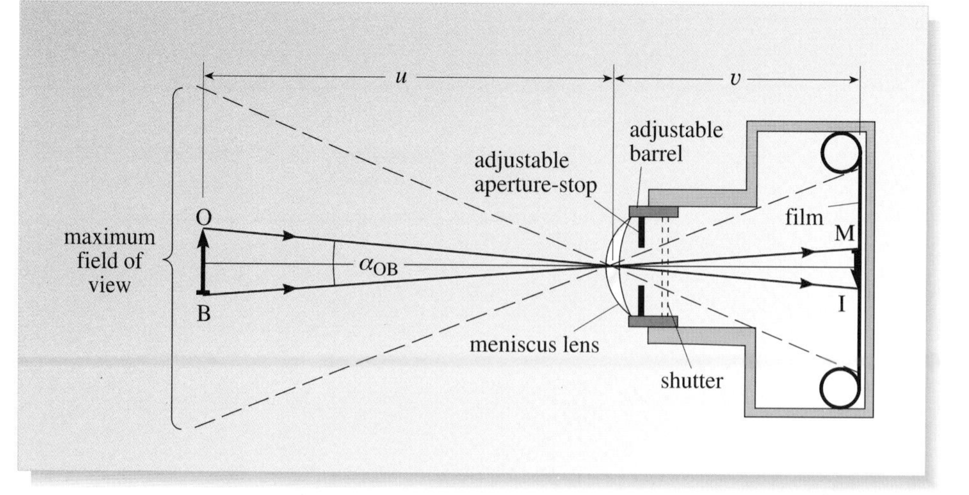

Figure 3.46 The basic elements of a simple photographic camera. Note that the maximum field of view of the camera (maximum value of α_{OB}) is determined by the size of the film frame.

The basic optics

In contrast to most of the optical systems looked at so far, it is important for the camera to have a large **field of view**. In other words, the object of interest to the photographer is likely to extend appreciably either side of the camera lens's optical axis: the visual angle α_{OB} subtended by the object could easily be as much as 30° or so. When the angle is as large as this, the various lens distortions are certainly not negligible. The cheapest cameras commonly use a single converging (convex) lens but, in an attempt to minimize the effects of these distortions, some cameras use the so-called 'convex meniscus' profile shown in Figure 3.46. In addition, the lens is usually *stopped down* so that, even with the adjustable iris diaphragm fully open, the rays from any one point on the object pass through only a small central part of the lens. To a large extent, what you are paying for in more expensive cameras is a much more complex lens system (made up of perhaps as many as six or seven separate lenses) specially designed to overcome these distortion problems.

Unlike the eye, which can change its focal length, the camera's basic lens system has a fixed focal length. Consequently, if it is always to produce focused images on the film of objects that may be at various distances from the camera lens, it is necessary to be able to adjust the image distance v by moving the lens position relative to the film plane. This is usually achieved by mounting the lens in a sliding or threaded barrel, as illustrated in Figure 3.46. The range of adjustment of this barrel must be sufficient to accommodate object distances ranging from infinity to perhaps just a metre or so.

Question 3.18 A particular camera is required to be able to sharply focus objects located anywhere between 1 m and infinity. If the adjustable distance between the lens and film plane can be reduced to a *minimum* value of 80 mm, what must be the focal length of the lens, and over what range should it be possible to adjust the lens position? ■

Calculating exposure

Photographic film consists of an *emulsion* deposited onto some kind of transparent base — glass for photographic plates, or a plastic polymer for roll-film. The emulsion itself is made up of microscopic crystals of a *silver halide* suspended in a transparent medium, usually gelatin. This emulsion is the photosensitive surface: where it is exposed to light, various photochemical reactions are initiated, which, as a consequence of the film being treated with a chemical *developer*, cause the exposed transparent crystals of silver halide to be converted into opaque black grains of metallic silver. The unexposed silver halide crystals are rendered soluble in water by treating the film with another chemical known as a *fixer*. When the film is washed in water, it can be seen to contain a *negative* image of the object; that is, where the object was bright, the image is black, and where the object was dark, the image is clear (for black and white film).

But, of course, black and white film does not produce images made up of only totally black and totally clear regions; it will also reproduce the whole range of intermediate shades of grey. And as you might expect, the greyness of any particular part of the film is determined by the amount of light energy per unit area falling on that part of the film: the greater this *energy density*, the darker the film. Now, there are two ways of changing the energy density: either by changing the *exposure time* or by changing the rate at which energy is channelled to the film. Put another way, the total *exposure* (energy density) to which the film is subjected will be given by the product of the exposure time and the intensity of the light. If the intensity is represented by the symbol I, and the exposure time interval by the symbol Δt, then

$$\text{exposure} = I\,\Delta t. \tag{3.27}$$

Changing the camera's exposure time is no problem: you just change the length of time for which the shutter is open. This is effected by adjusting the camera's *shutter-speed* control. But how can the intensity of light entering the camera be changed? Easy — you adjust the aperture stop and so change the size of the aperture. If the aperture is increased, more light will be admitted in a given time interval; but this light will still be focused down to the same-sized image. This is exactly analogous to increasing the light-gathering power of a telescope by increasing the diameter of the objective lens or mirror.

It is common practice with camera lenses to express their *effective* diameter (that is, the **aperture diameter** of the stop) as a fraction of the focal length of the lens. So, for instance, if the aperture diameter D were 1/8th of the focal length of the lens, then $D = f/8$. Not surprisingly, the number in the denominator of this expression has come to be known as the ***F*-number** of the lens (or more precisely of the lens–stop combination); that is

$$F\text{-number} = \frac{\text{focal length } f}{\text{aperture diameter } D}. \tag{3.28}$$

You will find this definition of F-number used extensively in many areas of optics.

So, the setting of the camera's aperture stop can be specified either by saying the aperture has a *diameter* of $f/8$ (spoken as 'f-eight'), or by saying its F-number is equal to 8; both these mean the same thing. Furthermore, the bigger the F-number, the smaller the diameter.

Now if you look at a typical (but non-automatic) camera, you will find that it gives you a selection of shutter-speed settings, ranging from very short to quite long exposure-time intervals. The important thing to note about these settings is that the normal increment between adjacent positions corresponds to a change of time interval by a factor of 2 (as near as makes no difference). Typical exposure-time settings might be, for instance, 1/500 s, 1/250 s, 1/125 s, 1/60 s, 1/30 s, 1/15 s, 1/8 s and 1/4 s. (On some cameras, only the denominator is used to indicate this time value; that is, '60' means 1/60 s.) By increasing the exposure time by a factor of two while keeping everything else constant, you double the amount of light (energy density) reaching the film. It would seem logical, therefore, to arrange things so that successive positions of the camera's 'intensity-admittance control' — its aperture-stop — would also change the amount of light reaching the film by a factor of two. In this way, a 'one-position' increase in the exposure time (by a factor of two) could be exactly counteracted by a 'one-position' decrease in the aperture (also by a factor of two). Such an arrangement greatly facilitates successful operation of the camera. But when you look at a camera's aperture-stop settings, you see that they specify the aperture diameter in terms of the F-number of the lens/aperture combination — and these F-number values might range from (say) $F = 2.8$, through $F = 4$, 5.6, 8, 11, 16, to (say) $F = 22$. But these numbers are *not* changing by a multiple of two each time!

Well true, and there is a very good reason for this. The light intensity at the film depends on the *area* of the aperture, and so it is the area that must change by a factor of two each time. And of course, the area of a circular aperture is given by

$\pi \times$ radius^2, or by $(\pi/4) \times$ diameter^2; that is, the aperture area is proportional to the *square* of the aperture's diameter. So if the area is to change by a factor of two, then the (diameter)^2 must change by a factor of 2, so that the diameter changes by a factor of $\sqrt{2}$. This is just what is represented by the sequence of *F*-numbers cited above (given that they are quoted to only two significant figures). The situation is summarized in Table 3.1; note that *f* is the focal length of the camera lens, and that all the aperture area values have been divided by $\pi/4$.

Table 3.1 Relationship between *F*-number and aperture size.

aperture diameter	$f/2.8$	$f/4$	$f/5.6$	$f/8$	$f/11$	$f/16$	$f/22$
F-number	2.8	4	5.6	8	11	16	22
(*F*-number)^2	8	16	32	64	128	256	512
$\dfrac{\text{aperture area}}{\pi/4}$	$f^2/8$	$f^2/16$	$f^2/32$	$f^2/64$	$f^2/128$	$f^2/256$	$f^2/512$

Question 3.19 A camera with a lens of focal length 48 mm is used to photograph a particular scene. The light levels in this scene are such that the optimum exposure, appropriate to the sensitivity of the film that is being used, can be achieved with aperture and shutter-speed settings of $f/8$ and 1/125 s, respectively.

(a) What is the aperture diameter of the camera at these settings?

(b) What would be an appropriate aperture-stop setting if the exposure time were to be reduced to 1/250 s?

(c) What is the *F*-number setting on the camera corresponding to a lens aperture diameter of 3 mm? What would be an appropriate shutter-speed setting to combine with this aperture-stop setting in order to photograph the same scene? ■

Depth of field and depth of focus

Being able to set the same exposure but with different combinations of aperture-stop and shutter-speed settings is a very useful facility. For example, if you want to photograph some rapidly moving object, you must 'freeze the motion'. In other words, you must use a very short exposure time; otherwise the image of the moving object will be blurred. It is important to be able to compensate for the potential reduction in total exposure in this case, by *increasing* the aperture (reducing the *F*-number). Conversely, if for some reason you deliberately *want* to integrate the light over a longer exposure-time period — perhaps you *want* a waterfall to blur and look sheet-like — then you can prevent over-exposure by reducing the aperture (increasing the *F*-number) to compensate.

But does changing the aperture diameter here have any effect on the image other than to change its intensity? The answer to this question is a very definite yes — changing the aperture also changes the available *depth of focus* and the corresponding *depth of field*. These two concepts are illustrated in Figure 3.47. A camera's **depth of focus** (Figure 3.47a) is defined as the range of distance over which the film plane can be moved relative to the lens without the image of a fixed object becoming unacceptably blurred. (For a given object distance, there is strictly only one image-plane distance for which the image is *sharply* focused. Note also that in practice, the distance between lens and film is varied by moving the lens rather than the film.) The complementary idea of **depth of field** (Figure 3.47b) is defined as the range of distance over which an *object* could be moved relative to the camera without its image in the film plane becoming unacceptably blurred. (For a given lens-to-film distance, there is strictly only one object distance that will give a truly sharp-focused image on the film.) In both these cases, the degree of blurring of the image can be reduced by reducing the diameter of the aperture; this is because reducing the aperture diameter reduces the angle of the cone of light producing the image. Putting this round the other way, for a given degree of blurring, the range of satisfactory image distances (for a fixed object), or object distances (for a fixed film plane), becomes larger as the diameter of the aperture is reduced. Study Figure 3.47 carefully until you are convinced that these statements are true.

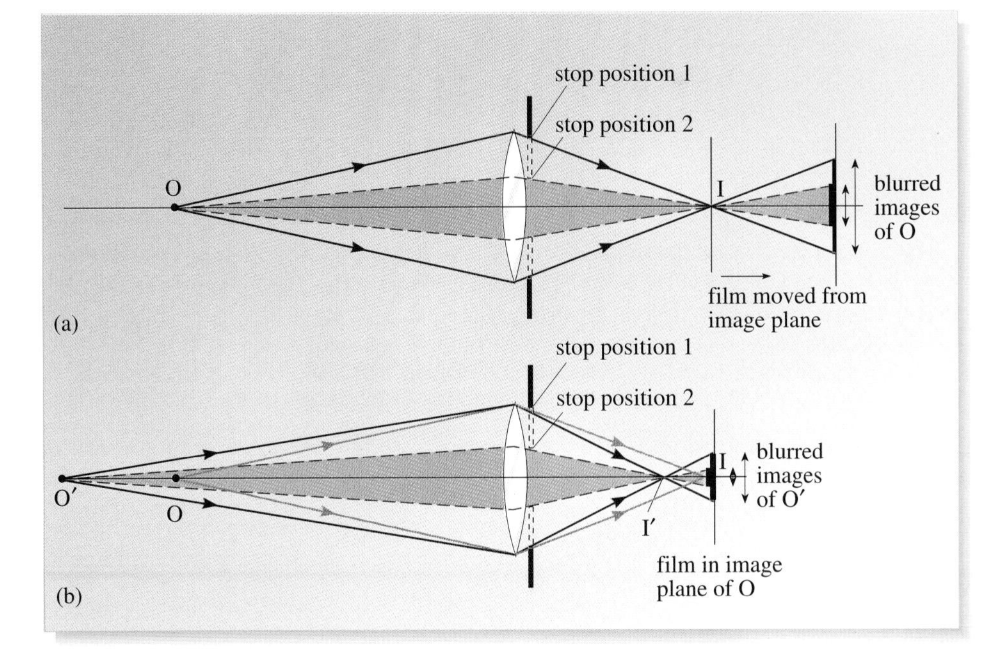

Figure 3.47 (a) Depth of focus; (b) depth of field (which is also, in practice, frequently but incorrectly called depth of focus). In both cases, the out-of-focus image is *less* blurred when the aperture-stop is in position 2.

So what are the practical consequences of this? Well, it means that when you have a *small* aperture, you have a *large* depth of field, which in turn means that objects over a range of distances from the camera will all appear to be in focus on the film. This is often desirable when taking general-purpose photographs. Conversely, if you set a large aperture, then the depth of field will be small: objects in front of, and behind, the subject on which you are focused will be blurred. This is more appropriate for portraits, or as a special effect.

● Why might it be a bit of a problem getting a satisfactory 'photo-finish' type photograph of 100 m sprinters crossing the finishing line on a dull day?

❍ Because such a photograph has mutually conflicting requirements. To 'freeze the motion' of the sprinters requires a short exposure time. To compensate for this, particularly on a dull day, you would have to open up the aperture. But a large aperture would give a small depth of field, so you'd have difficulty getting the whole of the finishing line in focus simultaneously. You'd have to settle for a sharply focused finishing line but with the motion of the runners blurred, or no blurring of the motion but with a blurred finishing line, or some compromise between these two extremes. ■

In all this discussion, the expression, 'without the image… becoming unacceptably blurred' has been used, without saying exactly what is meant by this. How blurred an image is depends on the ability of the eye to resolve small-scale structures. The **resolving power of the eye** is its ability to observe two closely spaced point-objects, and see them as two distinct points. For the typical eye, two such points tend to merge together into a single blur when the angular difference between them is less than about 1 arc minute ($= 1°/60 \approx 3 \times 10^{-4}$ radians). Now, two points on a photograph held at the normal near point (250 mm) will subtend an angle of 3×10^{-4} radians at the eye when they are $(250 \times 3 \times 10^{-4})$ mm = 0.08 mm apart. In other words, two points 0.08 mm apart on the photograph can only just be resolved, so any blurring of up to this amount will obviously go unnoticed. And if the photographic print you are holding at your near point has been enlarged by say a factor of four relative to the original film, then this means that, on the original film, blurring of up to about 0.02 mm would be acceptable. In practice, you'd probably be prepared to accept a considerably greater degree of blurring than even this.

Special camera systems

It is often the case that an observer looking at the image produced by a microscope or telescope desires a permanent record of that image. A possible solution, of course, would be to take a photograph of the image. This can be achieved in each case by placing the film in the plane of the real, inverted image produced by the objective lens or mirror. In effect, the microscope becomes a very short focal length camera, with a magnification determined by v_o/u_o, whereas the telescope becomes a very long focal length camera, producing an image, the size of which is proportional to the lens or mirror's focal length f_o.

Naturally, the idea of a camera that can magnify and photograph objects at a distance is very attractive. But it is not really practicable to implement it in the simplistic way described above; there is a world of difference between what you would expect from a telescope that can take photographs, and a camera that can photograph distant objects. Since the image is formed in the lens's focal plane, and because a long focal length lens is required to produce a reasonably sized image, a long-distance camera would necessarily have to have a long, and therefore cumbersome and unwieldy, lens barrel. The so-called **telephoto lens** constitutes an attempt to solve this problem.

In practice, commercially available telephoto lenses (Figure 3.48) contain a large number of lens elements, most of which help to reduce aberrations and other limitations. However, the basic principle behind most of them is a two-lens system, in which a *diverging* lens is placed a small distance behind the usual converging objective lens, in such a way that the *effective* or *equivalent* focal length of the system is much longer than the actual physical distance between the front lens and the system's focal plane. A fairly high-powered example of such a lens system is shown in Figure 3.49. The converging lens L_1, with a focal length f_1 = +50 mm, is placed 30 mm in front of a diverging lens L_2 of focal length f_2 = −25 mm. (Notice that the configuration is similar to that used by Galileo in his original telescope, referred to earlier.)

Figure 3.48 A telephoto lens.

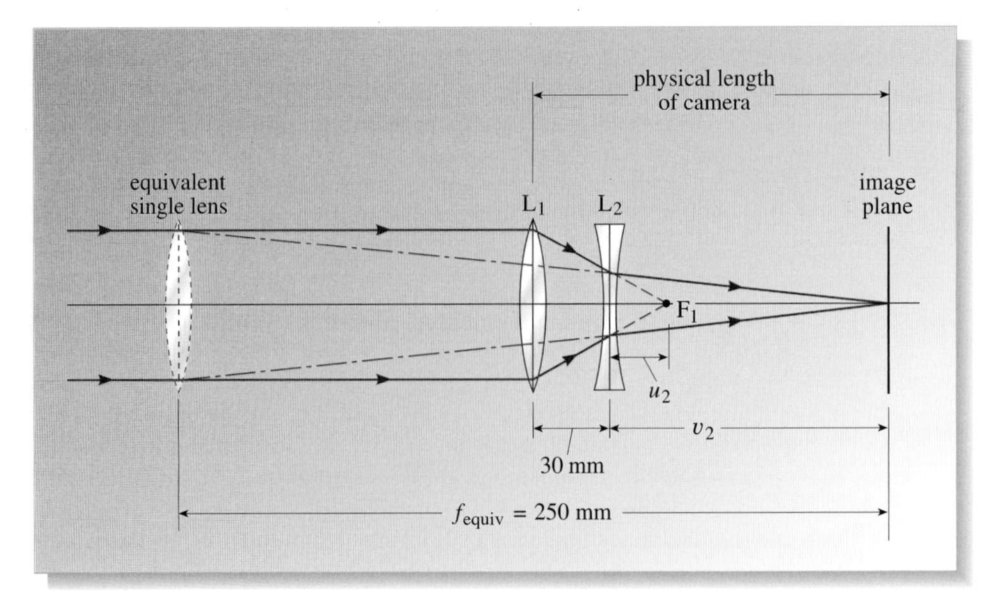

Figure 3.49 A fairly high-powered telephoto lens. By using a converging lens L_1 of focal length f_1 = 50 mm, positioned 30 mm in front of a diverging lens L_2 of focal length f_2 = −25 mm, it is possible to build a camera with a total *physical* length (first lens to film plane) of only 130 mm, but an *equivalent focal* length of 250 mm. To achieve the same effect with a single lens, that lens (shown dashed in the diagram) would have to be positioned 250 mm from the film plane.

- What is the value of the object distance u_2 for the diverging lens in Figure 3.49, and what value of image distance v_2 does this lead to?

○ The first lens tries to focus the parallel light to a point a distance 50 mm to its right. This point then constitutes a *virtual* object for the second lens; the object distance $u_2 = -(50 - 30)$ mm = −20 mm. For lens 2, the lens equation is $1/v_2 + 1/u_2 = 1/f_2$. Substituting values for u_2 and f_2,

$$\frac{1}{v_2} - \frac{1}{20\,\text{mm}} = -\frac{1}{25\,\text{mm}}$$

so

$$\frac{1}{v_2} = \frac{1}{20\,\text{mm}} - \frac{1}{25\,\text{mm}} = (1/100)\,\text{mm}^{-1}.$$

Hence the image distance is 100 mm beyond the second lens, and so the overall length of the camera would have to be 130 mm. ■

It is now possible to project backwards from the rays forming this final image, to locate the plane where they intersect the original parallel input rays. The distance between this plane and the image plane is the equivalent focal length of the two-lens system, which in Figure 3.49 is 250 mm.

Question 3.20 It's not a trivial matter to derive an equation that allows the equivalent focal length to be calculated from the focal lengths and separation of the two constituent lenses. However, you *should* be able to use straightforward geometrical (similar triangle) arguments to show that the value of f_{equiv} in Figure 3.49 really is 250 mm. Have a try. (*Hint* Let the diameter of the parallel input beam be D, and calculate the new diameter of the beam in the plane of the diverging lens L_2.) ■

5 Closing items

5.1 Chapter summary

1 Converging lenses bring incident parallel rays to a focus at the focal point on the optical axis, a distance of one focal length beyond the lens. Diverging lenses cause incident parallel rays to diverge *as if* from the focal point on the optical axis on the input side of the lens.

2 Converging mirrors reflect incident parallel rays back through the focal point on the optical axis on the input side of the mirror. Diverging mirrors reflect incident parallel rays so that they appear to diverge from the focal point on the optical axis behind the mirror. The distance from the focal point to the optical centre of the mirror is the focal length of the mirror, which for a spherical mirror is half its radius of curvature.

3 Lenses suffer from chromatic aberration such that the focal length of a lens is different for different frequencies. Both spherical lenses and mirrors suffer from spherical aberration such that rays are brought to a slightly different focus depending on their initial lateral displacement from the optical axis. The effect is small provided that only those rays close to the optical axis are considered.

4 The object and image positions, and the focal length of a lens or mirror, are related through the lens equation, $1/v + 1/u = 1/f$.

5 (a) A real image is one to which the rays converge. In the real-is-positive convention, a real image's distance from the lens (or mirror) is always assigned a *positive* value. A virtual image is one from which rays *appear* to diverge; its distance from the lens (or mirror) is always assigned a *negative* value. (b) A real object is one from which the rays diverge (or appear to diverge); its distance from the lens (or mirror) is assigned a *positive* value. A *virtual* object is one to which the rays would have converged had they not been intercepted by the lens (or mirror); its distance from the lens (or mirror) is assigned a *negative* value. (c) Converging lenses and mirrors are assigned positive focal lengths. Diverging lenses and mirrors are assigned negative focal lengths.

6 Converging lenses and mirrors convert plane wavefronts into converging spherical wavefronts. Diverging lenses and mirrors convert plane wavefronts into diverging spherical wavefronts. In addition, converging lenses and mirrors reduce, or even reverse, the curvature of a diverging spherical wavefront. Diverging lenses and mirrors increase the curvature of diverging spherical wavefronts.

7 The location, orientation and magnification of the image of an extended object can be found by tracing two or more key rays as they pass through a lens or are reflected by a spherical mirror. The key rays that are easy to trace through a lens or mirror are: (i) (lens only) the undeviated ray passing through the lens's optical centre, or (mirror only) the ray reflected back along its path after passing through the mirror's centre of curvature; (ii) a ray that passes through (or is heading towards, or directly away from) the focal point, and is then rendered parallel to the optical axis; (iii) a ray initially parallel to the optical axis, which is then bent so as to pass through (or appear to have diverged from) the focal point.

8 The (linear) magnification m of an extended object, defined as the ratio of the image 'height' to the object 'height', is given by $m = v/u$.

9 The human eye can be regarded as an adjustable focal-length converging lens system, which projects a real inverted image of the outside world onto the retina

on the inside wall of the eyeball. The 'normal' (idealized optically perfect) eye can focus on objects located anywhere between the normal near point (250 mm distant) and the normal far point (infinitely distant).

10 Short-sightedness can be corrected by inserting an appropriate focal-length diverging lens in front of the eye; long-sightedness requires the insertion of a converging lens. The loss of the ability to accommodate requires the prescription of bifocal lenses (or possibly just 'reading glasses').

11 A lens may be specified in terms of its power, $P = 1/f$, expressed in units of dioptres D, where $1\,\text{D} \equiv 1\,\text{m}^{-1}$.

12 It is common practice to specify an object's apparent size in terms of the visual angle α—that is, the angle the object subtends at the eye-lens. For small α, the retinal image size is then proportional to α. The magnifying power or angular magnification of an optical system is then defined as the angle subtended at the eye by the image divided by the largest angle that can be subtended by the object; that is, $M = \alpha_{IM}/\alpha_{OB}$.

13 A magnifying glass is a single converging lens, which enables an object to be brought closer to the eye than the near point and still remain in focus, and by so doing increases the effective visual angle of the object.

14 The magnified image that an optical system presents to the eye can be satisfactorily located anywhere between the eye's near point (where the instrument is in near-point adjustment) and far point (where the instrument is in far-point adjustment).

15 A compound microscope uses two converging lenses to produce greater magnification than is possible with a single lens. The overall angular magnification is the product of the *linear* magnification of the objective lens and the *angular* magnification provided by the eyepiece lens.

16 Two converging lenses can also be arranged to produce a refracting telescope. In far-point adjustment its length is given by $f_o + f_e$ and its angular magnification by f_o/f_e.

17 A converging objective mirror and a converging eyepiece lens can be used to construct a reflecting telescope. The Newtonian and Cassegrainian telescope designs use a secondary mirror to enable the image to be directed to the eye or other light-sensitive recording device. Parabolic mirrors or Schmidt correcting plates are used to overcome the problems of spherical aberration.

18 Telescopes also increase the apparent brightness of a point object by concentrating the light collected over a large area into an area less than or equal to the area of the pupil. Light-gathering power is maximized to a value of $(D_o/D_e)^2 = (f_o/f_e)^2$ when $D_e = D_p$.

19 A camera uses a converging lens to project a real inverted image of an object onto a film. The energy per unit area falling on the film; that is, the exposure, is equal to the product of the light intensity and the exposure time. The exposure time can be adjusted with the shutter-speed control; the intensity is controlled by adjusting the aperture-stop.

20 The aperture diameter of a lens is often specified as a fraction of the focal length of the lens; that is, $D = f/F$, where the denominator is the F-number. As the F-number is successively *increased* by a factor of $\sqrt{2}$, the *area* of the aperture (and hence the intensity) is *reduced* by a factor of 2. The larger the F-number, the smaller the aperture, and vice versa.

21 A camera's depth of focus is the range over which its lens-to-image-plane distance can be altered while still retaining a reasonably well-focused image of a fixed object. Its depth of field is the range of object distances that will produce a reasonably well-focused image without the need to change the camera's focus. Both these quantities are increased as the aperture is reduced in size, and vice versa.

22 A telephoto lens is a two-lens system (one lens converging, the other diverging), which together have an equivalent focal length considerably greater than the physical distance from the front lens to the system's focal plane.

5.2 Achievements

Now you have completed this chapter, you should be able to:

A1 Explain the meaning of all the newly defined (emboldened) terms introduced in this chapter.

A2 Explain how chromatic and spherical aberration arise, and under what circumstances they can be reduced or eliminated.

A3 Use the lens equation to calculate the object distance, image distance or focal length for single lenses or mirrors and simple combinations of lenses and mirrors, making use of the real-is-positive sign convention.

A4 Locate images of extended objects by drawing ray diagrams and using the concept of key rays to trace the ray paths through lenses and reflected from mirrors.

A5 Describe the operation of the human eye, explain how the conditions of short-sightedness and long-sightedness can be corrected, and use the concepts of near point and far point in calculations.

A6 Carry out calculations involving lens power specified in dioptres.

A7 Explain how a magnifying glass, compound microscope, refracting telescope, reflecting telescope, camera and telephoto lens each work, and be able to draw ray diagrams to illustrate their operation.

A8 Calculate the linear magnification or angular magnification of a range of simple optical systems and the overall magnification of a compound microscope or telescope.

A9 Calculate the light-gathering power of a telescope.

A10 Calculate the exposure of a photograph using the concepts of shutter speed, aperture diameter and F-number.

A11 Explain how a camera's depth of focus and depth of field are related to the aperture size.

5.3 End-of-chapter questions

Note to Open University students. It is recommended that you attempt at least half of the following 12 questions, and study the answers for the remainder.

Question 3.21 Figure 3.50 shows a combination of a converging mirror of focal length 200 mm and a diverging lens of focal length 50 mm. The parallel beam of light incident on the mirror is reflected back to pass through the lens and emerge again as a parallel beam of light which falls on the opaque screen. What must be the distance between the centre of the mirror and the centre of the lens?

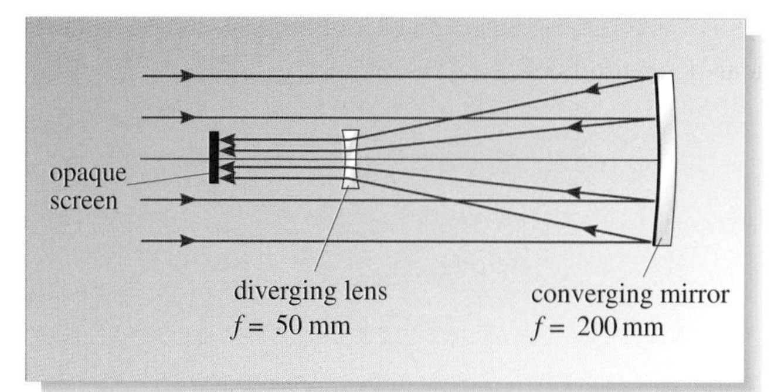

Figure 3.50 The broad parallel beam falling on the converging mirror emerges from the diverging lens as a narrow parallel beam (Question 3.21).

Question 3.22 A converging lens is set up, as in Figure 3.15a, to produce a real image, I, from a point object, O. The image, I, is now to be used as an object for a second converging lens placed to the right of the first lens. Sketch three diagrams showing where this second lens must be placed for I to be treated as:

(a) a *real* object producing a *real* image;

(b) a *real* object producing a *virtual* image;

(c) a *virtual* object producing a *real* image.

Can the image, I, ever behave as a *virtual* object and produce a *virtual* image?

Question 3.23 You are provided with a diverging lens and a converging lens, each of focal length 100 mm.

(a) A point-source object is placed on the optical axis 100 mm in front of the *diverging* lens. Where is the image located, and what is the nature of the image?

(b) The diverging lens is now replaced with the *converging* lens. Whereabouts in front of this lens must the object now be placed if it is to produce an image in exactly the same position as that produced by the diverging lens?

Question 3.24 An upright, extended object is placed in front of a diverging mirror. Sketch a ray diagram to determine (a) roughly where the image is located, (b) whether it is upright or inverted, (c) whether it is enlarged or reduced in size, and (d) in what way the image is changed as the object position is changed.

Question 3.25 If the mirror in the previous question has a radius of curvature ρ of 120 mm, and the object, which is 10 mm high, is placed 90 mm in front of the mirror, (a) how far behind the mirror is the image located, and (b) what is the height (in millimetres) of the image?

Question 3.26 A plane wave is incident on a converging mirror of focal length 100 mm, at an angle of 5° to the optical axis. By repeating the same sort of arguments as used in Question 3.9, demonstrate that this mirror focuses down the inclined plane wave to a point located in the focal plane of the mirror, but at a distance of 8.7 mm from the optical axis.

Question 3.27 A particular person suffering from myopia has an eye with the following properties: near-point distance d_{np} = 100 mm; far-point distance d_{fp} = 167 mm; distance from fovea to optical centre of eye-lens = 16.7 mm; minimum pupil diameter (i.e. in bright light) = 2.0 mm; maximum pupil diameter (i.e. when dark adapted) = 8.0 mm.

(a) Calculate the optical power and corresponding focal length of the eye-lens system at its far-point and near-point limits of clear vision.

(b) Over what range of F-number can the eye-lens be adjusted (i) when viewing a brightly lit subject, and (ii) when it is dark adapted?

(c) What would be the focal length and corresponding power of the spectacle lens prescribed to shift the eye's far point to infinity?

Question 3.28 The alphabetic characters in a piece of microfilm-text are 0.1 mm high. The text is unreadable with the 'normal' unaided eye, even when it is held at the eye's near point (250 mm away). To be just readable at this distance, the text would have to be 0.35 mm high.

(a) Calculate the *longest* focal length lens that, when used as a magnifying glass held very close to the eye, would render the microfilm-text readable. Where, in this case, would the virtual image of the text be located?

(b) What would be the maximum focal length lens that could be used to create a readable virtual image (of the text) anywhere between the eye's 'normal' near point (at 250 mm) and infinity?

(*Hint* Think carefully about what *adjustment* you should use in each of the above cases.)

Question 3.29 A (compound) microscope is constructed from two converging lenses — an objective lens of focal length 10.0 mm and an eyepiece lens of focal length 100 mm. The distance between the two lenses is *fixed* at 160 mm. Calculate the magnifying power of the microscope in both far-point and near-point adjustment, and determine how far in front of the objective lens the object must be placed in each case. Assume a standard near-point distance of 250 mm.

Question 3.30 To convince yourself that you fully understand the basic principles of operation of the telescope, try showing that:

(a) the magnifying power of a telescope's objective lens *alone* is

$$M_o = \frac{f_o}{d_{np}};$$

(b) the magnifying power of a telescope *in near-point adjustment* is

$$M(\text{near point}) = \frac{f_o}{f_e} + \frac{f_o}{d_{np}} = \frac{f_o}{d_{np}}\left(\frac{d_{np}}{f_e} + 1\right)$$

where d_{np} is the eye's near-point distance.

Hence demonstrate that, in both far-point and near-point adjustment, the *total* magnifying power of the telescope is given by the magnifying power of the objective alone multiplied by the magnifying power of the eyepiece alone. Assume throughout this question that the telescope is focused on objects that are *infinitely* distant.

(*Hint* The magnifying power is *always* given by α_{IM}/α_{OB}.)

Question 3.31 An amateur astronomer designs and builds a telescope specifically for observing the night sky. The objective lens has a diameter of 100 mm and a focal length of 500 mm. The eyepiece has a focal length of 50 mm and a diameter of 8 mm. (This value of diameter was chosen deliberately so as to match the pupil diameter of a 'normal' dark-adapted eye.) By calculating the magnifying power and light-gathering power of this telescope in far-point adjustment, show why it is not particularly well designed, and say what changes you would recommend to improve the design.

Question 3.32 A particular camera has exposure time and aperture-stop settings ranging from 1/250 s to 1/15 s and *f*/2.8 to *f*/22, respectively. Assume that the brightness of the scene that is to be photographed remains constant.

(a) By approximately what factor does the intensity at the film-plane change when the aperture diameter is altered from *f*/2.8 to *f*/22? Is this an increase or decrease in intensity at the film plane?

(b) For the film being used, the exposure for this particular scene is optimized when the camera is set to *f*/8 and 1/60 s. A waterfall forms part of the foreground of the scene. Deduce what camera settings should be used, to achieve the same optimum exposure, when it is desired (i) to 'freeze the motion' sufficiently to be able to see the individual drops of water in the waterfall, and (ii) to photograph the scene with the largest possible depth of field. ■

Chapter 4 Special relativity

1 Why do fast-moving muons live so long?

There is a type of fundamental particle called a **muon**, which in many ways resembles an electron. The main differences are the fact that muons are about two hundred times more massive than electrons, and that they decay into their less massive counterparts after a mean lifetime of a few microseconds. Muons can be created in the Earth's upper atmosphere when high-energy protons from outer space, known as *cosmic rays*, collide with the atoms they encounter in the upper atmosphere. The muons so generated travel predominantly in a downwards direction, towards the Earth's surface, at a speed close to the speed of light.

Experiments show the following result, which at first sight is quite puzzling. At the top of a mountain (Figure 4.1) the number of muons arriving per hour can be measured and recorded. At the bottom of the mountain, say a couple of thousand metres lower down, the number of muons arriving per hour can also be measured and recorded. Since the muons take several microseconds to travel from the top of the mountain to the bottom, a certain fraction of them should have decayed along the way, meaning that less will be detected at the bottom than at the top. In fact, knowing the mean lifetime of a muon (from measurements in the laboratory), one can predict how many muons should be recorded at the bottom of the mountain as a percentage of those recorded at its top. And here's the puzzle: in every case when such an experiment is performed, the number of muons recorded at the bottom of the mountain is far higher than predicted. In other words, far fewer muons have decayed than might be expected.

Figure 4.1 More muons are recorded arriving at the bottom of a mountain than are predicted on the basis of the number of muons arriving at the top.

So what's going on? The crucial fact here is that the muons are travelling close to the speed of light. Einstein's **special theory of relativity** implies that, according to the observer measuring the rate of arrival of the muons, time is passing more slowly for the muons. Consequently, the journey from the top of the mountain to the bottom takes the muons less time than measured by the observer, and more of the muons survive the trip. This result is an example of one of the key concepts of Albert Einstein's special theory of relativity. It may be paraphrased by the statement 'moving clocks run slow'.

Albert Einstein (1879–1955)

Albert Einstein (Figure 4.2) is widely agreed to have been the greatest physicist of the twentieth century. In 1905 he published four of the most influential papers in the history of physics. Two of these explained Brownian motion and the photoelectric effect, and the third and fourth outlined his special theory of relativity. Over the next few years it became clear to Einstein that an extension of his earlier work, a **general theory of relativity**, would also be a new theory of gravitation. The general theory of relativity, one of the greatest intellectual achievements of the twentieth century and a cornerstone of modern cosmology, was finally published in 1916. By the early 1920s, Einstein's best scientific work was done. He was none the less extremely influential in the physics community, and he did much to sow the seeds of the later developments of quantum mechanics. In his later years he worked to produce a unified theory of gravity and electromagnetism. Although this work was unsuccessful, it did nothing to diminish his extraordinary achievements.

Figure 4.2 Albert Einstein.

When most people hear the words 'Einstein' and 'special relativity', if they think of anything it is probably the most famous equation in all of physics: $E = mc^2$, which describes the equivalence between mass and energy. So you may be wondering why this book concludes with a chapter about special relativity, when the rest of it has been about *Dynamic fields and waves* such as are used to describe electricity and magnetism. The reason is that Einstein's first scientific paper describing his theory, in 1905, was entitled 'Zur Electrodynamik bewegter Körper' ('On the electrodynamics of moving bodies', *Annalen der Physik,* **17**, p. 891; Figure 4.3). In it he showed that Maxwell's equations obeyed the so-called principle of relativity, provided that a new way of transforming between different *frames of reference* was used and that the concepts and laws of mechanics are suitably altered. Maxwell's equations of electromagnetism are therefore already in accord with Einstein's special theory of relativity, and the demonstration of this will form a suitable finale to this book.

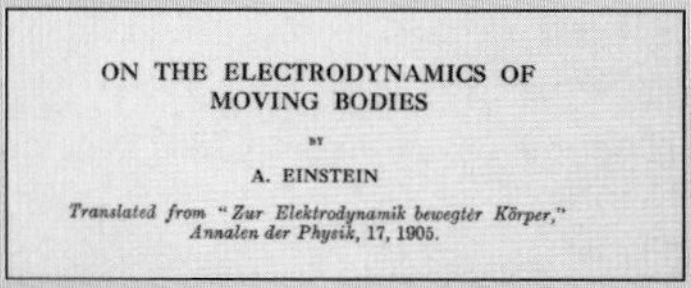

ON THE ELECTRODYNAMICS OF MOVING BODIES

BY

A. EINSTEIN

Translated from "Zur Elektrodynamik bewegtër Körper," Annalen der Physik, 17, 1905.

Figure 4.3 The front page of a translation of Einstein's 1905 paper 'Zur Electrodynamik bewegter Körper'.

Einstein's 1905 paper on special relativity

We owe a debt of gratitude to Max Planck for immediately accepting Einstein's paper for publication in *Annalen der Physik*, on his own authority. Planck read nearly all papers submitted for publication, and accepted anything he found interesting. Notably, he did not worry overmuch about the correctness of results if they seemed reasonable to him, as he believed this to be the responsibility of the author.

How could Planck know this was not the paper of a crackpot, like those every journal editor receives from time to time? Because there were three essential features in it that the work of crackpots does not have:

1. Einstein pointed out exactly where the other theories were wrong and what general principles to put in their place, using normal physics terminology.
2. Einstein followed up his principles by standard mathematical methods that all physicists knew.
3. Results derived from this theory could be tested by experiment.

Although this chapter does not follow Einstein's work in detail, it does have the same overall approach, being based on two postulates. Sections 2 and 3 explore the first postulate of relativity, and Section 4 examines the second postulate, relating to the constancy of the speed of light. In Sections 5 and 6, these two principles are brought together, and their implications for time and space are investigated. Then, in Section 7, the implications of special relativity for other areas of physics are explored, and the link between Einstein's special theory of relativity and Maxwell's theory of electromagnetism will conclude this book about dynamic fields and waves.

Special relativity has the reputation of being difficult to grasp, and yet it is fair to say that, alone among the topics of twentieth-century physics, it can be explained honestly and simply using mathematics that involve only basic algebra and geometry. The key to a straightforward (but quite rigorous) development of the basic ideas of special relativity is to adopt a careful and thoroughly logical approach to the experimental evidence and the implications that follow from it. By proceeding methodically, you may come to appreciate that, for many physicists, the fascination of Einstein's theory comes not only from its extraordinary predictions, but also from the irresistible, unrelenting force of the arguments that lead to these predictions.

2 The principle of relativity

Special relativity is based on two postulates — or fundamental principles — both of which are founded on experimental observation. This section will discuss the first of these, which embodies ideas that are familiar to you from your everyday experience. Indeed, it had already been formulated by Galileo in the seventeenth century. The second principle (which will be discussed in Section 4) is not so directly observable, and was firmly established only after some very careful experimental work in the late nineteenth century.

The first principle is to do with the universality of the laws of physics, and is exactly what should be expected from classical physics. It embodies our experience that everyday actions (such as eating a meal, or carrying out simple physical exercises) are no different if we cruise at high speed in an airliner than if we perform the same functions in a room at home. The process of formulating a precise general principle which is firmly based on everyday experience will require us to look closely at Newton's laws of motion. But before doing this, it is necessary to spend a little time on some basic definitions and technical terms that are needed to describe physical processes accurately.

2.1 Events, coordinates and clocks

An **event** is defined as a physical occurrence of very short duration that occurs in a very small region in space. Ideally, it occurs at a point in space and occupies an instant of time. There are two basic questions that can be asked about an event: 'where did it take place?' and 'when did it take place?'.

The simplest way to fix the position of an event without any ambiguity is to use a three-dimensional **coordinate system**. You have already used such a system several times. Figure 4.4 summarizes the way to use a coordinate system to specify the position of an event.

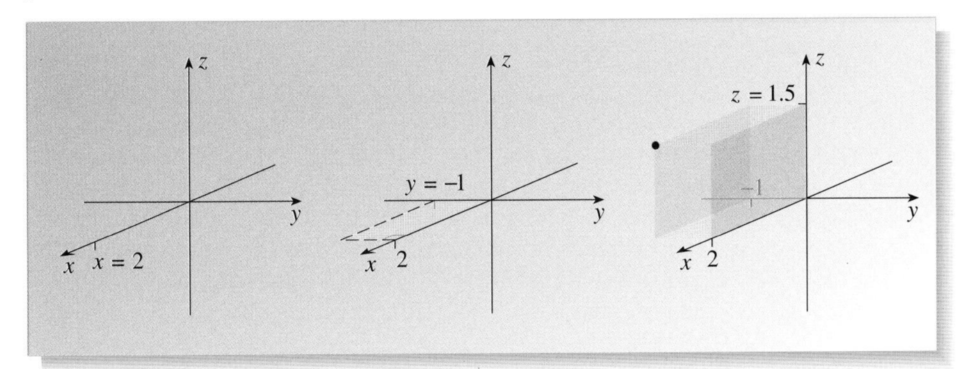

Figure 4.4 Fixing the location of a point in space by means of three position coordinates ($x = 2$ m, $y = -1$ m, $z = 1.5$ m) with respect to a given coordinate system.

To describe *when* an event happened, a *fourth* coordinate — a time coordinate — is needed. Time, which is measured in seconds, is simply the coordinate used to distinguish between different events that occur at a single position. However, fixing the time of an event turns out to be not as easy as fixing its position.

Imagine that you are standing at some particular point in a system of position coordinates watching other events that take place right in front of you. To enable you to answer questions about the times at which particular events occurred, all you need to do is to look at your watch and record the time of each event as it happened. Now suppose that you suddenly see a flash of light from a distant explosion. The explosion is an event, so you want to record its time. But what is its time? The time at which you *see* the flash is *not* the time of the explosion itself, but rather the time at which the light from the explosion entered your eye. *Your* watch is only useful for fixing the time of events that occur at *your* particular location. How can the time of the explosion itself be fixed, as opposed to the time at which the flash reached your position?

You could enlist the help of a friend, equipped with a watch that was synchronized with your own, and who was positioned at the point where the explosion is due to occur (rather him than you!). Your friend could then record the time of events as they happened at *his* position, including, at least in theory, the time of the explosion (Figure 4.5). This method would, indeed, enable you to fix the time of events at different locations, but the argument has lightly passed over one point of crucial importance. It was stated that the two watches had to be synchronized. This is obviously essential; there would be no point in your making a list of times of events at your position while your friend did the same thing for his position if you had no way of relating the two lists.

Figure 4.5 A rather dangerous way of assigning the time coordinate to an explosion.

How is it possible to make sure that two watches, of identical construction, but at different locations, are synchronized? This apparently simple requirement turns out to be full of pitfalls, and a discussion of a reliable synchronization procedure will be left until later. You might think, for instance, that before taking up your assigned positions, you and your friend could simply meet at some point and synchronize watches. But unfortunately this attractive solution has a flaw. It assumes that *after* the watches have been synchronized in one place, the process of moving them elsewhere will not affect their synchronization. Experiments with very accurate atomic clocks show that this assumption is not valid. When clocks are moved around after synchronization and then brought back together again, their relative motion may well have caused a loss of synchronization.

For the moment, therefore, please accept that *it is possible* to devise a straightforward and sound method of synchronizing clocks that are at rest relative to one another at different locations in space. (A way of doing this will be described in Section 5.) Once you have synchronized your clocks, you and your friend can easily assign a meaningful time coordinate to the explosion, even though *you* could not be present at the explosion with your own watch.

Having established a way of fixing the time of a distant explosion means that the original problem — that of assigning a time coordinate to *any* event no matter where it occurs — can now be solved. What is needed is not just two synchronized clocks, but (in principle at any rate) an infinite number of them.

If identical clocks were positioned at *every* point in space and *all* of the clocks were synchronized, then it would be a simple matter to fix the time of any event; all that would be required would be an observation of the time shown by the clock *at the location of the event*, at the moment that the event occurred.

There is one further important point that should be mentioned while talking about events in general. If one observer claims that two events occur together in space *and* time, then so will *all* other observers. This is sometimes referred to as the concept of **space–time conjunction**, and a simple illustration will clarify why this concept must always hold. Suppose that, to one observer, two objects arrive at the same place at the same time. For example, two speedboats that are travelling in opposite directions on a lake might crash into each other by arriving at the same place *and* time together. The point is that at such space–time conjunctions, physical effects can arise; in this example the speedboats could be damaged. Clearly, it would not make sense if while one observer claimed that both were at the same place at the same time (and one was damaged and sank as a consequence), another observer said that the speedboats did not occupy the same place and time! It will be helpful to invoke this idea later on when dealing with more complicated and potentially confusing situations.

2.2 Frames of reference

As you have seen, by using a system of coordinates and an infinite network of synchronized clocks, it is possible to specify uniquely the time and position of any event. Any system of coordinates and clocks that makes it possible to do this is called a **frame of reference**, a term you first met in *Predicting motion* Chapter 1. Once a frame of reference has been established, it is possible to give an accurate description of where and when events happened. Such a frame of reference even acts as a conventional standard of rest relative to which speeds can be measured. The coordinate system belonging to a particular frame of reference can be thought of as a kind of rigid scaffolding, providing the reference information that is needed when describing the motion of an object.

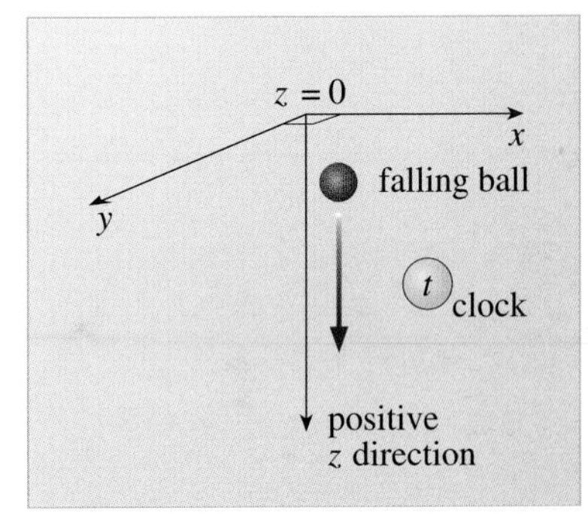

Figure 4.6 One frame of reference that can be conveniently used to describe a falling ball.

When setting out to describe a specific physical process, any frame of reference may be used. For example, to describe the motion of a ball from the moment it is dropped until it hits the ground, a frame of reference like that shown in Figure 4.6 could be used. In this frame of reference the ball starts at the point $z = 0$ and falls along the z-axis, so that the value of z corresponding to the position of the ball *increases* with time. On the other hand, the same process could equally well be described by using the different frame of reference shown in Figure 4.7, in which the positive z-axis points vertically upwards and the ball falls from a position where $z > 0$ towards $z = 0$. In such a frame of reference, the z-coordinate of the ball *decreases* with time. Changing the frame of reference does not change the physical process at all; it simply alters the way in which that process is described.

Most of relativity theory is more or less directly concerned with the consequences of describing the same physical processes from different frames of reference.

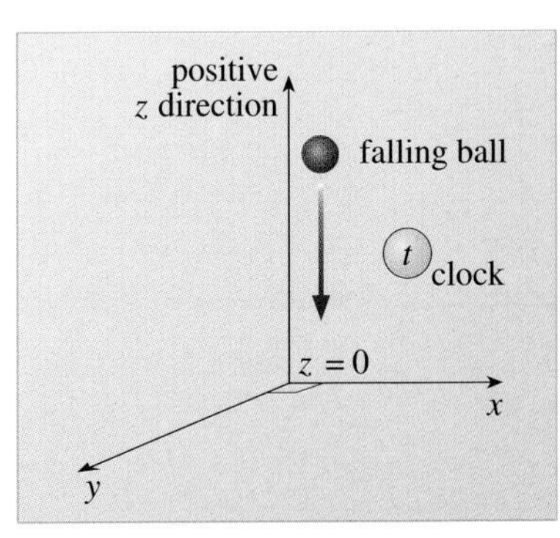

Figure 4.7 Another convenient frame of reference for describing the motion of a falling ball.

The final ingredient of any frame of reference is the **observer** — whom you can think of as the experimenter committed to using a particular frame of reference. Different observers may use different frames of reference, and every frame of reference has an observer associated with it. When an event occurs, all those observers who detect it will give it three spatial coordinates to fix its position, and one temporal coordinate to specify the time at which it occurred. The set of four coordinates used by one observer to define a particular event will, in general, be different from the set of coordinates given to the same event by a different observer. For instance, the observer using the frame of reference shown in Figure 4.6 might describe the starting position of the ball as $z = 0$; but the observer using the different frame of reference shown in Figure 4.7 would describe the same starting point by a different value of z.

It is important to understand the point about synchronization made earlier. When an observer detects an event, the time he or she assigns to it is the time shown on *one* of the synchronized clocks in his or her frame of reference — the one *at the location* of the event *when* it happens. Note also that the *observation* of an event by an observer (involving precise measurements with coordinate systems and synchronized clocks) is not the same as *seeing* it. Seeing an event could imply that there is a light-travel time involved between the time that light is emitted by the event and the time that light enters a person's eyes.

2.3 Newton's laws of motion revisited

The previous section spent some time discussing the concept of a frame of reference because frames of reference play a vital role in analysing physical processes. This section will demonstrate how Einstein was able to formulate the central question posed by special relativity in terms of so-called *inertial* frames of reference. To see why the word 'inertial' is so important in this context, it is necessary to take a closer look at the connection between Newton's laws and frames of reference.

Newton's first law (often called the law of inertia) states that neither the direction of motion nor the speed of an object changes unless it is acted on by an unbalanced force.

An important point to note about Newton's first law is that there are many situations in which this law does not appear to hold true, as discussed in *Predicting motion*. For instance, imagine someone sitting in the rear seat of a car that is acceler-ating forwards along a straight horizontal road. A ball on a horizontal tray placed on her lap is being held at rest by a string (Figure 4.8). If the string is cut, the ball does not remain stationary, but accelerates towards her. But if the observer regards the accelerating car as her frame of reference, the accelerating ball certainly disobeys Newton's first law in her frame of reference. Although in this frame of reference there are clearly no unbalanced forces acting on the ball, it *accelerates* towards her.

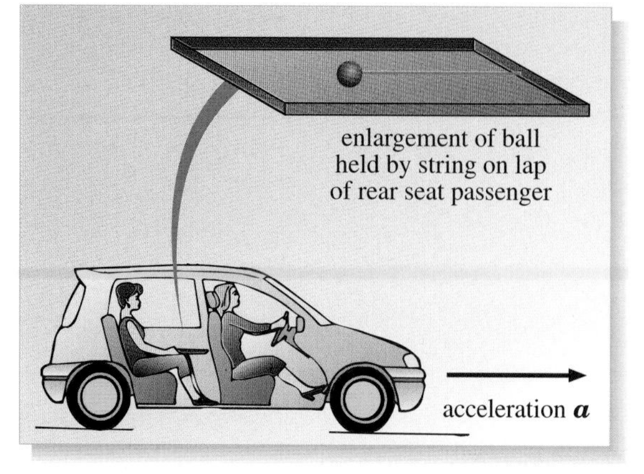

Figure 4.8 A woman sitting in the rear seat of an accelerating car holds a tray horizontally on her lap. When the string holding the ball is cut, the ball accelerates towards her, even though, in the frame of reference that is accelerating with the car, there appear to be no forces acting on it. In this frame of refer-ence, Newton's first law does not appear to hold true.

It is clear, therefore, that Newton's first law does not hold if tested by arbitrary observers: it applies only in some special frames of reference. Nevertheless, Newton believed that there existed such special frames of reference in which this law applies. As you saw in *Predicting motion*, such frames of reference are given a special name; they are called *inertial frames*.

> An inertial frame is a frame of reference in which a body that is not acted on by any unbalanced force moves with constant speed along a straight line.

Or, more succinctly;

> An **inertial frame** is a frame of reference in which Newton's first law is valid.

So Newton's first law effectively *defines* an inertial frame: it is a frame of reference which has the special significance that Newton's first law holds when physical events are observed from it.

- A particle is moving with constant speed v along the x-axis of some particular inertial frame. Suppose you are also travelling along the x-axis of the same inertial frame with constant speed u, in pursuit of the particle (Figure 4.9a). Show that your *personal* frame of reference is also an inertial frame.

○ Both you and the particle have constant velocity relative to the original inertial frame, so it is clear that the particle will have constant velocity relative to you. Thus, if you think of yourself as the origin of your own personal frame of reference that you carry with you (Figure 4.9b), you can see that the particle has constant velocity relative to your personal frame of reference. This means that as long as your personal frame of reference moves with constant velocity relative to an inertial frame, it too is an inertial frame. In fact, every one of the infinite number of frames of reference that have a constant velocity relative to an inertial frame is also an inertial frame. ■

It may be concluded that:

> Any frame of reference that moves with constant velocity relative to an inertial frame is also an inertial frame.

So, if just one inertial frame can be found, then it follows that an infinite number of them can be found. But is it possible to find one? The most obvious place to start is to ask if a frame of reference at rest with respect to the surface of the Earth is an inertial frame.

- The daily rotation of the Earth about its axis gives rise to a small acceleration in a laboratory fixed on the Earth's surface. What is the magnitude of that acceleration at the Equator? (*Hint* What is the angular speed of an object at rest on the Equator?) Take the Earth's radius, R_E, as 6.4×10^6 m.

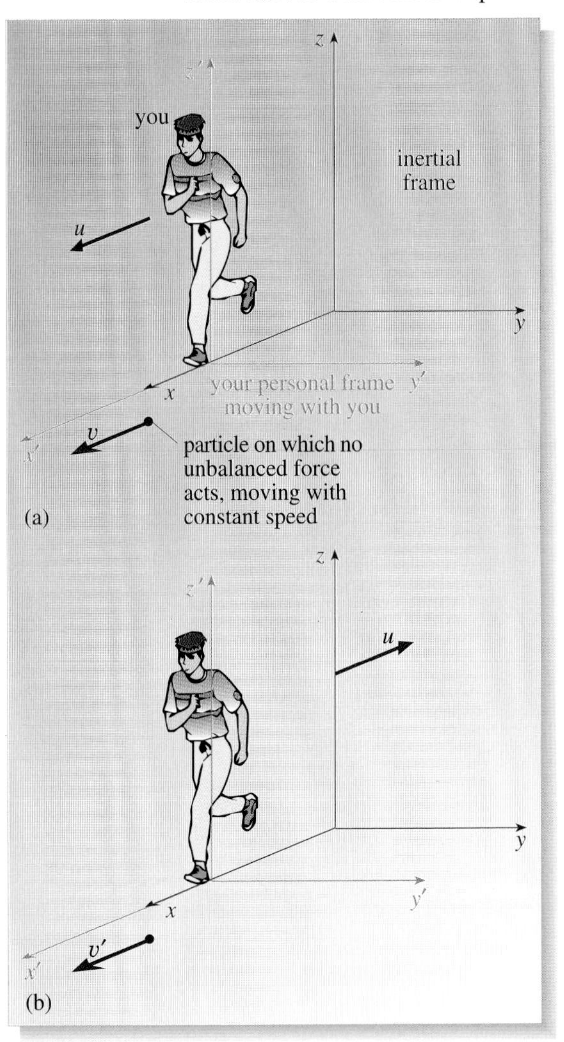

Figure 4.9 Viewing the motion of a particle that is not acted on by an unbalanced force from two different inertial frames. (a) The view from the original inertial frame. Both you and the particle move along the x-axis with constant speed. (b) The view from your frame of reference. The particle moves with a constant speed v', and the original inertial frame moves with constant speed u.

❍ An object at rest on the Earth's surface at the Equator experiences an acceleration of magnitude $a = v^2/R_E = \omega^2 R_E$ (*Describing motion*, Equations 3.31 and 3.32), where ω is the angular speed in radians per second.

$$\omega = \frac{2\pi}{1\text{ day}} = \frac{2\pi}{24 \times 60 \times 60\text{ s}} = 7.3 \times 10^{-5}\text{ s}^{-1},$$

giving $a = (7.3 \times 10^{-5})^2\text{ s}^{-2} \times 6.4 \times 10^6\text{ m} = 0.034\text{ m s}^{-2}$. ■

So your earthbound personal frame of reference is not, strictly speaking, inertial, nor is any other frame of reference fixed on the Earth. However, the acceleration involved is very small compared with those arising from everyday forces, and its effect is negligible for most everyday observations. Other terrestrial and solar motions involve even smaller accelerations. The Earth's acceleration due to its orbital motion around the Sun is about an order of magnitude smaller than that due to its axial spinning, whereas the Sun's acceleration towards the centre of our galaxy is about $3 \times 10^{-10}\text{ m s}^{-2}$. Thus, for most purposes a frame of reference fixed on the Earth provides a very good approximation to an inertial frame. Although a truly inertial frame is something of an idealization, the non-existence of an ideal inertial frame is not really a problem because there are plenty of frames of reference that are very nearly inertial (i.e. there are plenty of frames of reference in which Newton's first law is very nearly correct).

Now recall that Newton's second law states that, in order to make a body of mass m undergo an acceleration $\boldsymbol{a}$, a *force* is required that is equal to the product of mass and acceleration:

$$\boldsymbol{F} = m\boldsymbol{a}.$$

It is not difficult to see that Newton's second law is also *not* applicable in every possible frame of reference. Consider again the woman sitting in the accelerating car. She will observe that the ball she initially placed at rest on the horizontal tray surface is *accelerating* towards her (Figure 4.8), yet in the frame of reference in which she is at rest (i.e. a frame of reference that is accelerating forwards with the car), there are no unbalanced forces acting to account for this acceleration.

In general, in order for Newton's second law to hold true, a frame of reference must be used in which a body that is not acted on by an unbalanced force moves with constant velocity. But this is just how an *inertial frame* was defined above. So, as noted in *Predicting motion*, Newton's first law serves to define a special class of frame of reference — *inertial frames* — as particularly important and significant. Their significance is that it is in those frames of reference that Newton's second and third laws can be used.

Inertial frames play a central role in special relativity. Einstein's special theory is entirely concerned with observations that are made with respect to inertial frames. Indeed, that is why the theory is called the *special* theory of relativity: it only applies to a special class of frames of reference — the inertial frames. Extending this to consider observations made with respect to *non-inertial* frames would enter into the realm of the *general* theory of relativity. Because of the equivalence between the effects of gravity and acceleration, such a theory is also a theory of gravity. A detailed discussion of this is a subject left for other courses.

Before leaving the subject of inertial frames, there is one further point that needs clarification. You have seen that an accelerating frame of reference cannot be an inertial frame, but what exactly does the term 'accelerating' mean in this context? Surely, when saying that a frame of reference is accelerating this really means that it is accelerating *relative* to something else. What is that something else? Strange as it may seem, the answer appears to be 'the rest of the Universe'. There seems to be a link between the extent to which a frame of reference is inertial, and its acceleration relative to the distant stars and galaxies. This peculiar relationship between the accuracy of Newton's first law and the apparent motion of the large-scale distribution of matter in the Universe is not properly understood, and continues to be one of the mysteries of physics.

Question 4.1 A workman in a lift at the top of a skyscraper is idling away his time by pushing a cork to the bottom of a bucket of water with his finger, and watching the cork bob upwards when his finger is removed (Figure 4.10). Suddenly the rope holding the lift breaks and the lift and its occupant fall freely towards the ground. (Assume no resistance of any kind.) Taking advantage of the situation (and being a physicist at heart), the workman repeats his experiment with the cork, water and bucket. When he removes his finger this time, does the cork:

(a) sink to the bottom of the bucket?

(b) stay where it is in the water?

(c) rise to the surface as before, but more rapidly?

(d) do something else?

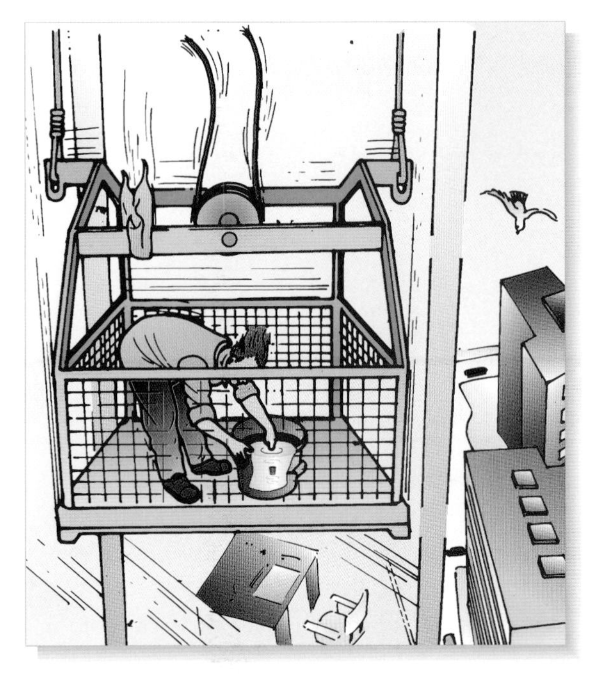

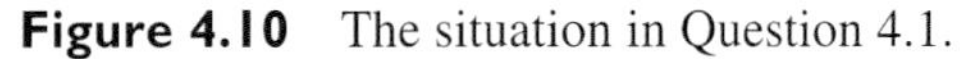

Figure 4.10 The situation in Question 4.1.

Explain your answer in terms of inertial and non-inertial frames. (*Hint* Archimedes' principle states that the upward force on a body immersed in a liquid is equal to the weight of the displaced liquid.)

Fortunately, due to a safety device, the lift slowed down and stopped before reaching the ground, so the workman survived to relate his findings to his friends! ■

2.4 The laws of physics and the principle of relativity

You have just seen that inertial frames play a vital role in the formulation of Newton's laws of motion. The same ideas will now be discussed in a more formal way to show that inertial frames are crucial to the process of discovering and *writing down* physical laws in general.

Suppose you set out to study the behaviour of a body that is not acted on by any forces. You find that the body continues with uniform velocity in a straight line along the x-axis, and this can be expressed mathematically as

$$\frac{dx}{dt} = \text{constant}. \tag{4.1}$$

(You have naturally expressed this law in terms of the coordinates x, y, z, t of *your* inertial frame.) If you carried out many such observations, all yielding this result, you might feel that this result is worthy of being called a 'law of physics'. (What you would have discovered, of course, is Newton's first law of motion.) But before elevating any statement to the status of a law, it would be sensible to check that other observers, in other frames of reference agree with your findings.

One of the most basic assumptions of special relativity is that any law of physics that is true in one inertial frame will also be true in all other inertial frames. This is called the *principle of relativity*. It implies that if you make a discovery in your inertial frame, you are only entitled to call it a law of physics if it can be shown to be true in every other inertial frame. One way of expressing the principle of relativity is to say that the laws of physics do not allow you to distinguish between inertial frames. Since laws of physics are usually expressed in mathematical terms, the most useful way of stating the principle of relativity is in a form that relates to mathematical equations:

The principle of relativity

The laws of physics can be written in the same form in all inertial frames.

Just what is meant by 'in the same form' can be explained with reference to the earlier example. If 'dx/dt = constant for a force-free particle' really is a law of physics, then an observer in a different inertial frame with coordinates x', y', z', t' should find that in her frame of reference the motion of a force-free particle is given by an equation of the same form:

$$\frac{dx'}{dt'} = \text{constant}'. \tag{4.2}$$

Of course, if the new observer fails to find any experimental support for Equation 4.2, you would have to conclude that your discovery is *not* a law of physics, even though it may be an accurate statement in your own frame of reference.

The principle of relativity is one of the foundation stones on which the theory of relativity is built. It is a remarkably powerful and far-reaching principle because it applies to *all* the laws of physics, even the ones that have not yet been discovered! In view of its importance, it is only proper to ask what evidence there is in support of the principle of relativity.

You will see in later sections that once the form of an equation is known in one inertial frame, it is possible to use a mathematical technique to work out the form of that equation in any other inertial frame. According to the principle of relativity, any equations that cannot be written in the same form in all inertial frames cannot be laws of physics. So, without doing any experiments, this technique makes it possible to pick out equations that *cannot* be laws of physics. By using this method, Einstein discovered that many of the equations that were thought to be laws of physics in his day could not be laws at all, even though they agreed with all existing experiments.

Einstein proposed modifications or alternatives to many of these false laws — alternatives that could be written in the same form in all inertial frames, in accordance with the principle of relativity. Subsequent experiments have shown that the old 'laws', though adequate for their own time, do not describe the wide range of results that are now available. On the other hand, the modified laws that do take the same form in all inertial frames have been supported by experiment. It is this *experimental* support for the consequences of Einstein's theory that provides convincing evidence that the basic assumptions of the theory, including the principle of relativity, must be correct.

Question 4.2 Suppose that while observing the motion of a particle in your inertial frame you come to the conclusion that the following equation describes the x-component of the acceleration of the particle:

$$\frac{d^2x}{dt^2} = \frac{k}{(x - X)^2},$$

where k is a constant and X is the x-coordinate of a second particle in your frame of reference. What relationship must an observer in a different inertial frame — with coordinate axes x', y', z' and t' — find to be experimentally valid if it is even worth giving further consideration to your formula as a possible law of physics? (You should, of course, *assume* that the principle of relativity is correct in answering this question.)

Question 4.3 Imagine that you are standing beside a road looking at a bridge that, naturally enough, is at rest relative to you. A friend is driving past in his car, which is moving with constant velocity relative to you. As he passes, he shouts out of the window that the principle of relativity is false because he can distinguish his frame of reference from yours simply by observing that the bridge is not at rest in his frame of reference. Is his claim justified? Give reasons for your answer. (Assume that a frame of reference fixed on the Earth is an inertial frame.) ■

3 Coordinate transformations

3.1 The central problem of relativity

According to the principle of relativity, a law of physics can be written in the same form in *all* inertial frames. A stringent test for any equation that you propose as a possible law of physics is therefore to check that it can be written in the same form in another inertial frame that is *moving* relative to your frame of reference. If it can, then it *may* be a physical law; experiment can then decide whether it is or not. If it cannot be written in the same form in the moving frame of reference, then it cannot be a law of physics, and you need not waste any more time testing it.

Suppose an equation can be expressed in terms of the coordinates of your frame of reference (x, y, z and t). Then, according to the principle of relativity, if the equation represents a law of physics the corresponding equation in a frame of reference that is moving with respect to you has the same form, but can be written in terms of coordinates (x', y', z' and t'). What is clearly needed is the relationship that links the coordinates of an event in your frame of reference and those of the same event in the other frame of reference. Such a relationship is called a **coordinate transformation**; if you know the coordinate transformation between two inertial frames, you can then transform arbitrary equations and see if they take the same form or not.

Figure 4.11 shows the simplest possible arrangement of two inertial frames that are in relative motion. In this example, frame of reference A is fixed on the platform of a railway station, with the x-axis pointing along the track. The observer in frame of reference A measures time by means of clocks that are at rest in his frame of reference and which have been synchronized with the main station clock. (As usual, you should imagine that the observer in frame of reference A has a synchronized clock wherever he needs one.) Frame of reference B is fixed in a railway carriage that is moving relative to the station with constant speed V in the positive x-direction. The observer in frame of reference B — a passenger on the train — measures the time at which events occur in his frame of reference using clocks that have been synchronized with a clock fixed in his carriage.

The frames of reference A and B are also arranged so that the corresponding axes are parallel, and the origin of B moves along the x-axis of A.

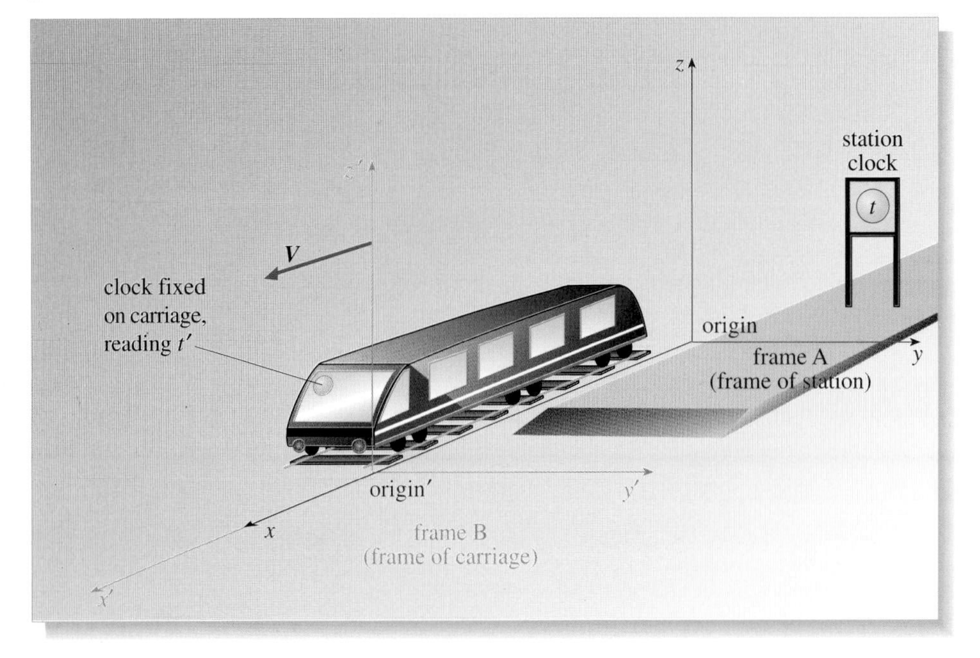

Figure 4.11 Two inertial frames in standard configuration. The two origins coincided at $t = t' = 0$.

One further condition is needed to ensure the simplest possible arrangement. Because the origin of B is moving along one of the axes of A, there will be an instant at which the origins of the two frames of reference coincide. This is an *event* or a *space–time conjunction* in the technical sense of Section 2.1; it happens at a point, and only lasts for an instant. The simplest situation will be if both observers A and B arrange that their respective clocks record zero at the time that both origins coincide.

This simplest possible arrangement of two inertial frames with constant velocity relative to each other (as specified above) is given the special name: **standard configuration**. The definition of two frames of reference in standard configuration is summarized in Figure 4.12.

Frames of reference in standard configuration will be used many times when discussing situations involving two inertial frames in relative motion.

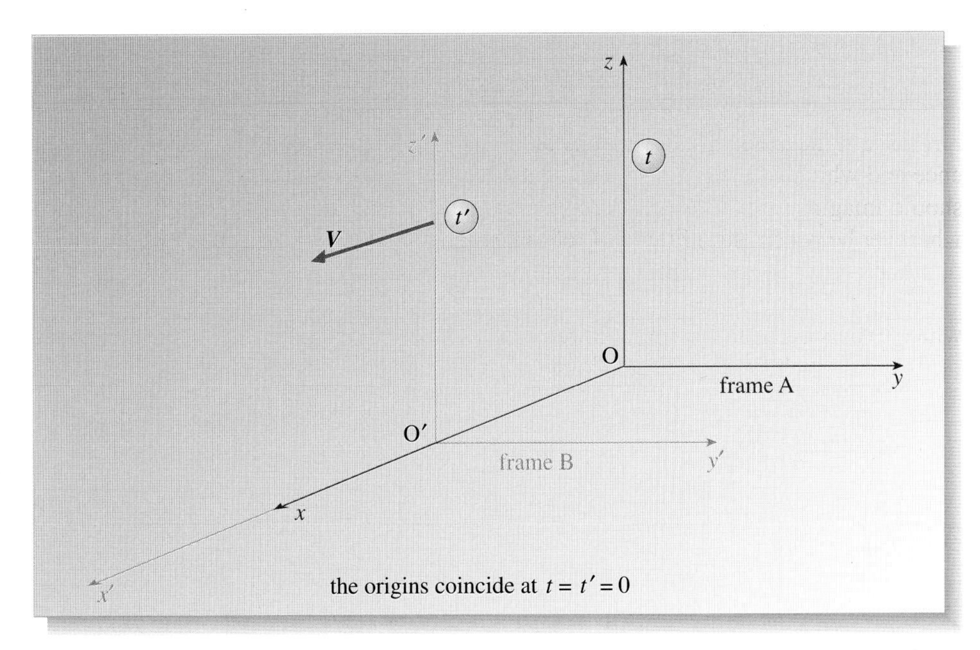

Figure 4.12 Standard configuration of two inertial frames A and B. The origins of A and B have been called O and O′, respectively, for simplicity, and coincide at $t = t' = 0$. Frame of reference B moves along the x-axis in the positive direction with speed V relative to frame of reference A (i.e. V is the magnitude of a velocity component V_x). The two frames of reference maintain the same relative orientation (y parallel to y', etc.) while moving.

Einstein was the first person to realize that finding the correct coordinate transformation that links two inertial frames in standard configuration held the key to an entirely new view of space and time. This is the problem that must be solved in order that the full power of the principle of relativity can be used to assess the viability of equations that are proposed as possible laws of physics.

The central problem of relativity

What is the correct transformation that links the coordinates of an event in inertial frame A to those of the same event in inertial frame B, when A and B are in standard configuration?

This central problem of relativity will be answered in Section 6, when all the components of Einstein's special theory of relativity will be drawn together. At this point, though, a different coordinate transformation will be examined — one that, before 1900, most physicists would have regarded as the 'common-sense' solution to the central problem of relativity.

3.2 The Galilean coordinate transformation

Look at Figure 4.13, which is a schematic version of two frames of reference A and B in standard configuration. An intuitive feeling for the relationship between the coordinates of an event in inertial frame A (x, y, z and t) to those of the same event in inertial frame B (x', y', z' and t') gives rise to the **Galilean coordinate transformation** equations:

$$\begin{aligned} x' &= x - Vt \\ y' &= y \\ z' &= z \\ t' &= t. \end{aligned} \tag{4.3}$$

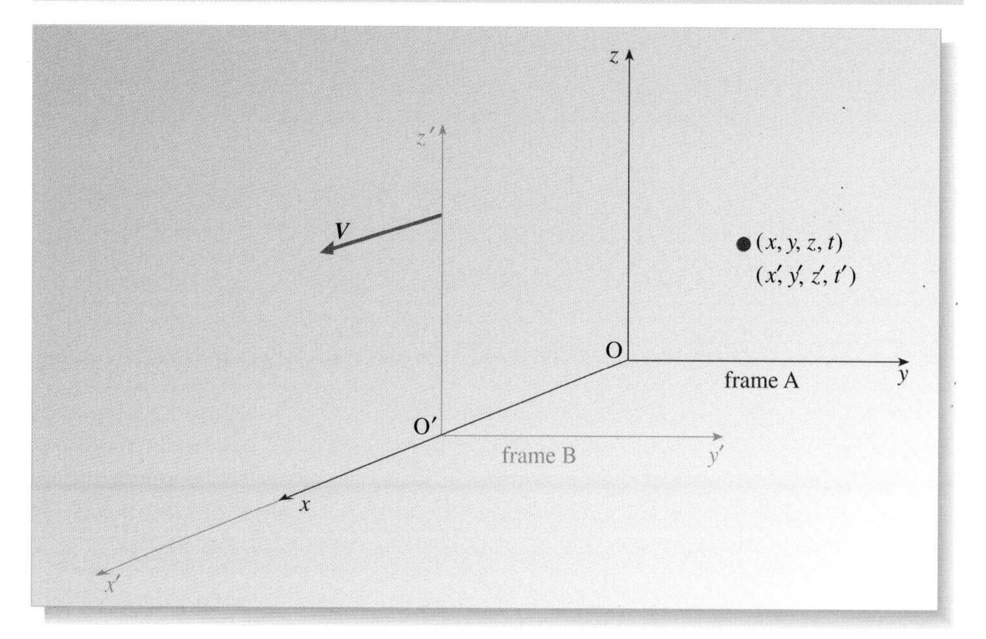

Figure 4.13 An event takes place at (x, y, z, t) in frame of reference A, and at (x', y', z', t') in frame of reference B. Frames of reference A and B are in standard configuration.

These equations indicate that an event observed in frame of reference B will be at exactly the same place and time as it would be observed in frame of reference A, except for the x-coordinate, which allows for the fact that the origin of frame of reference B has moved away from the origin of frame of reference A by a distance Vt after time t.

Note *Although these equations are a perfectly plausible solution to the central problem of relativity, they are* **not** *the correct solution.*

In order to appreciate why the Galilean transformation would have been regarded as the common-sense solution before Einstein, it is necessary to examine more closely what it indicates about space and time.

The first point to notice about the Galilean transformation is that a time interval $(t_2 - t_1)$ measured by an observer in frame of reference A will be identical to the same time interval $(t'_2 - t'_1)$ measured from B, since it is assumed that $t_2 = t'_2$ and $t_1 = t'_1$. This is exactly what most physicists would have expected up to 1900, and confirms Newton's view of time:

> ...absolute, true, and mathematical time, of itself, and from its own nature, flows equably, without relation to anything external...
>
> Newton's *Principia*.

The next point to notice about the Galilean transformation is that any length interval is the same in both frames of reference A and B.

- The two ends of a rod that is lying at rest along the x'-axis in frame of reference B are measured by an observer in B to have coordinates x_1' and x_2', when measured at the *same* time t_m'. In frame of reference A, the rod is moving, and its ends have coordinates x_1 and x_2 when measured at the same time t_m by an observer in A. Show that the Galilean transformation predicts that the length of the rod measured by both observers is the same.

- ❍ Since $t_m = t_m'$, the measurements of both ends of the rod occur at the same time in both frames of reference. To the observer in frame of reference A (in which the rod is moving), the ends of the rod have coordinates x_1 and x_2, and the length of the rod is $(x_2 - x_1)$. To the observer in frame of reference B (in which the rod is at rest), the length of the rod will be $(x_2' - x_1')$. From the Galilean transformations,

$$x_2' - x_1' = (x_2 - Vt_m) - (x_1 - Vt_m) = x_2 - x_1 \quad \blacksquare$$

In other words, according to the Galilean transformation, the two observers obtain identical results for the length of the rod. It is important to note that *two* measurements, x_1' and x_2', or x_1 and x_2, are needed, and it is also assumed that they are made *at the same time*. A moment's thought will confirm the requirement that the measurement of the position of both ends be made at the same time; this is what is understood by the length of a moving object. Measuring the position of the two ends at different times would certainly produce an erroneous result.

Perhaps the most surprising thing about the result you have just deduced is that it is *not* correct. You will see later that an observer in frame of reference A will obtain a different result from an observer in frame of reference B.

Length measurements in the y- and z-directions are unaffected by the relative motion of the two frames of reference, because the motion is perpendicular to both the y- and z-directions. Following on from this result, it is not difficult to show that even if the rod is pointing in some arbitrary direction with respect to the x-, y- and z-axes, then according to the Galilean transformation, its length would still be the same in the two inertial frames A and B.

3.3 The Galilean velocity transformation

Imagine you are standing beside a road observing two cars. One of the cars is travelling away from you at 110 km h^{-1}, and the other car travels in the same direction at 80 km h^{-1}, as shown in Figure 4.14. What is the speed of the faster car as measured by the driver of the slower one?

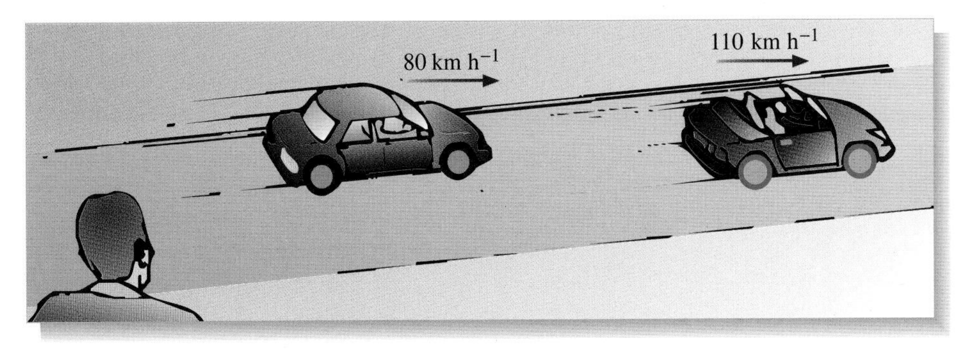

Figure 4.14 Relative speeds in an everyday situation.

You will have no difficulty in answering this question from your everyday experience. The relative difference in speed between the two cars is $30\,\text{km}\,\text{h}^{-1}$, so that the faster car appears to the driver of the slower car to be moving away from it at $30\,\text{km}\,\text{h}^{-1}$. What you have done, perhaps without realizing it, is to transform the velocity of the faster car from one inertial frame (in which you are at rest on the side of the road) to another inertial frame (in which the slower car is at rest).

It is possible to derive this result in a more formal way from the Galilean transformation. Consider two inertial frames A and B in standard configuration having relative speed V. An object that is moving with velocity $\boldsymbol{v}$ in frame of reference A has components v_x, v_y and v_z, along the x-, y- and z-axis directions, respectively. Its velocity as measured in frame of reference B is $\boldsymbol{v}'$ (with components v'_x, v'_y and v'_z). From the Galilean transformation (Equation 4.3):

$$v'_x = \frac{\mathrm{d}x'}{\mathrm{d}t'} = \frac{\mathrm{d}(x - Vt)}{\mathrm{d}t} = \frac{\mathrm{d}x}{\mathrm{d}t} - \frac{\mathrm{d}}{\mathrm{d}t}(Vt) = \frac{\mathrm{d}x}{\mathrm{d}t} - V = v_x - V$$

$$v'_y = \frac{\mathrm{d}y'}{\mathrm{d}t'} = \frac{\mathrm{d}y}{\mathrm{d}t} = v_y$$

$$v'_z = \frac{\mathrm{d}z'}{\mathrm{d}t'} = \frac{\mathrm{d}z}{\mathrm{d}t} = v_z.$$

So,
$$v'_x = v_x - V$$
$$v'_y = v_y \qquad (4.4)$$
$$v'_z = v_z$$

As you can see, the **Galilean velocity transformation** implies that v_y and v_z are the same in frames of reference A and B, whereas v'_x differs from v_x by the *relative* speed of the two frames of reference, V. This result agrees with the intuitive result for the two cars travelling along a road.

- A man is travelling in a car moving from left to right at $80\,\text{km}\,\text{h}^{-1}$ relative to an observer in frame of reference A, as shown in Figure 4.15. The man in the car throws two bricks out of the car at a speed of $10\,\text{km}\,\text{h}^{-1}$ relative to the frame of reference in which he is at rest, B. One of the bricks is thrown towards the front of the car and the other towards the rear. Use the Galilean velocity transformation to calculate the speed and direction of the two bricks as measured by the observer in frame of reference A immediately after they have been thrown.

Figure 4.15 Bricks thrown from a moving car travel at different speeds relative to an observer at the side of the road.

❍ Using the Galilean velocity transformation, for the brick thrown in the forward direction (i.e. in the same direction as the motion of the car) as seen by the observer in frame of reference A,

$$10\,\mathrm{km\,h^{-1}} = v_x - 80\,\mathrm{km\,h^{-1}}, \text{ so } v_x = 90\,\mathrm{km\,h^{-1}}.$$

For the brick thrown in the backward direction,

$$-10\,\mathrm{km\,h^{-1}} = v_x - 80\,\mathrm{km\,h^{-1}}, \text{ so } v_x = 70\,\mathrm{km\,h^{-1}}.$$

So in both cases the brick travels in the forward direction in frame of reference A, but at different speeds. As the motion is occurring only in the x-direction in this situation, v_y, v'_y, v_z, v'_z are all zero. ■

You could probably have quoted these answers without resort to the Galilean velocity transformation, based on your everyday experience. Again, the surprising thing is that these answers are *not* correct!

You have seen that in the Galilean transformation, velocities add in a direct way. It is interesting to examine this result in the light of the principle of relativity which requires that the laws of physics can be written in the same form in all inertial frames. If the Galilean transformation is the correct solution to the central problem of relativity, then the result just derived must be supported by experiment. The converse also holds; if it is found experimentally that the Galilean velocity transformation is *not* supported by experiment, it *must* be concluded that the Galilean velocity transformation is *not* the correct solution.

At first glance, it appears that all is well. The Galilean velocity transformation certainly agrees with everyday ideas about relative velocity, and everyday experience of moving objects like cars and trains. However, it is important to realize that these examples, which so strongly influence our views, all involve fairly low speeds — typically not more than $10^3\,\mathrm{m\,s^{-1}}$. Such a limited range of everyday speeds cannot provide sufficient evidence to ascertain the validity of the Galilean velocity transformation in all situations, because physicists frequently have to deal with much greater speeds. In fact, the predictions of the Galilean velocity transformation equations are in *direct conflict* with experiment at very high speeds.

Experimental observations on the behaviour of the velocity of light at the end of the nineteenth century were to radically change these long established views on space and time, and were to show conclusively that the Galilean transformation is NOT the correct coordinate transformation linking inertial frames. This is the subject of the next section.

4 The constancy of the speed of light

It was through experimental observations on the speed of light at the end of the nineteenth century that it became clear that the previously held views on space and time were incorrect. At first, the experimental work was closely related to furthering the understanding of the theory of light and electromagnetism, which was one of the pinnacles of nineteenth-century classical physics. It was Albert Einstein who made the connection between the very unusual and unexpected results from measurements on the speed of light, and the nature of space and time.

As you saw in Chapter 1, Maxwell's electromagnetic theory predicts that light travels through a vacuum at a constant speed c, given by

$$c = \frac{1}{\sqrt{\varepsilon_0\mu_0}} = 3.0\times10^8\,\mathrm{m\,s^{-1}}. \qquad \text{(Eqn 1.45)}$$

But what does Maxwell's prediction really mean? What reference point, or frame of reference, should be used when trying to interpret Maxwell's prediction? As you will see shortly, Maxwell's own answer to this question was quite wrong, though his method of calculating c remains valid.

Suppose that light from an explosion on the surface of a distant star is detected by an observer on Earth. Any light signal that is seen must have originated somewhere: it must have travelled away from its source and eventually have fallen on a detector (such as a photographic film); otherwise it would not have been observed. This simple analysis suggests three possible answers to the question of the reference point relative to which the light may be moving with speed c:

1 *c could be the speed of light relative to its source.* This seems a rather natural and attractive possibility.

2 *c could be the speed of light relative to the space through which it propagates.* This probably seems a rather strange alternative since the vacuum of space does not appear to provide any obvious reference points.

3 *c could be the speed of light relative to the detector.* This may not seem very likely on the ground that the light does not encounter the detector until the very end of its journey.

The three possibilities listed above are the simplest and most plausible answers to the question. The rest of this section will be devoted to trying to decide which, if any, of the three possible interpretations is the correct one.

4.1 Is c the speed of light relative to the source?

In Section 3, you have already looked at several examples where the speed of an object is relative to its source. This is the familiar situation of *relative speeds*, and the Galilean velocity transformation provides a way of working out the object's speed with respect to different inertial frames (Equation 4.4).

Imagine that light consists of particles that obey the laws of Newtonian mechanics. It would be very reasonable to interpret c as the speed of light relative to its source. You could imagine that the particles of light, emerging with speed c from a luminous object, were like bullets leaving the barrel of a gun with fixed velocity. If the gun were moving towards the target when fired, then the bullets themselves would move more rapidly towards the target than if the gun had been stationary. Thus, if this 'ballistic' picture of light is correct, light emitted from a source that is approaching a detector should have a higher speed relative to the detector than light from a source that is either stationary or moving away from the detector. This view is rather easy to picture because it is a kind of mechanical view of light. Newton might have been quite happy with this interpretation of the speed of light, but is it correct? An *experimental* test is needed to help us decide.

In principle, such an experiment is simple. All that is needed is a device that can measure the speed of light, and a moving source of light. Nature provides the required kind of source in the form of a *multiple star*. About a half of all the points of light seen as stars in the night sky are, in fact, systems consisting of two or more stars orbiting about their common centre of mass. Indeed, the closest stars to the Sun are those forming the triple system of Alpha Centauri (see Figure 4.16). Binary systems, composed of just two stars, are the most common of the multiple star systems.

Figure 4.16 The region of the sky containing the constellation of Centaurus. The triple star system Alpha Centauri is marked. Because the component stars of the system are so close together, they appear as a single point of light in this wide-field image. (Photograph taken by Sven Kohle.)

Imagine a binary system in which one of the stars is very massive, whereas the other is much less massive and performs circular orbits about its more massive companion, as shown in Figure 4.17. Suppose that the speed of the orbiting star (relative to its more massive companion) is v, and, in order to avoid unnecessary complications, think of the Earth as being at rest relative to the more massive component of the binary system. Now consider an astronomer on Earth making observations of the orbiting star.

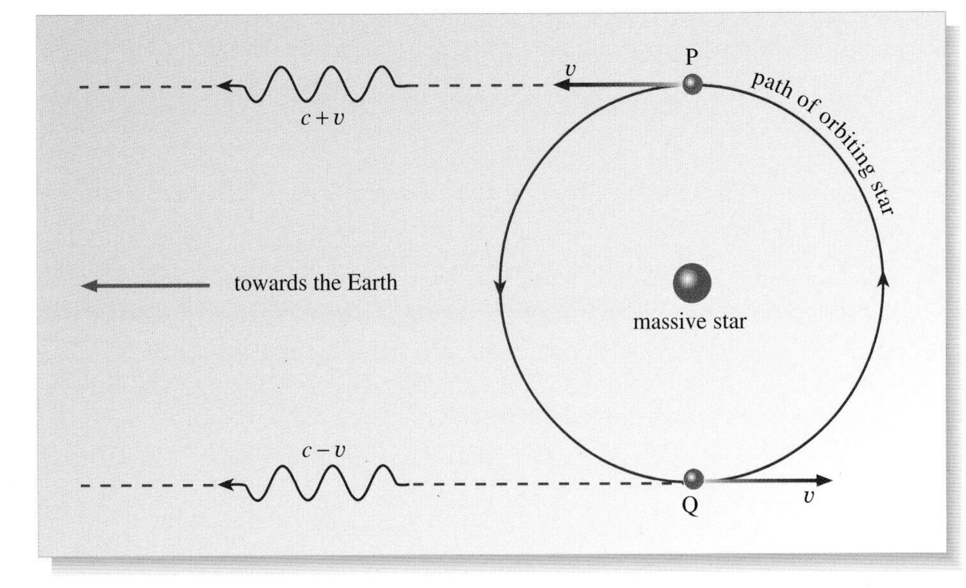

Figure 4.17 In the simple bullet-like model, light from a distant star has a speed, relative to the Earth, that depends on the motion of the star at the time of emission. Thus, relative to the Earth, the speed of the light emitted when the orbiting star is at P should be greater than when the star is at Q.

When the orbiting star is at position P, it is moving towards the Earth with speed v, and any light that it emits should, *if the 'ballistic' theory is correct*, have speed $c + v$ when observed from the Earth. Similarly, half an orbit later, when the star is at Q, it is moving away from the Earth with speed v, and the light that it emits should have speed $c - v$ relative to the Earth. Thus, relative to the Earth, the speed of light coming from the orbiting star should depend on the point in the orbit at which it was emitted.

In practice, observations of binary stars show no evidence for such an effect, and hence indicate that the speed of light from the orbiting star *always has the same value*, irrespective of the point in the orbit at which it was emitted.

The fact is that there is simply no experimental evidence to support the idea that the speed of light depends on the motion of its source. Based firmly on the findings of experiment, therefore, we must rule out the idea that light consists of particles that obey the laws of Newtonian mechanics and have speed c relative to their source.

Question 4.4 Two spacecraft, each carrying a laser, are travelling directly towards the surface of a planet. The speed of one of the craft relative to the planet is $2 \times 10^5\,\mathrm{m\,s^{-1}}$, and the speed of the other is $4 \times 10^5\,\mathrm{m\,s^{-1}}$. If both craft fired their laser beams simultaneously from a range of $10^9\,\mathrm{m}$, which of the two beams will strike the planet first, and what delay would you expect before the other beam strikes? ■

4.2 Is c the speed of light relative to space?

This section deals with the interpretation of c that Maxwell himself would have given, though he would not have posed the question given in the title in terms of 'space'. Maxwell would have used a term accepted at the time — the **ether** — which as you saw in *The restless Universe* was supposed to be an all-pervading, unusual form of matter.

Apart from electromagnetic radiation, the waves encountered most frequently in physics are those of sound propagating through the air, which you met in Chapter 2. These travel at about $340\,\mathrm{m\,s^{-1}}$ *with respect to the air*. It will help in the discussion of light propagation to look more closely at the role that the propagating medium plays in sound measurements. The following example explains this point.

Example 4.1 The relative speed of sound

(a) Two observers are at opposite ends of a long *enclosed* railway carriage, and are set up to measure the speed of sound inside the carriage. Sound is generated by an electric bell placed inside the carriage at its midpoint (equally distant from both observers — Figure 4.18a). Will the two observers obtain the same measurements for the speed of sound:

(i) when the carriage is stationary?

(ii) when the carriage is moving along the track at a speed of $50\,\mathrm{m\,s^{-1}}$?

(b) In a separate experiment, the observers are in an *open* railway carriage, which also has a bell at its midpoint (Figure 4.18b). How will their results for the speed of sound compare with those of part (a) above under the same two conditions:

(i) when the carriage is stationary?

(ii) when the carriage is moving along the track at a speed of $50\,\mathrm{m\,s^{-1}}$?

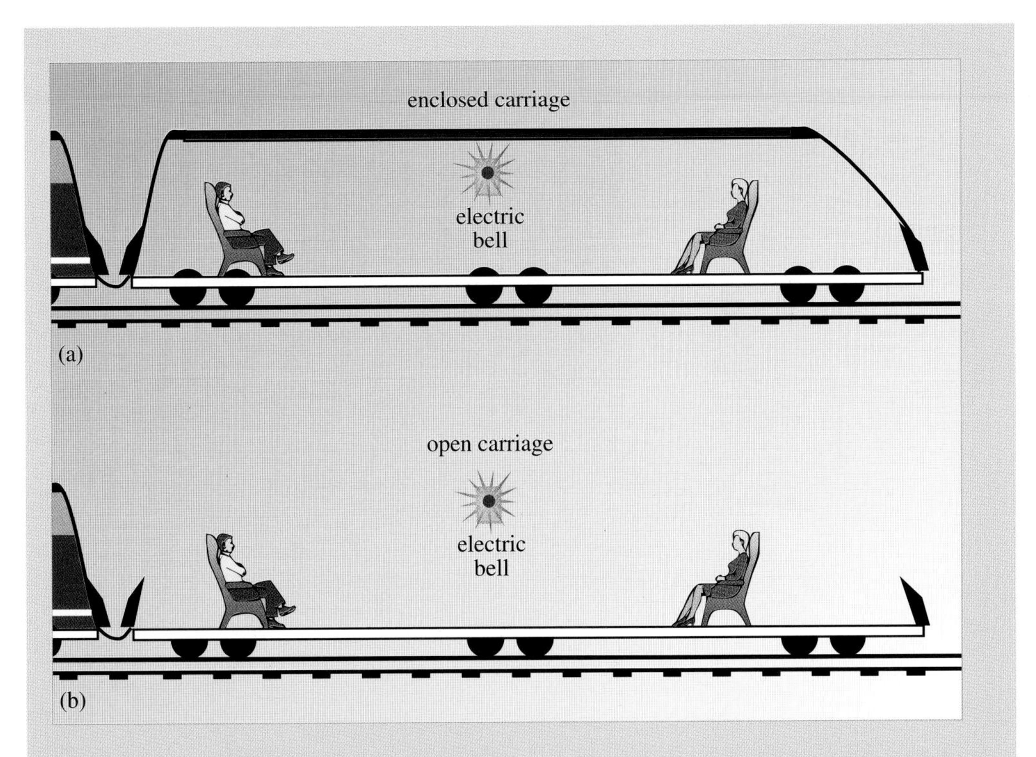

Figure 4.18 (a) Two observers are at opposite ends of a long *enclosed* railway carriage; they measure the speed of sound generated when a bell situated midway between them is sounded first when the carriage is stationary, and second when it is moving along the track at $50\,\mathrm{m\,s^{-1}}$ (Example 4.1a). (b) Two observers in an *open* carriage measure the speed of sound from a bell fixed at the midpoint when the carriage is stationary, and when it is moving at $50\,\mathrm{m\,s^{-1}}$ (Example 4.1b).

Solution

(a) The speed of sound is always relative to the propagating medium — in this case air. In the case of the *enclosed* carriage, the air inside the carriage travels along with the carriage, and is therefore effectively at rest with respect to the electric bell and the two observers.

In these circumstances, the speed of the carriage past the air *outside* the carriage is irrelevant. In case (i) and case (ii), both observers will obtain the same value for the speed of sound namely $340\,\mathrm{m\,s^{-1}}$.

(b) (i) If the open carriage is stationary, there is no relative motion between the observer and the air, and the speed of sound will again be $340\,\mathrm{m\,s^{-1}}$.

(ii) The situation is quite different if the open carriage is moving with respect to the outside air carrying the sound waves. Now it is necessary to take into account the relative motion between the observer and the air through which the sound propagates. For the observer facing the direction of motion of the carriage, the measured speed of sound will be $390\,\mathrm{m\,s^{-1}}$, whereas the observer with his back to the direction of travel will measure a speed of $290\,\mathrm{m\,s^{-1}}$.

The answer to Example 4.1 indicates that it is the relative speed of the observer *with respect to the air* (the propagating medium) that matters in all cases of sound propagation.

Throughout the nineteenth century, it was felt that electromagnetic effects also needed some kind of medium for their propagation, and it was thought that light would therefore travel at speed c relative to this propagating medium. The realization that light, which travels through a vacuum, is an electromagnetic wave, forced those who believed in an electromagnetic medium to endow it with rather strange properties. For example, the medium had to fill space (otherwise the stars would not have been visible), yet astronomical bodies such as the planets had to move through the electromagnetic medium without friction, or else their orbits would have changed. This peculiar all-pervading medium, which propagated electromagnetic effects but apparently did little else, was called the *ether*.

You can see now why asking 'Is c the speed of light relative to space?' does not really make sense. Space does not supply any natural reference points that can be used when measuring speeds. But it is perfectly sensible to ask 'Is c the speed relative to the ether?', in direct analogy with the speed of sound relative to the air through which it travels.

Having formulated a sensible question about the speed of light in terms of the ether, an obvious goal for nineteenth-century physicists was to test the ether interpretation experimentally.

In 1879, Maxwell put forward the idea that, since the Earth's orbital motion around the Sun ensured that the Earth was moving relative to the ether, then this relative motion should influence the speed of light relative to a detector fixed on Earth. (The situation is analogous to the sound measurements in the moving open carriage in Example 4.1.) In general, the amount of time required for a beam of light to travel from a point A to a mirror at point B and then to return to point A (Figure 4.19) should depend on:

1. the speed of the line AB relative to the ether;
2. the orientation of the line AB.

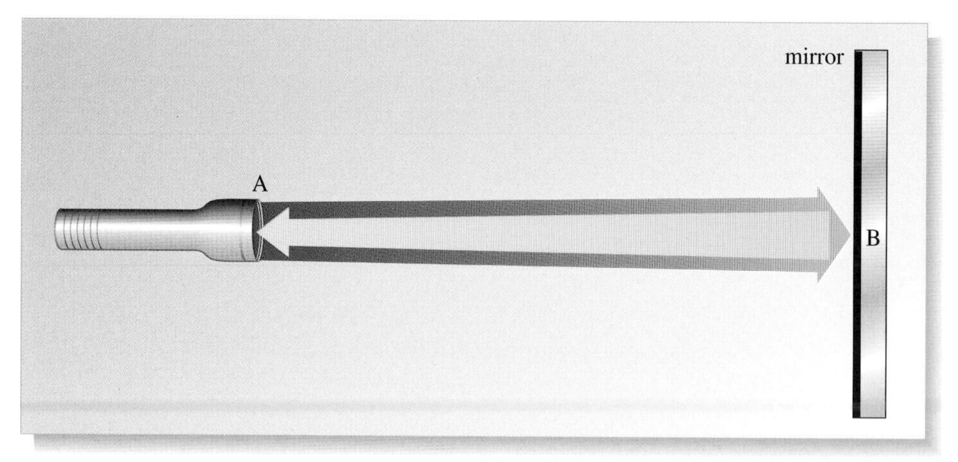

Figure 4.19 A beam of light travels from point A to a mirror at point B, and is then reflected back to A. The distance between A and B is fixed.

In the case of light, these two factors would have only a very small effect on the round trip time; but if the effect could be measured it would provide valuable information about the ether. It would even be possible to work out the speed of the Earth relative to the ether.

The first physicist to devise a sufficiently precise experiment to test the implications of the ether hypothesis was an American scientist, A. A. Michelson (Figure 4.20a). Figure 4.21 provides a simplified plan of Michelson's apparatus. A narrow monochromatic beam of light was shone onto a semi-silvered mirror, which reflects part of the beam at point A while allowing the rest to pass through. The mirror split the original beam into two parts travelling at right-angles to each other. The half of the original beam that had passed through the semi-silvered mirror was reflected by a mirror at point B, at a distance L from point A. The other half of the original beam — the half that had been reflected at A — was reflected again, by a mirror at point C, which was also at a distance L from A. After being reflected at either B or C, the two parts of the original beam returned to A, where they recombined. The recombined beam was then observed on the screen.

(a)

(b)

Figure 4.20 (a) Albert Abraham Michelson (1852–1931) was born in Germany but later moved to America and became a professor of physics at the University of Chicago. In 1907 he became the first American scientist to win a Nobel Prize, for his experiments that led to the demise of the theory of the ether. (b) Edward Williams Morley (1838–1923) collaborated with Michelson on the second, improved, version of his experiment, which failed to detect the Earth's motion through the ether.

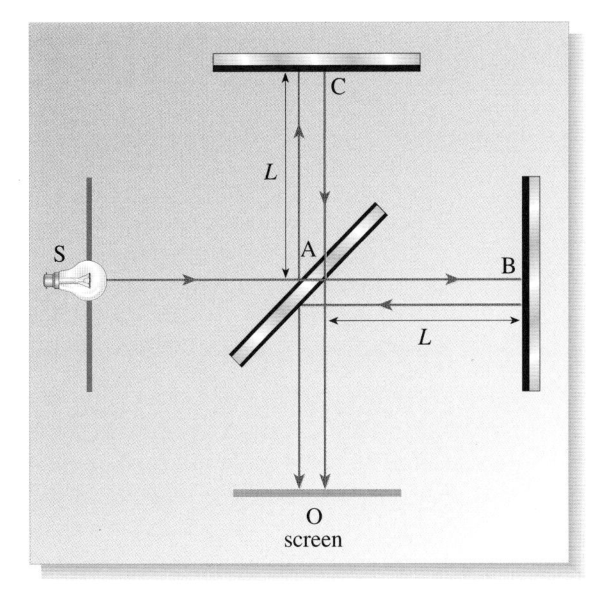

Figure 4.21 The apparatus used by Michelson in his attempt to detect the effect of the ether on the time required for light to make a round trip between *two* points separated by a distance L (the only rays shown are those that relate to the theory being tested in the experiment). The motion of the Earth moves the entire apparatus relative to the ether.

From Chapter 2, you should realize that when the two parts of the original beam recombine at A they will *interfere*. If the time taken for light to travel from A to B and back differs from the time required to go via C by a whole-number multiple of the period of oscillation of the light wave, then the interference will be constructive, the two waves will reinforce one another and the screen will be brightly illuminated. On the other hand, if the time difference is an odd number of *half*-periods, destructive interference occurs and the screen will not be illuminated. So the existence of a time difference should influence the illumination of the screen. Furthermore, rotating the whole apparatus in the plane of the points A, B and C should change the orientation of both AB and AC relative to the direction of the Earth's motion through the ether. Such a rotation would therefore alter the time delay along each path, so changing the illumination of the screen. It was this *change* in illumination, as a result of rotation, that Michelson tried to detect.

Michelson first performed his experiment in 1881, and found no sign of the predicted time difference. His results were challenged, so he repeated the experiment in an improved form in 1887, with the aid of E. W. Morley (Figure 4.20b). Once again there was no evidence of a time difference. The result of this **Michelson–Morley experiment** has been described as the most famous negative result in the history of physics. As far as Michelson himself was concerned, its implication was clear, as he wrote at the time:

> The result of the hypothesis of a stationary ether is thus shown to be incorrect.

There were several attempts at explaining the null result by invoking various 'mobile-ether' theories, in which the Earth dragged part of the ether around with it, but eventually the growth in complexity of such arguments (none of which were supported by experiment) led to a willingness to consider more radical alternatives.

It was into this atmosphere that Albert Einstein introduced his special theory of relativity. It presented a new way of interpreting electromagnetism that retained the successes of Maxwell's theory but did away with the superfluous speculations about the ether that Maxwell had employed. Thus, the ether theory was finally laid to rest by a combination of experimental discovery and theoretical innovation.

4.3 Is c the speed of light relative to the detector?

> When you have eliminated the impossible, whatever remains, however improbable, must be the truth.
>
> Sherlock Holmes in *The Sign of Four* by Sir Arthur Conan Doyle.

Three possible interpretations of the phrase 'the speed of light' were given at the beginning of this section. Experimental evidence has dismissed two of them, so only one remains. Is the third possibility correct? Is c the speed of light relative to the detector? If so, some surprising consequences will have to be accepted, as the following imaginary experiment shows.

Consider two spaceships A and B, each having a light beacon on its nose-cone. Suppose that spaceship A is at rest with respect to an observer, whereas spaceship B is travelling very fast, at a constant speed of $c/3$ (one-third of the speed of light) towards the observer, as in Figure 4.22. Now imagine that just at the moment when spaceship B passes spaceship A, the observer measures the speed of light from each spaceship.

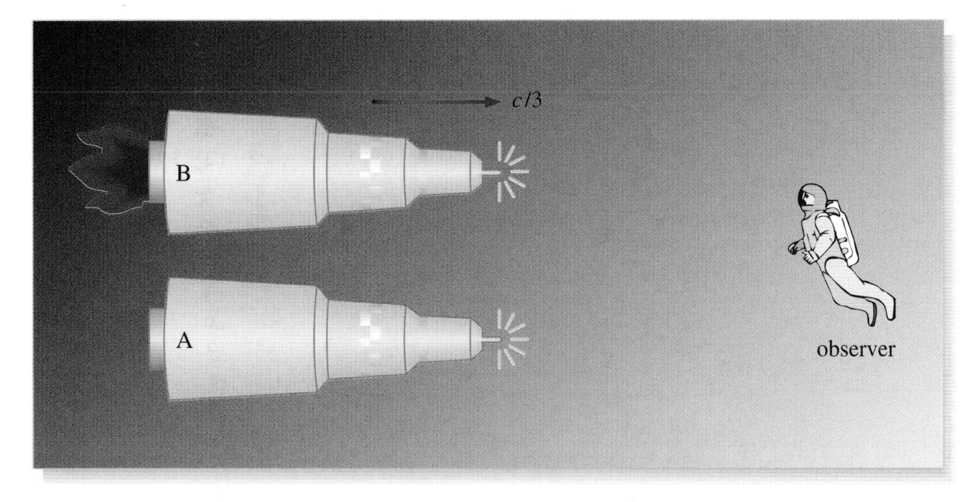

Figure 4.22 Spaceship A is at rest with respect to an observer, whereas spaceship B is travelling towards the observer at a speed of $c/3$. (*Note* We are indicating that a spaceship is moving by the flames from its rocket motor. However, it is important to recognize that special relativity *only* operates in inertial (i.e. non-accelerating) frames. So, in reality these spaceships are 'coasting along' with the rocket motor off.)

On the basis of intuition (with light behaving like bullets fired from a gun muzzle), you would expect that the speed of the light emanating from the nose cone of the spaceship B, as measured by the observer, would be $c + c/3 = 4c/3$. But it has just been established that the speed of light is not relative to its source (Section 4.1). Nor is it relative to some mysterious ether (Section 4.2). The fact is that the observer will obtain the *same* result, namely c, for the speed of light from both spaceships, even though one is moving at high speed towards him, and the other is stationary. In other words, the speed of light is always c relative to its detector. Moreover, this is true for *any* detector, so that *all* observers always measure the speed of light as c. You should find this prediction very surprising; it conflicts with usual ideas about relative speed — in particular, the different behaviour of light compared with, say, objects thrown from cars appear to defy 'common sense'. Later you will see that these apparently contradictory observations are in fact quite consistent.

This interpretation of c is consistent with the binary star experiment and with the Michelson–Morley experiment. Moreover, no experiment has ever provided any convincing evidence to the contrary. The fact that the constancy of the speed of light relative to its detector (with all observers measuring the speed of light as c) seems at odds with everyday experience simply shows how limited and unreliable everyday experience is!

The fact that all observers measure the speed of light as c is a surprising discovery. However, because it is supported by experiment it will be adopted as a trustworthy principle and used throughout the rest of this chapter. It is the basis of the second great postulate of special relativity. This discovery about the behaviour of light, when taken together with the principle of relativity, has some very surprising consequences, as you will find out in the next section.

Question 4.5 Suppose that you are on board a spacecraft travelling towards the Earth at a speed of $1.5 \times 10^8\,\mathrm{m\,s^{-1}}$. You are equipped with a laser that you are using to send signals to a detector orbiting the Earth. When your signals are received, it is found that their speed relative to the detector is $3 \times 10^8\,\mathrm{m\,s^{-1}}$. If you were to fix a device to the spacecraft to measure the speed of this laser light, so that you were measuring the speed of the laser signal relative to its source, what result would you find? ■

From the answer to Question 4.5 you can see that we require that c is the speed of light relative to *every* detector, including one moving at the same speed as the source. Thus, possibility 1 (beginning of Section 4) is also true in a sense.

5 Some spectacular new predictions of special relativity

> The relativity theory arose from necessity, from serious and deep contradictions in the old theory from which there seemed no escape. The strength of the new theory lies in the consistency and simplicity with which it solves all the difficulties using only a few very convincing assumptions.
>
> *The Evolution of Physics*, by A. Einstein and L. Infeld, Cambridge University Press (1938).

5.1 Einstein's two postulates

We have it from Einstein that he had been thinking seriously about the question of space, time and motion since he was 16 years old. He would ask himself questions such as: what would happen if I were riding on a rod and the rod were travelling at the speed of light; what would I see? For years he felt that he almost saw through the problems, but something eluded him. He tells us that one day he was on a bus in Berne when it passed the town clock tower. As the bus moved past, Einstein imagined again: what time would he find the clock reading if the bus were travelling at the speed of light? Something clicked and he imaged that he would read no time at all, as the light from the clock carrying the information about the time could not catch up with him. So perhaps time was really a local affair, and depends on the relative state of motion of clock and observer.

By the end of the nineteenth century, it was well known that there was a serious problem with the electromagnetic theory of light. Maxwell's theory predicted the correct numerical value of c, but it failed to explain why c was always the speed of light *relative to its detector.* It was the 26-year-old Albert Einstein, who realized that the problem arose from a contradiction between the constancy of the speed of light and the Galilean coordinate transformation. Rather than trying to modify electromagnetic theory, Einstein adopted a different approach. He rejected the Galilean transformation and set out to find a replacement for it — a new coordinate transformation that would solve the central problem of relativity.

To ensure that the new coordinate transformation would be consistent with the constancy of the speed of light for all detectors, Einstein built the constant c into his theory from the very beginning. He achieved this by basing his mathematical deductions on the following two postulates:

Einstein's postulates

Postulate I — The principle of relativity

The laws of physics can be written in the same form in all inertial frames.

Postulate II — The principle of the constancy of the speed of light

The speed of light (in a vacuum) has the same constant value c in all inertial frames.

The first postulate and its implications were discussed in Section 2. The second postulate is a more precise restatement of the conclusion arrived at in Section 4. All that has been done is to replace the rather loose idea of measuring the speed of light relative to a detector by the more limited notion of measuring the speed of light relative to an inertial frame. If you think of the inertial frame as the one in which the detector is at rest, then it is clear that Postulate II is equivalent to the experimental observation that the speed of light is always relative to the detector. Since this postulate is true for *all* detectors, c has the same value in *all* inertial frames. Remember that both postulates are firmly based on experimental observations.

Some simple mathematics that follows from these two postulates led Einstein to the new correct coordinate transformation, and thus to a new theory of physics, now known as the *special theory of relativity.* Before establishing the new coordinate transformation that arises out of special relativity, it will be instructive to examine the implications of Einstein's postulates for time and length measurements.

5.2 Time dilation

The fact that light travels at the same speed in all inertial frames will lead to a very surprising conclusion about the behaviour of time. In Section 2.2, a promise was made that this section would present a simple but safe way of synchronizing any number of clocks located in a given frame of reference. Armed with Einstein's second postulate, this can now be done.

The key to the procedure is to make direct use of the fact that the speed of light is the same in all inertial frames. To synchronize clocks carried by A and B in Figure 4.23, a light source is placed at a point midway between them. The light source is then switched on, and observers A and B set their clocks to $t = 0$ at the instant when they first observe the light. The speed of light will be c relative to both observers. Clearly, this *is* a safe way of proceeding, based firmly on the second postulate; it does not involve any assumptions about the behaviour of moving clocks. Since any pair of clocks can be synchronized in this way, it is possible to synchronize clocks by this method throughout a given inertial frame.

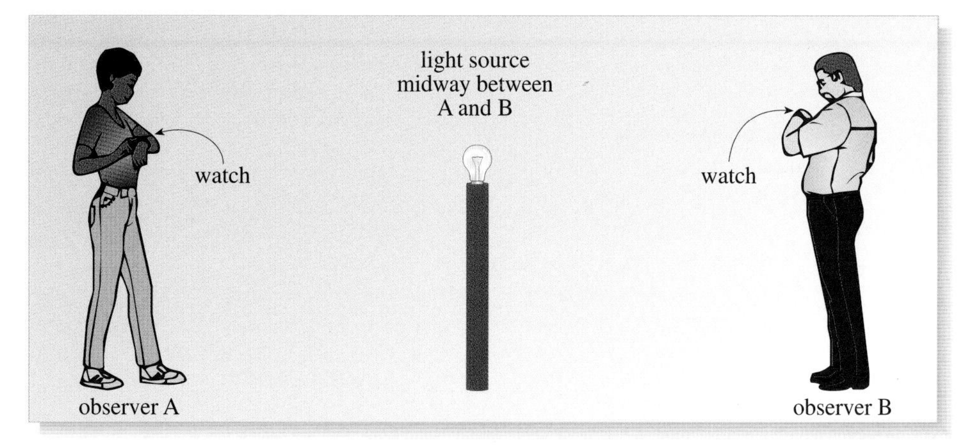

Figure 4.23 A way to synchronize watches for two observers A and B at rest relative to each other. The light source is at rest in their inertial frame and is midway between them.

To check that you have clearly understood the principle of the second postulate that is fundamental to the synchronization procedure described above, try the following.

- A and B are two astronauts in the same inertial frame, separated by a distance of 3×10^{10} m, who wish to synchronize the digital clocks on their respective spacecraft. Suppose that, by arrangement with B, A starts her clock at the same instant that she transmits a light signal. If it takes B one second to adjust his clock to any desired reading, at what value of t should B start his clock when he receives the light signal from A?

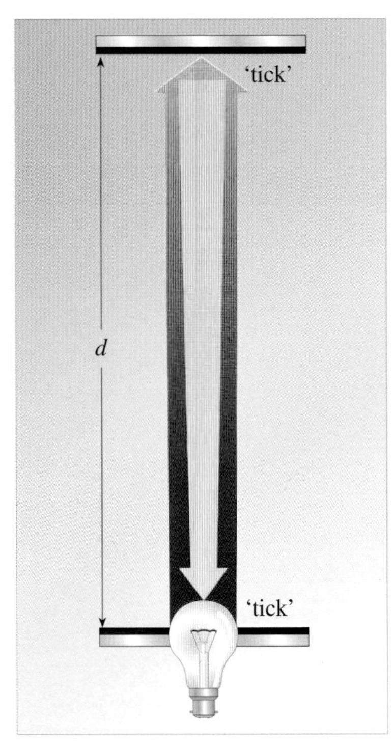

Figure 4.24 A light clock. A short pulse of light bounces back and forth between two parallel mirrors separated by a distance d, with the clock 'ticking' every time the light pulse hits one of the mirrors.

❍ The light signal will take 100 seconds to travel from A to B. When B sees the signal, he knows that the time according to A is t = 100 s. As B takes one second to adjust his clock, he should start it at 101 seconds. ■

We have happily talked about clocks and watches up to this point, without worrying about the *mechanism* used in these clocks to measure time intervals. In order to be absolutely confident that these time measurements are based on a principle that is simple and foolproof, the idea of a **light clock** will now be introduced. Figure 4.24 shows two parallel mirrors at a fixed distance apart. The light clock mechanism consists of a short pulse of light, which bounces back and forth between two parallel mirrors with the clock 'ticking' every time the light pulse hits one of the mirrors. Although this would be a pretty impractical clock in real life, it has the advantage that the time intervals, as measured by a light clock, depend only on the principle of the constancy of the speed of light, and can therefore be interpreted unambiguously using Einstein's second postulate.

The next thing to consider is a series of time measurements made by two observers who use a light clock to make measurements from two inertial frames in relative uniform motion (Figure 4.25a). This sort of exercise is often referred to as a **thought experiment**: it is an experiment that we can *imagine* taking place (although carrying it out in practice may be rather difficult), and for which the outcomes can be *predicted* based on underlying physical principles.

In this particular thought experiment, observer A is standing at the side of a railway track, and can be thought of as being at rest in inertial frame A. Observer B and the light clock are in the railway carriage. They are at rest in inertial frame B, which is moving past the frame of reference of observer A at a constant speed V. The direction of the relative motion of observers A and B is perpendicular to the light path as observed by observer B. The task is to measure the time interval between two consecutive ticks as measured by observers A and B, respectively.

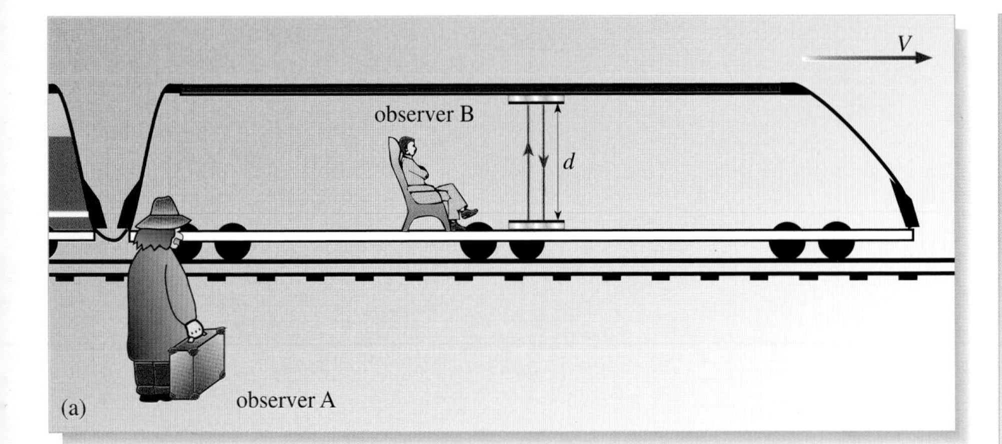

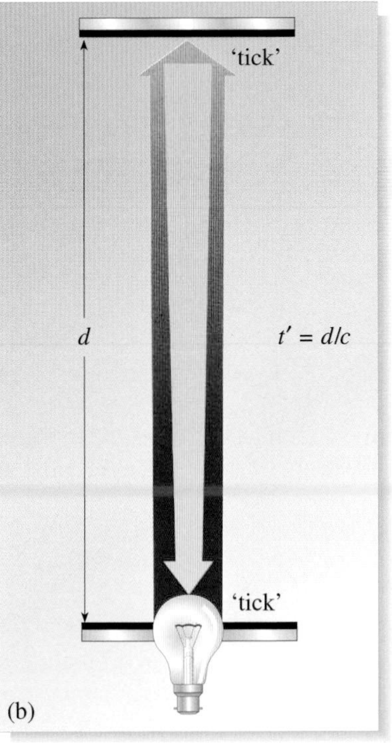

Figure 4.25 (a) Observer A measures the time interval between ticks as t, as the carriage carrying observer B and the light clock moves past him at speed V. Observer B measures the interval between ticks as t'; (b) the light path as observed and measured by B.

Let t' be the time between ticks as measured by observer B. To B, the light pulse bounces back and forth along a parallel path, as shown in Figure 4.25b, so that

$$t' = d/c. \tag{4.5}$$

To observer A, however, the light has to move a larger distance between ticks. As observed by A, in the time taken for the light pulse to travel from the bottom mirror to the top mirror, the top mirror will have moved along the direction of motion of the carriage (i.e. the motion of frame of reference B relative to frame of reference A). When the clock ticks, the light clock will be in the centre position PY shown in Figure 4.26a. As measured by observer A in frame of reference A, it is clear that the light has to move a longer distance between ticks (namely XY) compared with the observations of observer B who is travelling with the clock. The geometry of the light path between successive ticks, as observed by A, is illustrated in Figure 4.26b. Look carefully at how the lengths of the sides of the triangle PXY are labelled. Side PY, which is the perpendicular distance between the mirrors, can be obtained by rearranging Equation 4.5 to get

$$d = ct'.$$

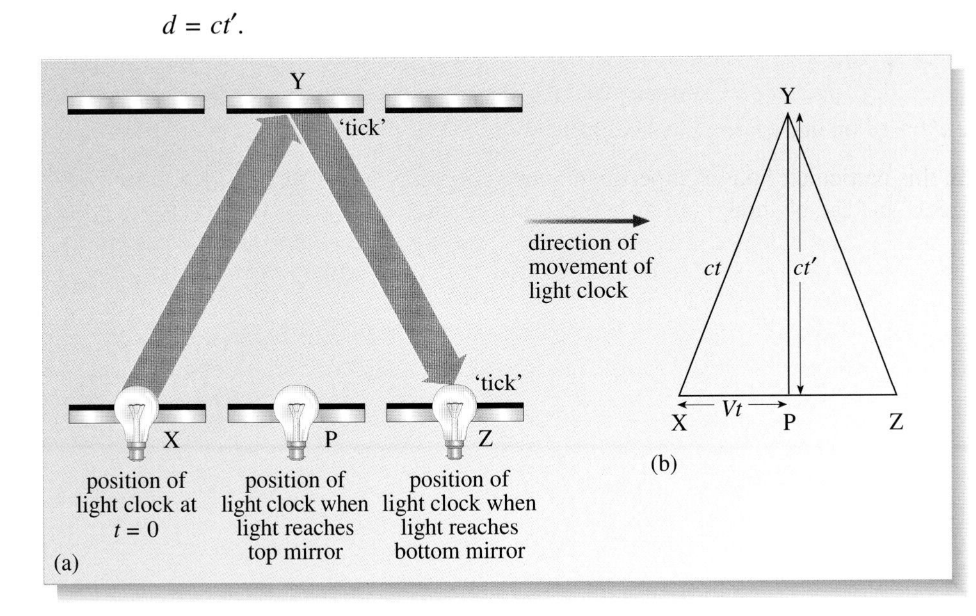

Figure 4.26 (a) The light path between successive ticks, from observer A's viewpoint; (b) the geometry of the light path illustrated in Figure 4.26a.

What about side XP? This is the distance moved by the top mirror during the time t (as measured by observer A) it takes for the light to travel from the bottom to the top mirror. With the carriage moving at a constant speed V in frame of reference A, the length of XP is thus Vt.

Finally, there is the length of XY. This length can be evaluated by making use of the second postulate, which states that the speed of light is *the same in all inertial frames*. So it is clear that the length of XY is ct, the length travelled by light (at constant speed c) in the time interval t measured by A for the light to travel between mirrors (i.e. the time interval measured by A between successive ticks).

We now have a right-angled triangle PXY with all the lengths known, and Pythagoras's theorem can be used to obtain a relationship between the sides:

$$XP^2 + PY^2 = XY^2. \tag{4.6}$$

Substituting into Equation 4.6 with the values indicated in Figure 4.26b,

$$V^2t^2 + c^2t'^2 = c^2t^2$$

or, rearranging,

$$c^2t'^2 = t^2(c^2 - V^2).$$

Then dividing both sides by c^2 gives

$$t'^2 = t^2\left(1 - \frac{V^2}{c^2}\right).$$

and finally

$$t' = t\sqrt{1 - \frac{V^2}{c^2}}. \qquad (4.7)$$

You should think carefully about the meaning of Equation 4.7. Depending on the relative size of V compared with the speed of light, the factor $\sqrt{1 - V^2/c^2}$ will range from a value close to one (when $V \ll c$), to values approaching zero if V is comparable in size to c. Notice therefore that t' is always less than t. This means that the time interval t' between the two reflections as measured by observer B is *less than* the time interval t between the same two events as measured by observer A. This is a startling result. It is given the name **time dilation** and is more colloquially expressed as 'moving clocks run slow'. The time dilation formula is often expressed in a different form by writing $t = \Delta T$ and $t' = \Delta T_0$. Then the dilated time ΔT is given by

$$\Delta T = \frac{\Delta T_0}{\sqrt{1 - V^2/c^2}}. \qquad (4.8)$$

This equation has the advantage of showing that time *intervals* are involved (hence the Δ signs). The subscript 0 is used to indicate the time interval in the frame of reference in which the clock is at rest, which may be easier to remember than the choice of primed and unprimed frames of reference in Equation 4.7. You will see shortly that a similar convention is used to describe the effects of special relativity on length measurements.

No matter how it is written, this is not an easy result to understand in an intuitive way, and it is likely to take some time before you feel confident in discussing the implications of how measurements of time can behave in this way. Before looking more closely at some consequences of time dilation, it is useful to emphasize exactly how this surprising result has arisen. The key ingredient in the analysis was that the second postulate was used in working out the lengths of the right-angled triangle PXY. It was the fact that the speed of light *is the same in all inertial frames* that forced us to assign the value ct and ct' to sides XY and PY. This is the crucial feature that distinguishes special relativity from Newtonian mechanics. It is then inevitable that the time intervals t and t' are different, and are related according to Equation 4.7.

Question 4.6 A friend who has heard vaguely that 'moving clocks run more slowly' attempts to explain this effect to you as a *mechanical* phenomenon. She argues that the rapid movement of a watch has somehow affected the mechanism of the watch, causing it to 'really run more slowly'. How would you convince her that her explanation was false? (*Hint* Think about Einstein's first postulate.) ■

You should be convinced now that time dilation is not some kind of mechanical defect that happens to moving clocks. It is, rather, a direct consequence of the two postulates of special relativity. Observer B who travelled with the clock in frame of reference B would not see anything strange happen to a wrist watch on his arm. What has been discussed does not involve some mysterious process happening inside the atoms of a moving object; the effect has arisen out of the measurement process. The fact that observers in relative motion will generally disagree about the time interval between any two events is to do with what is meant by a 'time interval', and the way it is measured, and not with the workings of clocks. A light clock (or any other kind of clock) is simply a device that is used to make this clear.

Moreover, since time dilation is independent of the method of time measurement, then presumably there will be a time dilation effect produced by a light clock even if the direction of relative motion between two observers is parallel to the light path in the light clock.

This is indeed the case, as you can see by considering the situation in Figure 4.27, where two identical light clocks are at right-angles. It is arranged that the two light beams return to essentially the same point P in space as the light clocks tick.

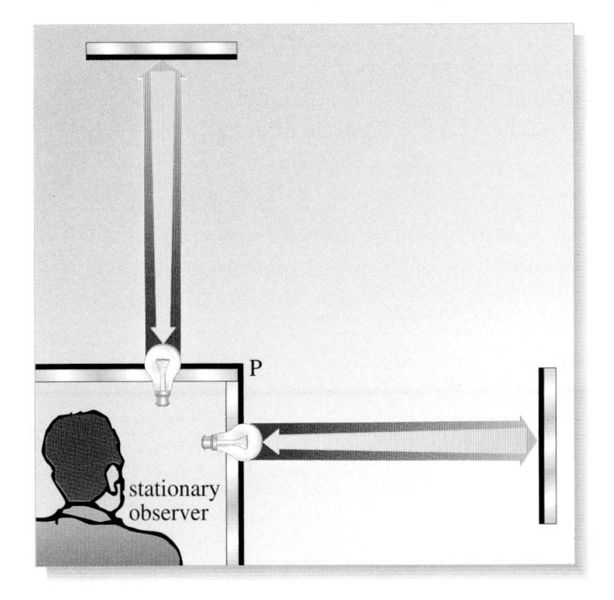

Figure 4.27 Two identical light clocks at right-angles, such that the two light flashes return to the same point P as they tick. We consider the situation when the light clocks are stationary with respect to an observer, and also when they move with uniform velocity V (relative to the observer) in a direction parallel to one of the light beams.

Imagine two light flashes leaving point P, one travelling upwards and the other sideways. When the two light clocks are at rest with respect to the observer, the two light flashes will both return to P at the same time, and then repeat this cycle indefinitely. There is no doubt that in these circumstances, both clocks will keep the same time.

Now suppose that the same set-up is moving in a direction parallel to *one* of the light clocks with constant speed V, relative to the observer. It has previously been established that there is a time dilation effect for the clock where the light flashes are moving perpendicular to the direction of motion. But what will be the effect on the light clock where the flashes are moving parallel to the motion of the set-up? This question can be answered by invoking the concept of space–time conjunction. Since the two light flashes keep on returning to the *same* point P at the *same* time when the set-up is stationary relative to the observer, then they must always do so. In particular, the light flashes must also return to P at the same time when the frame of reference in which the clocks are stationary is moving relative to the observer. This can only be the case if *both* light clocks are subject to the same time dilation effect.

The rather imprecise expression '*moving clocks run slow*' causes much misunderstanding. What has been discussed are the conclusions of time interval measurements made by two observers, moving relative to each other, who examined clocks and applied the laws of nature to their findings. The central conclusion should be that time has lost its absolute nature (as understood in Newtonian mechanics). Clocks do not tick in some relentless sense, measuring some absolute quantity called time. Thus, time dilation is a property of time itself; time has become a *relative* quantity. It is a quantity that is *personal* to the particular observer who is measuring it. It must be concluded that both the clocks and all life processes of the observer in frame of reference B would slow down *relative* to the clocks of the observer in frame of reference A.

Question 4.7 The distance to the triple star system Alpha Centauri is such that it takes light 4.2 years to travel from that system to the Earth. If an astronaut left the Earth and travelled at a constant speed of $0.9c$ relative to the Earth towards Alpha Centauri, how long would it take, according to an observer on Earth and according to the astronaut, for him to reach there? ■

5.3 Lorentz contraction

You've just seen that, according to special relativity, time is a relative quantity, and its measurement is personal to the particular observer who is measuring it. Now, since time has lost its absolute character, what about length measurements? In fact, was the derivation of the time dilation formula careful enough with regard to measuring lengths? After all, in the derivation, it was assumed that the length between the mirrors in the light clock, which was perpendicular to the motion, was unaffected by the relative motion between the observers. To check if this was indeed a safe assumption, look at the new thought experiment in Figure 4.28.

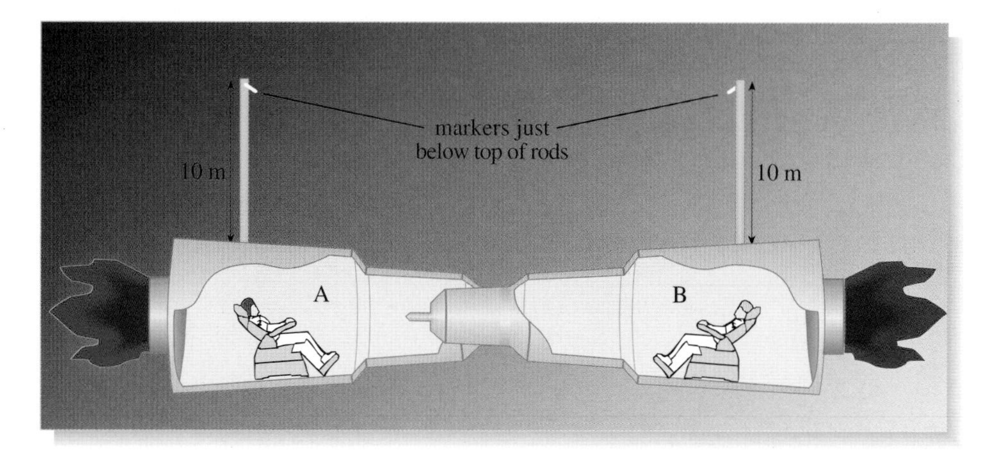

Figure 4.28 A thought experiment on length contraction.

The figure shows two spaceships A and B travelling past each other at high speed. Each is equipped with a vertical rod 10 m long, fitted with a marker at its top end; the marker can just scratch a mark on the end of the other rod as they almost touch each other in passing. If there is any change in the length of spaceship B's rod, as seen from spaceship A, this will obviously have an effect on where the marks will or will not appear, and, of course, one can also think of events from the point of view of spaceship B.

The way to resolve what happens in this sort of situation is to invoke the principle of relativity. Think first about how things appear from spaceship A's standpoint. If spaceship B's rod contracts compared with ship A, then B's marker will scratch a mark below the top of rod A. In this case, rod A's marker would be above the top of rod B, so there would be no mark at all on rod B. Now, the whole thing is symmetrical, yet if the above were to happen, there would be an asymmetric result. And remember, what has happened is a real physical effect — a mark scratched on rod A and no mark on rod B — a real effect that cannot depend on whether it is observed from spaceship A or B.

Now for the crucial point: according to the principle of relativity, there is no reason for preferring one inertial frame over another. The laws of physics must be the same for an observer in spaceship A as they are in spaceship B. So anything that happens to spaceship A's rod must also happen to the rod on spaceship B. This can only be the case if the rod marks are exactly on top of each other, implying that to either observer the moving and static rods are of the same length, and there is no contraction (and, for that matter, no expansion either).

So, lengths are not affected by motion perpendicular to their measured length, but what about the effect on length measurement of motion in the direction of measurement? To investigate this, look at the modified light clock experiment in Figure 4.29. This time the parallel mirrors are positioned so that the light bouncing back and forth between them is orientated exactly along the line of motion of the spaceship carrying the light clock. The mirrors are a distance L_0 apart, as measured in the inertial frame B, in which the mirrors are stationary. In this example it is more convenient to have the clock 'tick' once every time the light pulse returns to the light source.

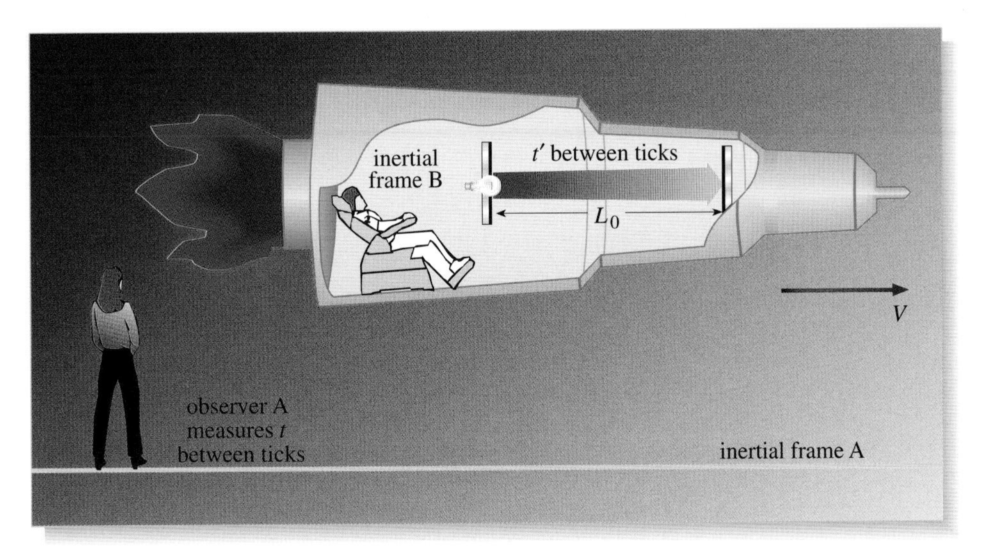

Figure 4.29 A modified light clock experiment. The light clock 'ticks' every time the light pulse returns to the light source.

Clearly, in inertial frame B, the time interval between ticks is simply

$$t' = \frac{2L_0}{c}. \tag{4.9}$$

But what about from the viewpoint of inertial frame A, where the light clock moves to the right at speed V? This calculation can be carried out with the help of Figure 4.30.

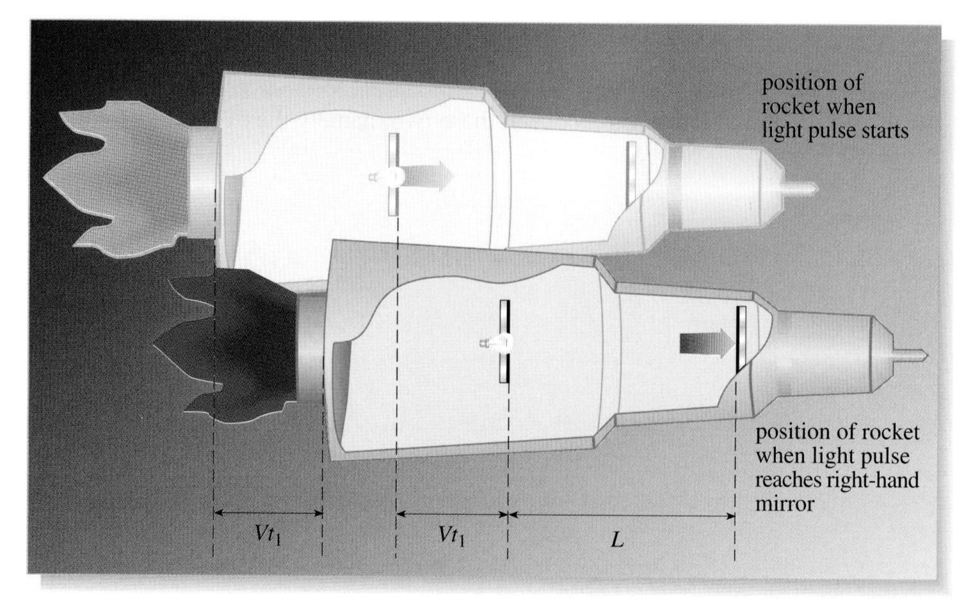

Figure 4.30 The time interval between ticks as observed from inertial frame A.

The distance between the mirrors as measured in inertial frame A is L, and this may (or may not) be the same as L_0. The first thing to work out is the time for light to reach the right-hand mirror, t_1. Since light travels at speed c in all inertial frames, the distance travelled by light in time t_1 is therefore ct_1. But this is the distance between where the left-hand mirror was *when the light set out* and where the right-hand mirror is *when the light arrives*, as shown in Figure 4.30. Since time t_1 has elapsed between the two events, and the mirror moves to the right with speed V, the total distance travelled by the light is therefore $L + Vt_1$. So clearly

$$ct_1 = L + Vt_1$$

and hence
$$t_1 = \frac{L}{c - V}. \tag{4.10}$$

- Using a similar argument to that above, obtain an expression for the time t_2, the time that the light takes to travel back from the right-hand mirror to the left-hand mirror.

○ Since the light is now travelling in the opposite direction to the spaceship, in this case the total distance travelled by the light is $L - Vt_2$. So $ct_2 = L - Vt_2$, or rearranging

$$t_2 = \frac{L}{c + V}. \quad ■ \tag{4.11}$$

Clearly, the total time between ticks, as measured by the observer in inertial frame A, is just the sum of $(t_1 + t_2)$. So

$$t = \frac{L}{c - V} + \frac{L}{c + V} = \frac{L(c + V)}{(c - V)(c + V)} + \frac{L(c - V)}{(c + V)(c - V)}$$

$$t = \frac{2Lc}{c^2 - V^2}.$$

Dividing both numerator and denominator by c^2 this becomes

$$t = \frac{2L}{c(1 - V^2/c^2)}. \qquad (4.12)$$

Now, Equation 4.9 was an expression for t', the time measured in inertial frame B, and Equation 4.12 is an expression for t, the time measured in inertial frame A. But you already know a relationship between these two, namely the expression for time dilation, Equation 4.7. So, substituting for t and t' in Equation 4.7 yields:

$$\frac{2L_0}{c} = \frac{2L}{c(1 - V^2/c^2)} \times \sqrt{(1 - V^2/c^2)}.$$

The factor of $2/c$ cancels from both sides of the equation, leaving

$$L_0 = \frac{L}{1 - V^2/c^2} \times \sqrt{1 - V^2/c^2}.$$

Then dividing both numerator and denominator by the square-root term gives

$$L_0 = \frac{L}{\sqrt{(1 - V^2/c^2)}}.$$

So, finally, rearranging this in terms of L, the length as measured in inertial frame A:

$$L = L_0\sqrt{1 - \frac{V^2}{c^2}}. \qquad (4.13)$$

This is an extremely important result, and will be used throughout the rest of this chapter. What Equation 4.13 implies is that an object whose measured length is L_0 when at rest, contracts by the factor $\sqrt{1 - V^2/c^2}$ when measured by an observer relative to whom it moves at a speed V in a direction parallel to its measured length. This startling effect is called the **Lorentz contraction** (or sometimes *Fitzgerald–Lorentz contraction*). The formula was originally derived by the Dutch physicist H. A. Lorentz (Figure 4.31a), and independently by the Irish physicist G. F. Fitzgerald (Figure 4.31b), to explain the null result of the Michelson–Morley experiment by a real contraction of a body along its direction of motion through the ether. In their explanation, the speed V was relative to the ether and the contraction was actually caused *by* the ether. It was Einstein's special theory of relativity which was able to show the correct explanation for this effect.

(a)

(b)

Figure 4.31 (a) Hendrik Antoon Lorentz (1853–1928) was a Dutch physicist who studied at the University of Leyden, where he subsequently became Professor of Mathematical Physics. By refining his own extension to Maxwell's theory of electromagnetism, he obtained in 1895 the space and time transformations later derived by Einstein. For later work, he was awarded the Nobel Prize for Physics in 1902, jointly with fellow countryman Pieter Zeeman. (b) George Francis Fitzgerald (1851–1901) was an Irish physicist who was a professor at Trinity College Dublin. In 1889 he suggested that the shrinkage of a body due to motion at speeds close to that of light would account for the results of the Michelson–Morley experiment.

Just as in the case of time dilation, there are many misunderstandings surrounding Lorentz contraction. For example, it's often said that a moving object 'appears to be contracted', as if Lorentz contraction were some kind of optical illusion. This is certainly not the case: the contraction is a real physical effect concerning measured lengths.

● Imagine two frames of reference in standard configuration. A spaceship containing a clock and a measuring rod aligned along its x'-axis is at rest with respect to inertial frame B. To an observer in inertial frame A, the spaceship is moving with a large velocity along his x-axis. The observer in inertial frame B (inside the spaceship) measures a time interval on his clock as ΔT_0 and the length of his measuring rod as L_0. Will the observer in inertial frame A measure the time interval on B's clock and the length of B's rod as greater than or less than ΔT_0 and L_0, respectively?

❍ From Equation 4.8, the observer in inertial frame A will measure the time interval ΔT as *greater* than ΔT_0: time is *dilated* relative to the measurement made in the frame of reference in which the clock is at rest. From Equation 4.13, the observer in inertial frame A will measure the length of the rod L as *less* than L_0: lengths are *contracted* relative to the measurement made in the frame of reference in which the rod is at rest. ■

5.4 Symmetry

In describing time dilation and Lorentz contraction, frames of reference moving relative to other frames of reference have been referred to. Since it's only *relative* speed that is important, it is perfectly reasonable to reverse the argument in Figure 4.29 and assume that inertial frame A is moving to the left at speed V with respect to inertial frame B, which contains the light clock. This situation is summarized in Figure 4.32. Einstein's first postulate says that it's quite all right to do this. But, in that case, wouldn't time dilation lead us to expect that it is the clock in inertial frame A that is running slow, as observed by B? The answer, you might be surprised to hear, is yes. Speed is relative, so from B's standpoint, inertial frame A can be thought of as moving, and clocks in inertial frame A would be expected to run slow compared to those in inertial frame B. So according to special relativity, the clock in inertial frame A runs slow from the standpoint of the observer in inertial frame B *and* the clock in inertial frame B runs slow from the standpoint of the observer in inertial frame A. This interesting conclusion highlights the care that must be taken in talking about measurements in special relativity.

However, you need never be perplexed by this situation. When A says that B's clock runs slow, these observations are valid for A, but not for B. Observer B can make his own measurements and his conclusion is that A's clock runs slow. Both are correct! There is no physical contradiction because each set of results refers to its own inertial frame. There is a remarkable symmetry between all observers who use inertial frames: they will all agree that 'moving clocks run slow' and that 'moving rods contract in the direction of motion'. There is no phenomenon that will permit an observer in an inertial frame to conclude that he is the one who is really moving, since that would be in conflict with the principle of relativity, Einstein's first postulate.

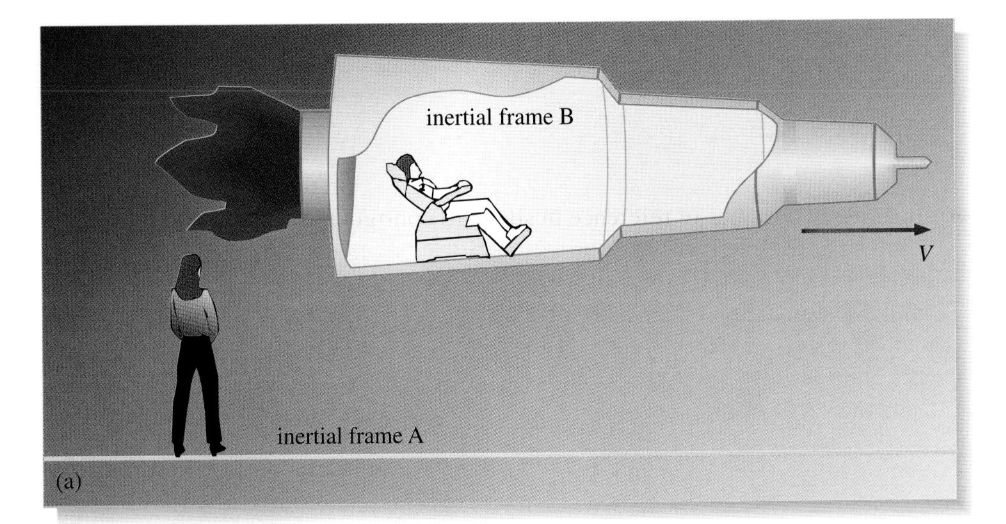

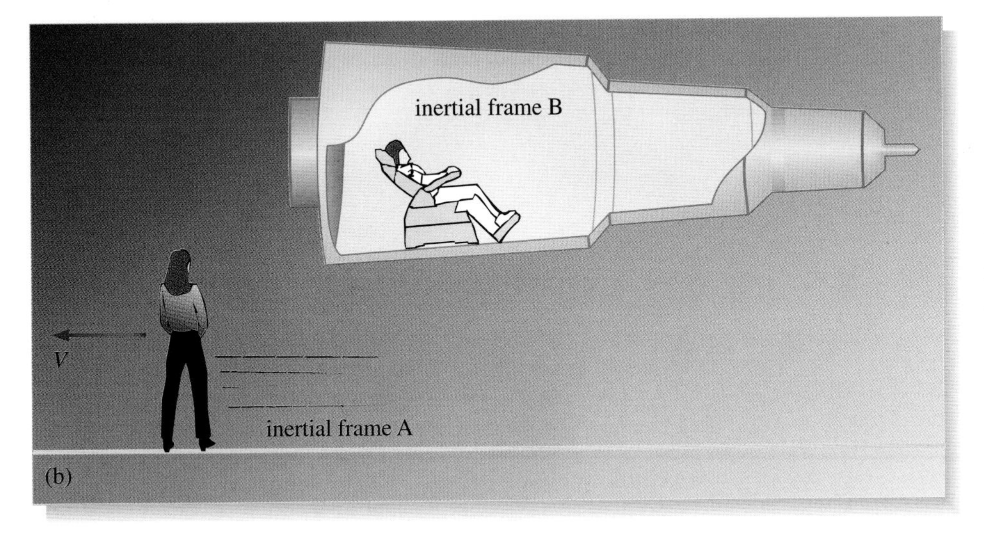

Figure 4.32 There is symmetry between all observers who use inertial frames. Viewed from inertial frame A, B's clock runs slow; whereas viewed from inertial frame B, A's clock runs slow. Both observers are correct in asserting that the other's clock is running slow and that the other's lengths contract in the direction of motion.

Box 4.1 The twin 'paradox'

By definition, there are no paradoxes in Nature. A theory in physics is meant to describe some of this reality, and so unless it is determined to be in conflict with the results of some experiment, it can be taken to be free of contradiction — and special relativity is in this position.

Then what is meant by a paradox in relativity? The answer is simply predictions of the theory that contradict common sense. But common sense is not a guide to the deeper parts of Nature. There is no point in defending relativity from such objections. None the less, some of the processes that vex our common sense, while free of contradiction, involve interesting physics. And in discussing why they are consistent, deeper insight into the nature of relativity can be found.

The twin 'paradox' concerns the application of the time dilation result to the problem of a pair of twins. One goes on a journey at a speed close to that of light. When the twin returns, the twin that remained at home will expect that the travelling twin will be younger by a certain factor, while the twin who travelled will expect that the twin who stayed home will be younger by the

You are not expected to be able to reproduce the reasoning outlined in this box.

same factor. Since relativity predicts these contradictory results, so the paradox goes, it must be inconsistent.

Evidently both results cannot be true, so what does relativity actually predict, how did the frame of reference symmetry manage to get broken, and is there a physical way to check on the answer?

Filling in some details of the account above, suppose that the twin who stays at home is called Xavier, and he is the reference body for the inertial frame H (for 'home'). The travelling twin, Yvonne, takes off from home in a spacecraft, and after a short initial period of acceleration, reaches cruising speed V. The journey continues at this speed until the spaceship reaches a distance L metres from home, as measured by Xavier. This is the *outward* leg of the journey, during which Yvonne is the reference body of the inertial frame to be denoted frame of reference O (for 'outward'; see Figure 4.33).

At the outermost point of the trip, the spaceship executes as sudden a reversal of velocity as it can, re-accelerates to the speed V, and continues back to home. During this return journey, Yvonne is the reference body of the inertial frame to be denoted frame of reference R (for 'return'). Note that frames of reference R and O are quite different: there is a relative speed between them of $2V$.

Arriving back home, the spaceship decelerates sharply and lands, after which the twins compare watches.

In order to understand what really happens, suppose that the twins beam flashes of light at each other at a regular frequency f, as observed locally. Then the real physical effect of the total journey is that Yvonne sends fewer flashes than Xavier sends. Since every flash that is sent by one twin is also received by the other, Xavier receives fewer flashes than Yvonne receives. When Yvonne gets home she is demonstrably younger than Xavier, by a mutually agreed number of flashes.

Now suppose they compare notes, on Yvonne's return. Yvonne shows Xavier her log of flashes received. This records that flashes were received at a rate $f_{\text{away}} < f$ when she was moving away from Xavier and at a rate $f_{\text{towards}} > f$ when she was returning towards Xavier. Inspection of Xavier's log shows the same pattern. The flashes sent by Yvonne when she was travelling away are received at a frequency f_{away} by Xavier, and those sent when she was returning are received at a frequency f_{towards} by Xavier.

So how is the symmetry broken? The answer is rather simple: inspection of Yvonne's log shows that the frequency received from Xavier increased when she reversed direction that is, half way through the trip (according to her). But poor Xavier had to wait to receive Yvonne's higher-frequency signals; he observed the increase some time *after* the midpoint of his log because of the finite time of travel of the signals. So it is clear, without using any equations, that Yvonne is younger than Xavier at the space–time conjunction of their reunion in Figure 4.34. It is a simple consequence of the fact that they *agree* on the values of f_{away} and f_{towards}, which depend only on their relative motion. Put like this, it is hard to see why some people regard it as paradoxical.

It would be different if Yvonne drove a car with a sound alarm blaring, while Xavier sounded an identical alarm at home. Using sound, they would disagree on the receding and approaching Doppler-shifted frequencies, since the formulae of Chapter 2 depend on whether the source or the receiver is moving relative to the air. This disagreement turns out to compensate for the postponement in increase in the frequency received by Xavier. In the case of sound, one has to do a rather complicated calculation to show that the twins agree on the total number of cycles received. In the case of light, it is immediately obvious that they cannot receive the same number of flashes, since the principle of relativity requires them to agree on the formula for the Doppler effect, and the finite speed of light makes Xavier's log change at a point later than its midpoint.

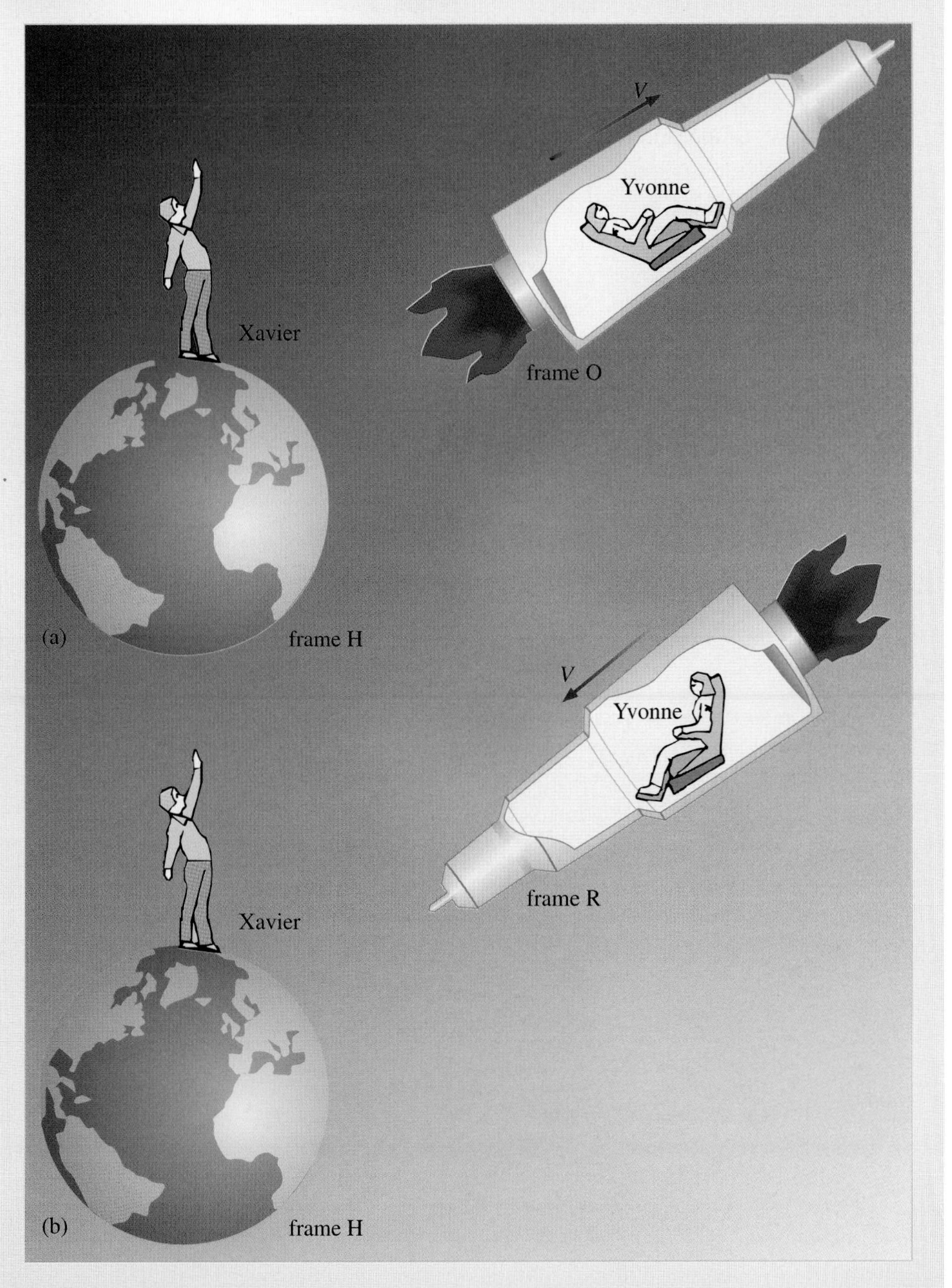

Figure 4.33 (a) Xavier stays at home tied to inertial frame H, while Yvonne travels outwards in inertial frame O at speed V, before (b) returning at the same speed in inertial frame R.

Figure 4.34 When Yvonne returns from her rocket trip, she is substantially younger than her stay-at-home twin Xavier.

Incidentally, the Doppler formula for light is

$$\frac{f}{f_{\text{away}}} = \frac{f_{\text{towards}}}{f} = \sqrt{\frac{c+V}{c-V}}$$

where V is the speed of recession or approach.

It is often claimed that the twin 'paradox' cannot be solved using special relativity. During acceleration, deceleration and turnaround, Yvonne is not in anything like an inertial frame, and that motion is not covered by special relativity. The acceleration and deceleration may be neglected for numerical purposes, but the turnaround is crucial. Let us be clear about it: Yvonne knows that the turnaround is non-inertial because the ship's motive power has to be turned on, as a result of which measurable g-forces act. In contrast, Xavier experiences no such forces. So the turnaround beaks the symmetry: Yvonne and Xavier experience asymmetric states of motion due to the turnaround.

Symmetry between the twins is lost, therefore, but is it the g-forces that give the numerical difference in ages? No, and this is easy to see. For just suppose that Xavier and Yvonne were two of triplets, and the third sibling, Zebedee, sets off along with Yvonne in an identical ship. But when Yvonne turns around, Zebedee continues for an identical length again; then turns around using the identical manoeuvre that Yvonne did, and comes home. Now if only the g-forces gave the numerical differences in age, Yvonne and Zebedee would be the same age, but they are not, and Zebedee is the younger owing to his longer journey.

It is sometimes said that the *twin paradox* is really a problem in general relativity, since that is the theory that deals with non-inertial frames. Technically this is true, and no question of symmetry between frames of reference would ever arise if the problem were done this way. But an examination of the solution in general relativity confirms that the actual age change during the turnaround manoeuvre can be made small, and that most of the solution reduces to the special relativity formulation anyway. In other words, general relativity is necessary to justify strictly that the accelerations break the symmetry, and to calculate the age changes during those periods. But these changes can be made quite small, and special relativity gives essentially the correct numerical answer.

Question 4.8 In Question 4.7, you calculated the time taken (according to its astronaut) for a spaceship to travel to Alpha Centauri (4.2 light years distant from Earth) at a constant speed of $0.9c$ relative to the Earth. Now suppose that the same astronaut turns his spaceship around as soon as he arrives at Alpha Centauri and returns to Earth, travelling at $0.9c$ relative to the Earth throughout its journey. How long would the round trip take:

(a) according to the astronaut?

(b) according to the astronaut's twin sister, who stayed on Earth throughout the voyage? ■

5.5 Muon decay

In Section 1, you were introduced to the strange case of the muons for which time seemed to slow down. With the real physical effect of time dilation, we now have an explanation for this phenomenon.

In a frame of reference in which they are at rest, muons have a mean lifetime of 2.2×10^{-6} s against spontaneous decay. An observer on the surface of the Earth will find such particles moving downwards with a uniform speed of, say, $0.98c$. So by time dilation, the Earth-bound observer will find that, in his frame of reference, the moving muons have a mean lifetime that is five times longer — that is, 1.1×10^{-5} s. Consequently, on average, the muons will penetrate towards the surface a good bit further before decaying. A test of special relativity, therefore, is to count cosmic ray muons at various altitudes and see if the count agrees with the relativistic prediction. Indeed it does.

Question 4.9 A certain mountain is 4050 m high, as measured by an observer who is stationary with respect to the mountain. (a) What is the height of the mountain as measured in the frame of reference rushing earthwards with the cosmic ray muons travelling at $0.98c$? (b) How much time elapses for the muons in travelling from the top of the mountain to sea-level? ■

6 The Lorentz transformation

It is clear that the Galilean transformation of time and length is in disagreement with the startling predictions of time dilation and Lorentz contraction. This is not surprising, since the Galilean velocity transformation is in direct contradiction to the behaviour of light measurements in different inertial frames, and is therefore in conflict with Einstein's second postulate.

The time is now ripe to return to the central problem of special relativity posed in Section 3.1.

What is the correct coordinate transformation that links two inertial frames, A and B, in standard configuration?

The effects of time dilation and Lorentz contraction were deduced by making direct use of Einstein's postulates. Perhaps the correct coordinate transformation can be formulated using these results, thus ensuring that Einstein's postulates are 'built in' to the new coordinate transformation.

The correct coordinate transformation was deduced by Einstein in the famous 1905 paper, and is of such importance that it is given a special name — *the Lorentz transformation*. As noted earlier, the Lorentz transformation was first suggested in 1895 by H. A. Lorentz in connection with his attempt to modify the laws of Maxwell's electromagnetic theory. It incorporates three basic effects — Lorentz contraction and time dilation, which you have already met, and the relativity of simultaneity, which will be discussed shortly. The Lorentz transformation is the key to a correct understanding of the way length and time measurements transform from one inertial frame to another. It also leads to many other predictions of special relativity, extending beyond our views of space and time to change our views of mass and energy radically, and link electrical and magnetic phenomena.

6.1 The coordinate transformation of special relativity

Consider two frames of reference A and B in standard configuration. (Refer back to Figure 4.12 if necessary to make sure that you understand the meaning of standard configuration.) As usual, the two clocks at the origins are synchronized at $t' = t = 0$, at the point where the origins coincide. Shortly after the origins coincide, an event occurs at a coordinate $(x, 0, 0, t)$, as measured by observer A in frame of reference A, and $(x', 0, 0, t')$, as measured by observer B in frame of reference B. The task is to find the correct relationships between x and x', y and y', z and z', and t and t'.

Well, since there is no Lorentz contraction of length measurements in directions perpendicular to the relative motion of frames of reference A and B, the relationships between y and y' and between z and z' remain simple and straightforward:

$$y = y' \tag{4.15a}$$

$$z = z'. \tag{4.15b}$$

These results, of course, are also identical with the equivalent Galilean transformation equations.

Now, what about the x-coordinate? Figure 4.35 shows the situation after time t (measured in frame of reference A), when the event has coordinate x in frame of reference A, and x' in frame of reference B. According to the observer in frame of reference A, the origins have now separated by distance Vt.

To an observer in frame of reference A, the coordinate x' (this is the distance of the event from the origin, as measured by an observer B in frame of reference B) will appear contracted by the factor $\sqrt{1 - V^2/c^2}$. This is due to the Lorentz contraction effect discussed in Section 5.3. The position of the event along the x-axis, which A measures as x, can therefore be written as follows:

$$x = Vt + x'\sqrt{1 - \frac{V^2}{c^2}}.$$

Rearranging,

$$x'\sqrt{1 - \frac{V^2}{c^2}} = x - Vt$$

and so

$$x' = \frac{x - Vt}{\sqrt{1 - V^2/c^2}}. \tag{4.15c}$$

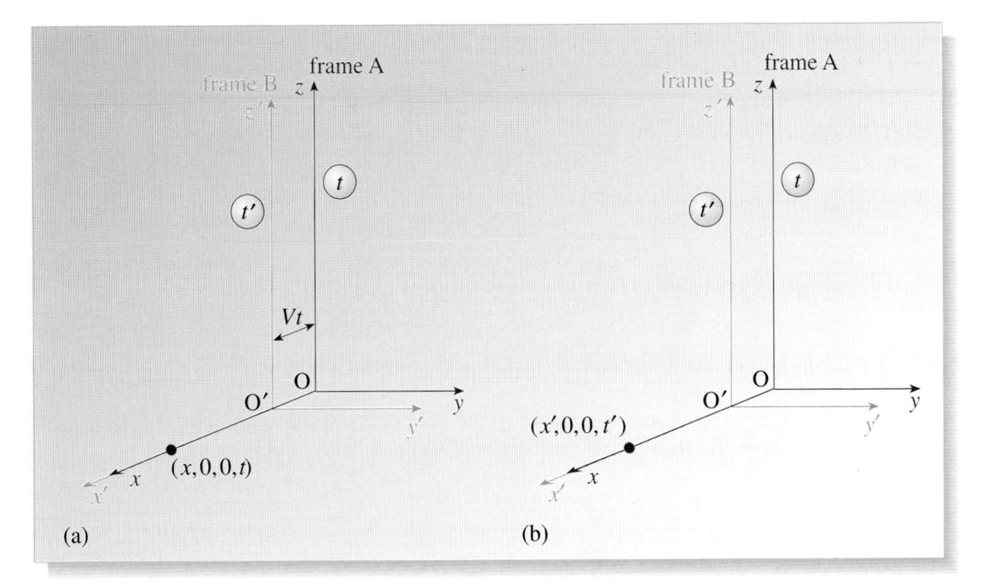

Figure 4.35 (a) The situation after time t (measured in frame of reference A), when the event $(x, 0, 0, t)$ occurs. The origins have separated by distance Vt. (b) The coordinate x' of the event from the origin, as measured in frame of reference B will appear contracted by the factor $\sqrt{1 - V^2/c^2}$ to an observer in A.

Note that Equation 4.15c is identical with the equivalent Galilean transformation (Equation 4.3), except for the factor $\sqrt{1 - V^2/c^2}$ in the denominator. This means that for velocities that are very small compared with c (i.e. $V \ll c$), the transformation equation becomes equivalent to the Galilean transformation of Newtonian mechanics.

In view of the time dilation effect, you would also expect a different result from the simple $t' = t$ relationship of the Galilean transformation. The algebra is a bit tedious and so will be omitted, but without using any new ideas it is not too difficult to show that the correct transformation is:

$$t' = \frac{t - Vx/c^2}{\sqrt{1 - V^2/c^2}} \tag{4.16}$$

which again reduces to the Galilean result $t = t'$ for $V \ll c$, as you will see in Section 6.5.

Bringing all these results together (Equations 4.15a, 4.15b, 4.15c and 4.16), the full Lorentz transformation that correctly links the coordinates of an event in frame A to those of the same event in frame B (with A and B in standard configuration) can now be written down. This provides the solution to the central problem of relativity.

The Lorentz transformation

$$\begin{aligned} x' &= \frac{x - Vt}{\sqrt{1 - V^2/c^2}} \\ y' &= y \\ z' &= z \\ t' &= \frac{t - Vx/c^2}{\sqrt{1 - V^2/c^2}}. \end{aligned} \tag{4.17a}$$

The factor $1/\sqrt{1-V^2/c^2}$ occurs so often in special relativity that it is often given the symbol γ (gamma), and is referred to as the **Lorentz factor**. Using this symbol, the algebra in special relativity can often be simplified. The Lorentz transformation equations are written out using the γ notation as follows:

$$\begin{aligned} x' &= \gamma(x - Vt) \\ y' &= y \\ z' &= z \\ t' &= \gamma\left(t - \frac{Vx}{c^2}\right) \end{aligned} \tag{4.17b}$$

It is clear that the Lorentz transformation is very different from the Galilean transformation of Section 3.2. The new coordinate transformation contains c, the speed of light — a factor that was absent from the Galilean coordinate transformation. The presence of c is not surprising since the postulate of the constancy of the speed of light plays a fundamental role in special relativity. However, there are other differences. In particular, note the greater complexity of the equations, linking x' and x, and t' and t, and the fact that $1/\sqrt{1-V^2/c^2}$ appears in both of them.

The great advantage of knowing the Lorentz coordinate transformation is that it can be used to provide the correct answers to questions involving length and time measurements in a convenient and straightforward manner. Effects like time dilation and length contraction are *built-in* to the Lorentz transformation, so one no longer has to worry about being misled by intuition or common sense!

6.2 Causality and simultaneity

One of the most dramatic consequences of the relative nature of time is that the *order* in which two events occur can, in certain circumstances, depend on the frame of reference of the observer. To see that this effect does indeed follow from the Lorentz transformation, try the following question.

Question 4.10 Suppose an observer in frame of reference A assigns the following coordinates to two events labelled 1 and 2:

Event 1: $x_1 = 12 \times 10^8\,\text{m}$, $y_1 = 0$, $z_1 = 0$, $t_1 = 7\,\text{s}$;

Event 2: $x_2 = 30 \times 10^8\,\text{m}$, $y_2 = 0$, $z_2 = 0$, $t_2 = 11\,\text{s}$.

Work out the coordinates of the same events as observed from a frame of reference B, in standard configuration with frame of reference A, when the relative speed of B with respect to A is $V = 4c/5$. ■

The result for Question 4.10 has shown that in frame of reference B, Event 1 occurs *after* Event 2, whereas in frame of reference A, Event 1 happens *before* Event 2. In other words, the order of the two events in frame of reference B is opposite to the order in which they occurred in frame of reference A.

You would be right in thinking that such a change in ordering might have some very disturbing consequences. Suppose, for instance that Event 1 represents pushing the plunger of a detonator, and Event 2 was the ensuing explosion. This would mean that an observer in frame of reference B would observe the explosion take place *before* the plunger had been pressed; effect would precede cause! Fortunately, such a situation can never arise. If you look at the coordinates of the two events in frame of reference A, you can see that the distance $x_2 - x_1$ between them is greater than the distance that light can travel during the period of time that separates them, $c(t_2 - t_1)$. In mathematical terms, $x_2 - x_1 > c(t_2 - t_1)$. An analogous result is true in frame of reference B.

Provided that nothing travels faster than light, it is not possible, *in either frame of reference*, for the two events to be connected by *any* kind of signal. Such a signal simply would not be able to travel fast enough to link the two events in either frame of reference. Therefore, with this proviso, Event 1 cannot be the *cause* of Event 2 in frame of reference A, nor can Event 2 have been the *cause* of Event 1 in frame of reference B. In fact, provided $V < c$, and provided no signal travels faster than light, it can be shown that Event 1 could not have caused Event 2 or vice versa *in any frame of reference*.

If, on the other hand, the separation of Event 1 and Event 2 had been such that a signal travelling at c or slower could have passed between them (so that Event 1 *could* have caused Event 2), the Lorentz transformation can be used to show that Event 1 will precede Event 2 in *all* inertial frames.

That cause should precede effect is called **causality**, and special relativity always preserves this logical relationship, provided it is assumed that c is the maximum speed at which a signal can travel.

There is plenty of experimental evidence that c is indeed a universal 'speed limit'. For instance, experiments in which electrons are accelerated through ever greater potential differences show that there is a limiting speed that the electrons can attain, however great the potential difference. Turning the argument around, if special relativity is true, then causality *requires* that the speed of light is the ultimate speed attainable.

There is a 'speed limit' in the Universe for the propagation of any material object or information: it is the speed of light in a vacuum, c.

Another potentially disturbing consequence of the relative nature of time is known as the **relativity of simultaneity**. This means that two spatially separated events which are simultaneous according to one observer (i.e. happening at the same time in his frame of reference) may occur at two different times according to another observer, provided the second observer is moving relative to the first. This is illustrated by the following question.

Question 4.11 Suppose that an observer in frame of reference A observes two *simultaneous*, but spatially separated, events to which she assigns the following coordinates:

$$\text{Event 1: } x_1 = 15 \times 10^8\,\text{m},\ y_1 = 0,\ z_1 = 0,\ t_1 = 8\,\text{s};$$

$$\text{Event 2: } x_2 = 21 \times 10^8\,\text{m},\ y_2 = 0,\ z_2 = 0,\ t_2 = 8\,\text{s}.$$

What are the times of occurrence of these two events as observed from frame of reference B, in standard configuration with frame of reference A, and moving at a speed of $c/2$ relative to A? ■

Even though Events 1 and 2 are simultaneous in time according to observer A, they are spatially separated in her frame of reference and so cannot be causally connected. Once again, causality is not violated.

6.3 The velocity transformation of special relativity

The Galilean coordinate transformation failed to solve the central problem of relativity because it led to incorrect predictions about the way in which two different observers would measure the velocity of an object. In particular, you saw in Section 2 that if the Galilean transformation were correct, then contrary to experiment, the speed of light would *not* be the same in all inertial frames. Since the postulate of the constancy of the speed of light is *built in*, the Lorentz transformations will not suffer from this difficulty. In this subsection, the correct rule for transforming velocity components between different frames of reference will be deduced. You will also see exactly how special relativity guarantees that any light beam will have the same speed relative to any observer in an inertial frame.

Suppose frames of reference A and B are in the standard configuration, with frame of reference B moving at speed V along the positive x-axis of frame of reference A. A particle travels at constant velocity v_x along the x-axis of frame of reference A. What is the velocity of this particle, as observed in frame of reference B, according to the special theory of relativity?

Clearly the constant velocity of the particle in frame of reference A is simply

$$v_x = \frac{\Delta x}{\Delta t}. \tag{4.18}$$

If the particle travels at constant velocity in frame of reference A, it experiences no forces. It must therefore travel at constant velocity in any inertial frame. So the velocity is also constant in frame of reference B and points along the x'-axis. It is given by

$$v'_x = \frac{\Delta x'}{\Delta t'}. \tag{4.19}$$

Now, the Lorentz transformations (Equation 4.17) imply that

$$\Delta x' = \gamma(\Delta x - V\Delta t)$$

$$\Delta t' = \gamma\left(\Delta t - \frac{V\Delta x}{c^2}\right)$$

where $\Delta x = x_2 - x_1$ and $\Delta t = t_2 - t_1$, and the algebra has been simplified by using γ. So Equation 4.19 becomes

$$v'_x = \frac{\gamma(\Delta x - V\Delta t)}{\gamma(\Delta t - V\Delta x / c^2)}$$

Cancelling γ, and dividing both the numerator and the denominator on the right-hand side by Δt, yields

$$v'_x = \frac{\left(\dfrac{\Delta x}{\Delta t}\right) - V}{\left(1 - \dfrac{V}{c^2} \times \dfrac{\Delta x}{\Delta t}\right)}.$$

Using the initial expression for v_x (Equation 4.18), gives the final result

$$v'_x = \frac{v_x - V}{1 - Vv_x/c^2} \qquad (4.20)$$

● (a) What is the result of setting $V = 0$ in Equation 4.20, and what does this imply?

(b) What is the result of setting $v_x = c$ in Equation 4.20, and what does this imply?

❍ (a) Setting $V = 0$ gives $v'_x = v_x$. So if the frames of reference are not in relative motion, the same velocity is measured in both.

(b) Setting $v_x = c$, we have

$$v'_x = \frac{c - V}{1 - V/c} = \frac{c - V}{(c - V)/c} = c\,.$$

So if a pulse of light moves with speed c in inertial frame of reference A, it must also move with speed c in inertial frame of reference B. (This is the second of Einstein's postulates of special relativity.) ■

An interesting application of the velocity transformation equation (Equation 4.20) is to demonstrate that no matter how fast two objects travel towards each other, the magnitude of their relative velocity (the speed of one measured by the other) cannot exceed c. The following question should convince you how this remarkable property comes about from the velocity transformation equations of special relativity.

Question 4.12 An astronaut sitting in the observation tower of a space station sees two spaceships X and Y approaching him from opposite directions at high speed (Figure 4.36). He measures the approach speed of both X and Y to be $3c/4$. What is the speed of spaceship Y as measured by X? (The trick here lies in being able to describe this situation in terms of two frames of reference in standard configuration. If you have any difficulty with this, read the *first* part of the answer. You should then be able to carry out the necessary calculation.) ■

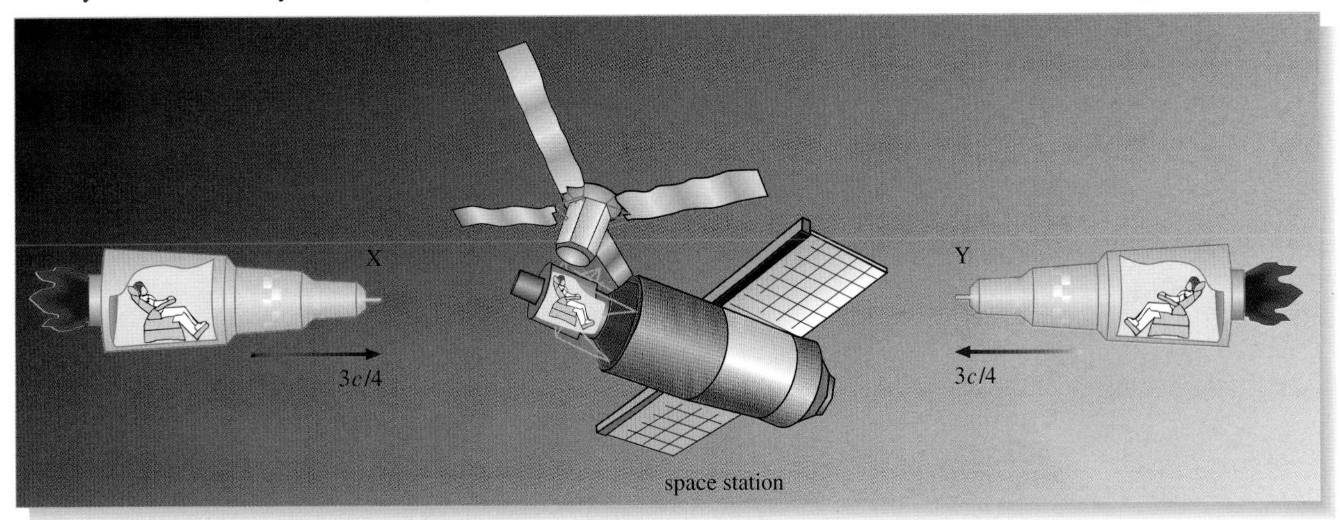

Figure 4.36 The situation described in Question 4.12.

6.5 Special relativity at low speeds

You have just seen that according to the velocity transformation law of special relativity, velocities are not *additive* in the simple Galilean sense. But if V and v_x are both much less than the speed of light, then the term Vv_x/c^2 in Equation 4.20 is very much less than one and therefore can be neglected. The velocity transformation expression then becomes indistinguishable from the Galilean rule (Equation 4.4) that everyday experience confirms:

$$v'_x = v_x - V.$$

This is exactly the relationship obtained in Section 2 when discussing the relative motion of a car travelling along a road. The important point to stress is that the velocity transformation law of special relativity is the *correct* form: it is true for *all* speeds from everyday speeds right up to the speed of light. The 'common-sense' Galilean velocity transformation is *incorrect*. It is merely a very good approximation for the correct law that can be safely used at low speeds.

In everyday life, you almost never encounter objects that move at speeds greater than $10^{-5}c$ relative to you. ($10^{-5}c$ is very roughly ten times the speed of sound.) Under such conditions, it is clearly reasonable to make the following approximations to the Lorentz transformation:

$$x' = \frac{x - Vt}{\sqrt{1 - V^2/c^2}} \approx x - Vt$$

$$y' = y$$

$$z' = z$$

$$t' = \frac{t - Vx/c^2}{\sqrt{1 - V^2/c^2}} \approx t - \frac{Vx}{c^2}.$$

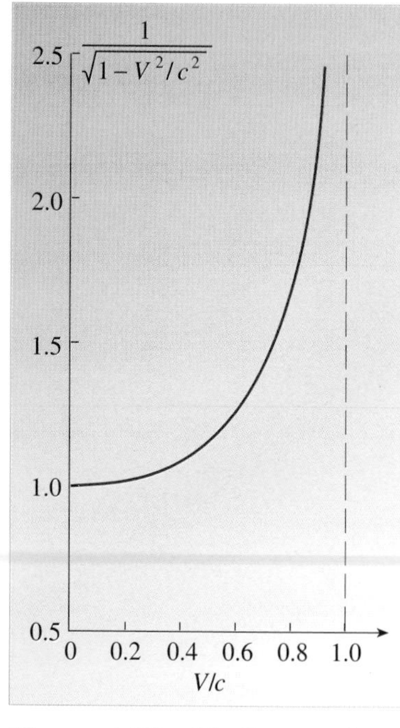

Figure 4.37 The Lorentz factor $1/\sqrt{1 - V^2/c^2}$ plotted against V/c.

As you can see, in this low-speed approximation, the three position equations immediately take the form of the first three equations in the Galilean transformation (Equation 4.3). As for the time equation, the distance x can be thought of as a typical distance involved in everyday observations, usually no more than a few kilometres. This implies that if $V \ll c$ then the term Vx/c^2 is negligible and t' becomes equal to t. So, for everyday speeds and distances, the Lorentz transformation equations become indistinguishable from the Galilean transformation equations. Of course, it is the Lorentz transformation that is correct at *all* speeds, including those involved in everyday experience. But in practice, since it approximates so closely to the simpler Galilean transformation, it is more convenient to use the latter in dealing with ordinary macroscopic phenomena.

It is only in situations that involve high speeds ($V \sim c$) that the ubiquitous Lorentz factor $\gamma = 1/\sqrt{1 - V^2/c^2}$ becomes significantly different from 1. Figure 4.37 shows the way in which the Lorentz factor increases as V is increased: as V tends to c, the Lorentz factor tend to ∞. You should be able to see that $1/\sqrt{1 - V^2/c^2}$ is only significantly different from 1 when V is an appreciable fraction of the speed of light.

7 Relativistic physics

Once Einstein had formulated the principle of relativity, and had shown that the Lorentz transformation was the correct coordinate transformation that linked inertial frames, it became possible to test all of the so-called 'laws' of physics, to see whether or not they took the same form in all inertial frames. Any so-called 'law' that did not retain its form when transformed by means of the Lorentz transformation could not really be a law of physics at all.

Using this simple test, Einstein found that Maxwell's laws of electrodynamics were basically sound. After reinterpreting some of Maxwell's ideas and disposing of the ether, Einstein found that the Maxwell equations did preserve their form under Lorentz transformation. Since the Maxwell equations did take the same form in all inertial frames, they were *possible* laws of physics, and this will be considered further in Section 7.2. However, other parts of physics did not fare so well.

Mechanics, in particular, had to be modified to a considerable extent to ensure that its laws took the same form in all inertial frames. Newton's second law, in the form $\boldsymbol{F} = m\boldsymbol{a}$, is not a law at all in the relativistic sense, nor is Newton's law of gravitation. Section 7.1 describes the changes required to bring mechanics into line with the demands of the principle of relativity.

7.1 Relativistic mechanics

You have already seen, in *Predicting motion* Chapter 3, that in relativistic mechanics, the **relativistic momentum $\boldsymbol{p}$** of a material particle of mass m and velocity $\boldsymbol{v}$ is defined by the following vector equation:

$$\boldsymbol{p} = \frac{m\boldsymbol{v}}{\sqrt{1 - v^2/c^2}} \qquad (PM\text{, Eqn 3.20})$$

The mass m that appears in this formula is just the usual mass of the particle — in other words the mass that would be determined in a low-speed experiment. However, just as great care had to be taken over the definition of length and time measurements in Section 5, so equal care should be taken about defining the mass of a moving particle. To be really precise, m is called the *rest mass* of the particle and is defined in the following way:

The **rest mass** m *is* the mass as measured by an observer *relative to whom the particle is at rest.*

Because of the way it is defined, the rest mass is totally independent of the speed of the particle. Although the particle may be moving with a very high speed relative to you, the rest mass of the particle is still the mass that would be found by some other observer relative to whom the particle was at rest. With this definition, the rest mass is *not* a relative quantity: it has the same value for all observers and is independent of the speed of the particle.

The definition of relativistic momentum given in *Predicting motion* Equation 3.20 can be used to help define the concept of relativistic force. As usual, things must be arranged so that when v/c is small the Newtonian results are recovered, since they are known to be accurate at low speed. Bearing this in mind, a reasonable definition of **relativistic force** is:

$$\boldsymbol{F} = \frac{\mathrm{d}\boldsymbol{p}}{\mathrm{d}t} = \frac{\mathrm{d}}{\mathrm{d}t}\left(\frac{m\boldsymbol{v}}{\sqrt{1 - v^2/c^2}}\right). \qquad (4.21)$$

This says that the force acting on a particle is equal to the rate of change of the (relativistic) momentum of the particle, an idea you should recall from *Predicting motion*. Equation 4.21 is the relativistic analogue of Newton's second law.

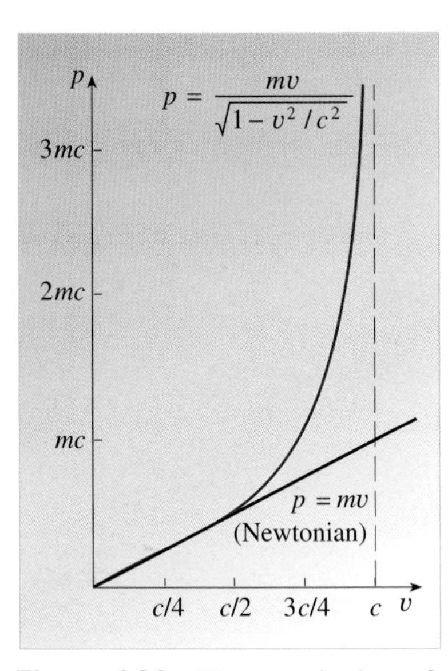

Figure 4.38 The magnitude p of the relativistic momentum plotted against the speed v. Near $v = c$, even large increases in the relativistic momentum have only a small effect on the speed. (The Newtonian momentum magnitude as a function of v is also included for comparison.)

It follows from Equation 4.21 that if a constant force acts on a particle, the (relativistic) momentum of the particle must increase at a constant rate. In Newtonian mechanics, this would mean that the velocity of the particle also increased at a constant rate, since velocity and momentum are directly proportional in Newtonian mechanics. However, *Predicting motion* Equation 3.20 implies that velocity and momentum are not directly proportional in relativistic mechanics.

Figure 4.38 shows the way in which the speed of a particle moving in a straight line and the magnitude of its momentum are related in relativistic mechanics. As you can see, increasing the momentum of the particle by a given amount has a diminishing influence on the speed of the particle as the speed increases. Applying a constant force to a particle for a very long time would certainly give it a large relativistic momentum, but that does not mean that the particle would ever attain the speed of light. As you can see from Figure 4.38, the speed of light is a limit that no material particle of non-zero rest mass can ever quite reach, however large its momentum may be.

In *Predicting motion* Chapter 3, it was stated that the (total) **relativistic energy** of a particle of mass m travelling at speed v is given by

$$E_{\text{tot}} = \frac{mc^2}{\sqrt{1-v^2/c^2}}. \qquad (PM\text{, Eqn 3.22})$$

As noted there, the total energy of a particle is the sum of its relativistic translational kinetic energy and its **mass energy**, where the mass energy is given by perhaps the most famous equation in all of physics (see Figure 4.39):

$$E_{\text{mass}} = mc^2. \qquad (PM\text{, Eqn 3.21})$$

Figure 4.39 The most famous equation in all of physics (Gary Larson/© Chronicle Features/ Far Works Inc.).

This leads to the definition of **relativistic translational kinetic energy** as

$$E_{\text{trans}} = \frac{mc^2}{\sqrt{1 - v^2/c^2}} - mc^2. \qquad (PM, \text{Eqn } 3.23)$$

Question 4.13 Using Equations 3.20 and 3.22 from *Predicting motion*, verify the following relationship between E_{tot} (the relativistic energy) and p (the magnitude of the relativistic momentum):

$$E_{\text{tot}}^2 = p^2c^2 + m^2c^4. \quad \blacksquare \qquad (4.22)$$

Equation 4.22 is more than just an interesting relationship between relativistic energy and relativistic momentum. It is an important result in its own right, because it enables us to extend the ideas of relativistic mechanics to cover particles such as photons ('particles' of light) that always travel at speed c. From *Predicting motion* Equation 3.20, you might think that such particles have infinite momentum, but all particles that travel at the speed of light have zero rest mass, and are not subject to this equation. However, from Equation 4.22, it is clear that the relativistic energy of such particles is simply related to the magnitude of their relativistic momentum by the formula

$$E = cp \qquad (4.23)$$

which is only true in the case $m = 0$.

Question 4.14 (a) What is the mass energy of a proton? ($m_{\text{p}} = 1.67 \times 10^{-27}$ kg)

(b) At what speed must a proton travel if its relativistic translational kinetic energy is equal to its mass energy?

(c) What is the relativistic energy of a proton that has a relativistic translational kinetic energy three times as great as its mass energy? ■

7.2 Electromagnetism

As noted earlier, electromagnetism is the real key to special relativity and this is where Einstein's first paper on the subject, in 1905, was focused. How should electromagnetism be considered, bearing in mind the results of special relativity?

As noted in Chapter 2, the electric and magnetic fields of electromagnetic radiation are perpendicular to each other. Consequently, it is best to use vector notation to understand what is happening. Suppose that we have two inertial frames, A and B, not necessarily in standard configuration. Suppose also that observers in frame of reference A find that frame of reference B is moving past it with constant velocity $\boldsymbol{v}$. In frame of reference A there is an electric field $\boldsymbol{\mathscr{E}}(x, y, z, t)$, and a magnetic field $\boldsymbol{B}(x, y, z, t)$ at each space–time point, and in frame of reference B there is also an electric field $\boldsymbol{\mathscr{E}}'(x', y', z', t')$, and a magnetic field $\boldsymbol{B}'(x', y', z', t')$ at each space–time point.

You might expect that the vector components will be mixed up by a Lorentz transformation, and indeed they are. The result is easiest to understand by referring to the fields in terms of vectors that are in directions parallel to (symbol '$\parallel$') and perpendicular to (symbol '$\perp$') the velocity vector $\boldsymbol{v}$ in a given frame of reference. A Lorentz transformation of the fields results in the equations

$$\mathscr{E}'_{\parallel} = \mathscr{E}_{\parallel}$$

$$\boldsymbol{B}'_{\parallel} = \boldsymbol{B}_{\parallel}$$

$$\mathscr{E}'_{\perp} = \gamma(\mathscr{E}_{\perp} + \boldsymbol{v} \times \boldsymbol{B}_{\perp})$$

$$\boldsymbol{B}'_{\perp} = \gamma\left(\boldsymbol{B}_{\perp} - \frac{\boldsymbol{v} \times \mathscr{E}_{\perp}}{c^2}\right). \qquad (4.24)$$

In spite of a formidable appearance, these equations are not difficult to understand. Parallel to the velocity, the components of electric and magnetic fields do not change. The components perpendicular to $\boldsymbol{v}$ are quite different. The most striking feature is that the electric and magnetic fields of the original frame of reference are mixed together to create the fields in the new frame of reference. In general, there is no natural separation of the electric and magnetic fields therefore. This is very important, and represents an important triumph of relativity theory.

The splitting of an electromagnetic field into electric and magnetic components is only a function of the frame of reference. Fundamentally, electric and magnetic fields have no independent existence.

● What does Equation 4.24 predict will be seen in a frame of reference (B) that is moving relative to a frame of reference (A) in which *only* a magnetic field exists?

❍ There will be both an electric *and* a magnetic field in frame of reference B. In particular, it predicts that $\mathscr{E}'_{\parallel} = \mathbf{0}$, $\mathscr{E}'_{\perp} = \gamma\boldsymbol{v} \times \boldsymbol{B}_{\perp}$, $\boldsymbol{B}'_{\parallel} = \boldsymbol{B}_{\parallel}$ and $\boldsymbol{B}'_{\perp} = \gamma\boldsymbol{B}_{\perp}$. ■

So, beginning with a static magnetic field, as from a permanent magnet, a Lorentz transformation to a moving frame of reference will yield an electric field in the new frame of reference. It is this field that drives the charges in a conductor, resulting in a current, and so induction by motion (Chapter 1) is a consequence of the Lorentz transformation. Moreover, relativity requires what was said in Chapter 1: moving a magnet past a conducting loop and moving a loop past a magnet describe the same physical phenomenon, but from the point of view of different frames of reference.

An important point to notice, though, is that the Lorentz transformation for electric and magnetic fields does not provide the *complete* solution to electromagnetic induction. You saw, in Chapter 1 Section 2.2, that an electric field can also be induced by a *changing* magnetic field, such as one whose strength varies with time. Although this effect is 'built in' to Faraday's law, it is *not* a consequence of special relativity alone. Faraday's law is notable in physics in describing *two* effects — induction by motion *and* induction by a changing field — which arise from different processes.

7.3 Conclusion

Einstein's special theory of relativity is one of the greatest achievements of physics. The theory holds enormous appeal for a wide audience, and many people who have never studied physics at least know of its existence. To physicists, the appeal of special relativity comes largely from its logic, internal consistency and agreement with experiment. The theory demands that we throw away our prejudices, and put our trust in a few simple experiments and our own powers of reasoning. Once we have accepted Einstein's postulates, all else follows inevitably; we are carried along on an irresistible tidal wave of logic. If we sometimes feel uncomfortable with the results, that's just too bad; they are unavoidable once the postulates have been accepted, and they are supported by a wealth of experimental evidence.

Quite apart from its intellectual appeal, relativity also has great practical value. The principle of relativity and the Lorentz transformation provide a kind of framework into which would-be physical theories must fit. This gives a way of 'testing' physical theories without recourse to experiment. For this reason, the special theory of relativity is one of the most useful tools available to the theoretical physicist.

Finally, for a wider audience, relativity has the appeal of the exotic. Its results are sufficiently strange to be intrinsically interesting, and yet not so bizarre as to be totally beyond belief. When coming to grips with relativity for the first time, you expect to have your mental powers stretched by the demands of Einstein's theory. In studying this chapter, you will, we hope, have experienced that mental extension, and benefited from it. Perhaps you have even enjoyed the experience!

> Nothing puzzles me more than time and space; and yet nothing troubles me less, as I never think about them.
>
> Charles Lamb.

8 Closing items

8.1 Summary

1. An event is a physical occurrence that occupies a point in space and an instant in time. If an observer finds that two events occur together in space *and* time, then so will *all* other observers.
2. A frame of reference is some system that makes it possible to assign to any event four numbers that distinguish it from all other events. Three of the numbers fix the position of the event relative to a point called the origin, and these numbers are called the position coordinates of the event. The fourth number is called the time coordinate, and fixes the time at which the event occurred in this frame of reference.
3. An inertial frame is a frame of reference in which Newton's first law is valid. Any frame of reference that moves with constant velocity relative to an inertial frame will also be an inertial frame.
4. An observer is someone who consistently uses a specific frame of reference to assign four coordinates to every event that he or she observes. Any observer who sets out to write down the laws of physics will naturally express those laws in terms of the coordinates of his/her own frame of reference.
5. The principle of relativity states that the laws of physics can be written in the same form in all inertial frames.
6. Given the coordinates of an event in one frame of reference, a coordinate transformation is a set of equations that makes it possible to work out the coordinates of that same event as viewed from another frame of reference. By using a coordinate transformation, it is possible to determine the results of experiments in one frame of reference simply by transforming results obtained in another frame of reference.
7. If the coordinate transformation between any two inertial frames is known, it is possible to check whether an equation expressed in terms of the coordinates of one inertial frame can be written in the same form in all other inertial frames. If it can, then, according to the principle of relativity, that equation is a possible law of physics.
8. The central problem of relativity is to determine the coordinate transformation that links two inertial frames, A and B, that are in standard configuration.

9 The Galilean coordinate transformation is a plausible solution to the central problem of relativity. However, it leads to a prediction about the transformation of velocities that is in conflict with experiment. Thus, although the Galilean transformation is intuitively appealing and agrees with everyday experience, it is NOT the correct solution to the central problem of relativity.

10 There is no experimental evidence to suggest that light behaves like particles that obey the laws of Newtonian mechanics with speed c relative to their source. Furthermore, the Michelson–Morley experiment, which investigated the existence of an ether through which light was supposed to propagate through space, produced a null result.

11 The velocity of light in a vacuum is always c relative to the detector, and so all observers measure the speed of light as c.

12 Einstein was able to use two postulates — the *principle of relativity* and the *principle of the constancy of the speed of light* — to formulate the theory of special relativity, which gave a correct method for relating observations made in different inertial frames.

13 According to special relativity, the duration of an interval of time is a relative quantity. The rate at which a clock ticks depends on the frame of reference in which it is measured ('moving clocks run slow'):

$$\Delta T = \frac{\Delta T_0}{\sqrt{1 - V^2/c^2}} \tag{4.8}$$

This effect is called *time dilation.*

14 Length is also a relative quantity: The length of a rod depends on the frame of reference in which it is measured. 'Moving rods contract in the direction of motion': $L = L_0\sqrt{1 - V^2/c^2}$ (Eqn 4.13). This effect is called *Lorentz contraction.*

15 For small values of V/c, time dilation and Lorentz contraction become negligible; that is, $\Delta T = \Delta T_0$ and $L = L_0$, as in the Galilean transformation of lengths and time (in other words, the expectations of Newtonian ideas of space and time).

16 Einstein used the two postulates of special relativity to deduce the *Lorentz coordinate transformation*, which is the correct transformation linking the space–time coordinates of an event in two inertial frames in standard configuration.

$$\begin{aligned} x' &= \frac{x - Vt}{\sqrt{1 - V^2/c^2}} \\ y' &= y \\ z' &= z \\ t' &= \frac{t - Vx/c^2}{\sqrt{1 - V^2/c^2}}. \end{aligned} \tag{4.17}$$

For small values of V/c the Lorentz transformation gives approximately the same results as the Galilean coordinate transformation.

17 The use of the Lorentz transformation leads to the following results.

(i) Simultaneity is relative: two spatially separated events that occur simultaneously in one inertial frame are not necessarily simultaneous in other inertial frames.

(ii) The order of events is relative: the order in which two events occur depends on the frame of reference in which they are observed, provided one event could not have caused the other.

18 The velocity transformation of special relativity is such that the speed of light is the same in all inertial frames. For small values of V/c the Lorentz velocity transformation gives approximately the same results as the Galilean velocity transformation.

19 The relativistic momentum of a particle of mass m and velocity $\boldsymbol{v}$ is defined by

$$\boldsymbol{p} = \frac{m\boldsymbol{v}}{\sqrt{1-(v^2/c^2)}} \qquad (PM\text{, Eqn 3.20})$$

and the relativistic energy of a particle of mass m and speed v, is defined by

$$E_{\text{tot}} = \frac{mc^2}{\sqrt{1-(v^2/c^2)}}. \qquad (PM\text{, Eqn 3.22})$$

20 The splitting of an electromagnetic field into electric and magnetic components is a function of the frame of reference. Fundamentally, electric and magnetic fields have no independent existence.

8.2 Achievements

Now that you have completed this chapter, you should be able to:

A1 Explain the meaning of all the newly defined (emboldened) terms introduced in this chapter.

A2 Explain and use the principle of relativity.

A3 Both recognize and use the Galilean coordinate transformation and the Galilean velocity transformation.

A4 Outline the Michelson–Morley experiment, and describe what it showed.

A5 State and use Einstein's two postulates.

A6 Write down and use formulae describing the time dilation effect and Lorentz contraction.

A7 Recognize and use the Lorentz transformation.

A8 Recognize and use the special relativistic rule for the transformation of the x-component of velocity.

A9 Discuss the concepts of the relativity of simultaneity and causality.

A10 Write down and use the formulae for the relativistic momentum and the relativistic energy of a particle of rest mass m and velocity $\boldsymbol{v}$.

A11 Describe how electromagnetic induction by motion arises naturally as a result of a transformation between frames of reference.

8.3 End-of-chapter questions

Question 4.15 Suppose that two inertial frames, A and B, are in standard configuration with $V = 3c/5$. What are the coordinates in frame of reference B of an event to which an observer in frame of reference A assigns the coordinates $x = 18 \times 10^8$ m, $y = -1$ m, $z = 0$, $t = 10$ s?

Question 4.16 (a) Consider an observer in a spacecraft that is travelling directly away from a distant space station at half the speed of light. Suppose that the space station fires a missile travelling at three-quarters the speed of light (relative to the space station) towards the spacecraft. If the observer in the spacecraft measures the speed of the approaching missile, what value would she obtain?

(b) Suppose that the observer in (a) is flying directly *towards* the space station at half the speed of light. Once again, the space station fires a missile travelling at three-quarters the speed of light towards the spacecraft. If the observer measured the speed of approach of the missile, what result would she obtain? (*Hint* Remember that velocity components can be positive or negative, according to their direction.)

Question 4.17 An elementary particle P of mass m_P moves to the right at a speed of $0.95c$, as measured in a laboratory frame of reference. It decays into a neutrino (a particle with zero rest mass) moving to the right and another particle Q of mass m_Q, moving to the left with speed v_Q, as measured in the inertial frame in which P is at rest. (a) How fast does the neutrino move in the laboratory frame of reference? (b) How fast does Q move in the laboratory frame of reference? (Express your answer in terms of v_Q.)

Question 4.18 A (super fast!) runner carrying a horizontal pole of length 20 m, in his frame of reference (B), races towards a tunnel. In the frame of reference (A) in which the tunnel is at rest, the length of the tunnel is 10 m. The speed of the runner is such that, in frame of reference A, the pole length is contracted to 10 m.

(a) Calculate the speed of the runner with respect to the tunnel.

(b) In frame of reference A the arrival of the pole at the far end of the tunnel (Event 1) is simultaneous with the rear end of the pole crossing the threshold of the tunnel (Event 2). Calculate the time interval between Event 1 and Event 2 in the runner's frame of reference (B).

(*Hint* Use the Lorentz transformation for time, and assume that *both* events occur at $t = 0$ in frame of reference A.)

Question 4.19 Triplets take part in the following space-travel experiment. Bob stays at home on Earth, while Bill travels to Barnard's star, 6 light-years away, and Ben travels to the star Tau Ceti, 12 light-years away. Both Bill and Ben travel at a speed $0.8c$, and when they reach their respective destinations, immediately turn around and come home at the same speed. When *all three* triplets are finally back on Earth, how much have they each aged since the experiment began?

Question 4.20 Referring to Equations 4.24, suppose that the two frames of reference are in standard configuration; that is, the relative velocity between the two frames of reference is parallel to the x- and x'-axes, and the velocity vector therefore has components $(v_x, 0, 0)$. Write the parallel and perpendicular $\mathscr{E}$ and $\boldsymbol{B}$ vectors in terms of components, and hence re-write the relationships in Equation 4.24 in terms of the x-, y- and z-components of the electric and magnetic fields. (*Hint* Use the component form of the cross product, and you should end up with six equations, one for each component of $\mathscr{E}$ and $\boldsymbol{B}$.) ■

Chapter 5 Consolidation and skills development

5.1 Introduction

This final chapter has two main aims: first to help you consolidate what you have learned from this book, and second to further develop your problem-solving skills, as introduced in *Predicting motion*.

Section 2 is mainly concerned with consolidation. It provides an overview of Chapters 1 to 4; drawing out some of the major themes that run through those chapters and linking them together.

The examples and problems contained in earlier chapters have given you plenty of opportunity to acquire problem-solving skills, but Section 3 aims to further develop these skills by revising the three-step problem-solving technique introduced in *Predicting motion*. As noted in that book, the technique is no substitute for knowledge and experience, but it does provide a useful framework, which can be used throughout the course and in your subsequent studies.

The remaining sections are devoted to consolidation activities. Section 4 contains a short test of basic skills and knowledge; Section 5 provides you with the opportunity to tackle the interactive questions for this book in the multimedia package; Section 6 invites you to try some of the longer questions contained in the *Physica* package. *Physica* will help you to develop some of the higher-level problem-solving skills that rely on strategy and insight, rather than simple manipulation.

5.2 Overview of Chapters 1 to 4

This section highlights some of the major themes that link the earlier chapters of this book. It is not intended to be a summary of those chapters, and does not seek to replace or repeat the material already contained in the summaries of those four chapters. Indeed, you may find it worthwhile to reread those chapter summaries before proceeding.

The subjects discussed in the first four chapters of this book probably appeared to be only tenuously connected with each other when you first glanced down the table of contents. The material should, however, have given you an appreciation for the underlying unification of the topics of electromagnetism, waves, optics and special relativity.

The first chapter explored the phenomenon of electromagnetic induction, building on the contents of *Static fields and potentials*. To discuss electromagnetic induction, we introduced a new quantity, the *magnetic flux*, ϕ, most concisely expressed as $\phi = \boldsymbol{A} \cdot \boldsymbol{B}$, where the vectors $\boldsymbol{A}$ and $\boldsymbol{B}$ represent, respectively, the area enclosed by a loop, and the magnetic field through the loop. Whenever the magnetic flux through an area changes, there will be an *induced EMF* around the boundary of that area, with magnitude proportional to $\mathrm{d}\phi/\mathrm{d}t$, directed so as to oppose the change that caused it. Since a current will generate a magnetic field, a changing current flow will generate a changing magnetic flux, and hence cause a *self-induced EMF. Inductors* exploit this phenomenon and are used in many everyday devices, such as telephones and radio-tuners.

Gathering together the material from *Static fields and potentials* with that in Chapter 1, leads to Maxwell's equations, which provide a concise description of the relationships between magnetic and electric fields, charge and current. Maxwell's equations predict electromagnetic radiation, and demand that in a vacuum it must always propagate with the fixed speed c. This reveals the basic nature of light: propagating electric and magnetic fields.

In Chapter 2 the properties of waves were explored. Waves are periodic, regularly repeating disturbances which propagate energy. There are two basic types of wave: *longitudinal waves* in which the disturbance is in the direction of propagation, and *transverse waves* in which the disturbance is perpendicular to the direction of propagation. Wave properties are described by the quantities: *period* (or *frequency* or *angular frequency*), *wavelength* (or *angular wavenumber*), *speed*, *amplitude* and *phase*. You should be able to define and interrelate these terms with ease. Waves propagating in two or three dimensions can be analysed using the concept of the *wavefront*. Everywhere along a wavefront the phase of the wave has the same value, and successive wavefronts are separated by one wavelength. The wavefront is always perpendicular to the direction of propagation. Wave phenomena such as *reflection*, *refraction*, *diffraction*, and *interference* can all be understood by applying the principle of *superposition* and the Huygens principle to derive the resultant disturbance when a wave encounters an obstacle or another wave.

Chapter 3 considered the application of some of the wave phenomena introduced in Chapter 2 to visible light. Reflection and refraction are used to manipulate light in optical instruments such as telescopes, microscopes and cameras. Since our vision is the most highly developed human sense, it is unsurprising that complex and sophisticated optical systems were invented long before the nature of light was properly understood. Nor is it surprising that we can use cameras, magnifying glasses and binoculars effectively without understanding how their optical properties result from *Snell's law* and the *lens equation,* though Chapter 3 aims to have provided this foundation.

The *Doppler effect* describes the change in frequency which occurs when there is motion of either the wave source or the observer. For sound waves, motion of either the source or the observer, relative to the air through which the wave propagates, leads to such a frequency shift, as derived in Chapter 2. The speed of light c arises naturally in Maxwell's equations, leading to the nineteenth-century idea that light propagates with speed c through a hypothetical medium called the *ether*. The Michelson–Morley experiment tried to measure the motion of the Earth relative to this ether, and arrived instead at the conclusion that there is no stationary ether through which light propagates. Contrary to nineteenth-century expectations, the speed of light is a fundamental constant: all observers will always measure the speed of light in a vacuum to have the value $c = 3.00 \times 10^8\,\text{m}\,\text{s}^{-1}$.

The fundamental empirical result of Michelson and Morley was followed by Einstein's special theory of relativity, which we derived in Chapter 4. Astonishingly, this truly revolutionary development is founded only upon the constancy of the speed of light, and the *principle of relativity,* which demands that the laws of physics must take the same form in all inertial frames. Perhaps even more amazing is the fact that the mathematical derivation of the *Lorentz transformation* uses only Pythagoras's theorem. This basis, coupled with daring and relentless logic, enabled Einstein to dispense with Newton's idea of absolute time. Since c is the fastest speed with which information can propagate, observations are necessarily affected by the light travel time between an event and the observer. This means that observers in different *frames of reference* might disagree over whether two events are simultaneous or the order in which they occur. In our everyday experience the light travel time is rarely very large, so the concept of simultaneity seems a valid one. However, if c were much smaller, we would be far more aware of special relativistic effects.

By deriving the Lorentz transformation equations, which relate the lengths and times measured by observers in two different inertial frames, Einstein was able to simplify and clarify several issues in different fields of physics. These transformation equations do not bear his name because they had, in fact, already been derived (in a style analogous to a student working backwards from a known answer without understanding the logic) to explain the apparently puzzling Michelson–Morley result. Time dilation and length contraction are now well-tested predictions: without special relativity we would be unable to understand many contemporary experimental results.

Perhaps the most impressive aspect of special relativity lies in the realization that the solution to the mysteries of electromagnetism was to overturn the hitherto unquestioned basis of Newtonian dynamics. A reinterpretation of the connections between superficially unrelated subjects proved to be an immensely powerful approach. Such *unifications* of apparently diverse phenomena are prized by scientists. If many empirical facts can be simply explained with a single logically self-consistent theory, then a genuine advance in understanding has probably been made. As a rule, scientists will seek the simplest possible explanation for phenomena, and will then use the deduced theoretical hypothesis to predict the results of a new experiment or observation. If the prediction is correct, then the theory is accepted provisionally; if the prediction is incorrect, then the theory has failed, and requires refinement. It is worth consciously adopting the policy of trying to identify the most concise underlying principles whenever you study science. It is far easier to remember and apply basic principles, deriving detailed results as required, than it is to memorize all the results you might need.

In this book you have read about one of the great revolutions of twentieth-century physics. In the subsequent books *Quantum physics: an introduction* and *Quantum physics of matter* you will learn about its other great achievement. Contemporary physicists seek to build on this impressive intellectual heritage by understanding how electromagnetism, nuclear forces and gravity might all be described within a single 'Theory of Everything'.

5.3 Problem-solving skills — the importance of diagrams

As noted in *Predicting motion*, problem-solving is an important skill in many subjects, but nowhere more so than in physics, where problems generally involve the formulation of a mathematical model to represent a particular physical situation, the analysis of that model, and the interpretation of the answers that it provides. This section revisits the problem-solving skills introduced in the earlier book, and concentrates in particular on the use of *diagrams* to help solve physics problems.

To recap, the stages in solving a problem are:

Stage I Preparation	Stage II Working	Stage III Checking
Summarize the information given	Work out a plan of attack	Check that the units are correct
Draw a diagram	Do the algebra	Check that the answer is sensible
Write down the equations you may need	Put in the values	Check how the answer would vary for different input

Here are some examples illustrating the problem-solving technique, each of which emphasizes the importance of using diagrams.

Example 5.1

Plane light waves from a laser of wavelength 600 nm are incident on a pair of narrow slits spaced 20 μm apart. A diffraction pattern is observed on a screen placed 2.0 m away from the slits. (a) How far away from the optical axis does the *third*-order maximum in the pattern occur? (b) What is the highest-order maximum theoretically visible in the diffraction pattern, and why might it be difficult in practice to actually see this order?

Solution

Preparation This is an example of a Young's slits type of experiment. We shall therefore need the diffraction equation $n\lambda = d\sin\theta_n$, where n is the order of the diffraction pattern, λ is the wavelength of the light, d is the spacing between the two slits and θ_n is the angle from the optical axis at which the order n appears in the diffraction pattern. A diagram is useful to clarify what is needed, such as that shown in Figure 5.1.

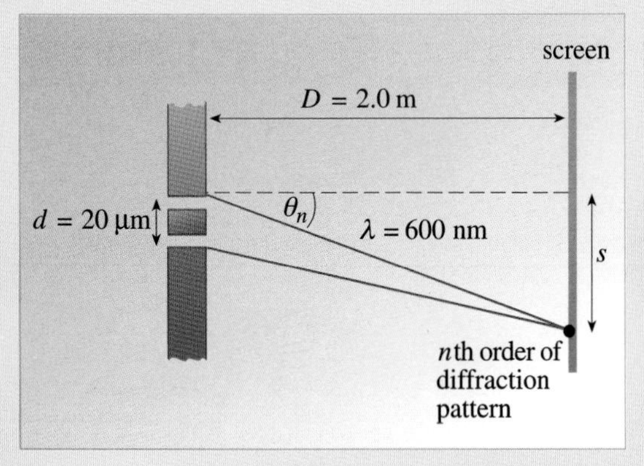

Figure 5.1 The situation described in Example 5.1.

Working (a) We shall first calculate the angle of diffraction, and then use trigonometry to determine the distance of the third maximum in the diffraction pattern from the optical axis.

It is important that all quantities are first converted to a common unit of length, such as the metre.

So, using $n\lambda = d\sin\theta_n$, in this case

$$\sin\theta_3 = (3 \times 600 \times 10^{-9}\,\text{m})/(20 \times 10^{-6}\,\text{m}) = 0.09,$$

and so $\theta_3 = 5.16°$.

Now, if the distance of the third-order maximum from the optical axis is s, and the distance from the slits to the screen is D, then Figure 5.1 shows that $\tan\theta_3 = s/D$, so

$$s = D\tan\theta_3 = 2.0\,\text{m} \times \tan 5.16° = 0.18\,\text{m}$$

Therefore, the third maximum in the diffraction pattern appears 18 cm from the optical axis.

(b) For a particular order to be visible in the diffraction pattern, it must be the case that $\sin\theta \leqslant 1.0$ (since 1.0 is the maximum value of the sine function). In other words, $(n \times 600 \times 10^{-9}\,\text{m})/(20 \times 10^{-6}\,\text{m}) \leqslant 1.0$ and therefore $n \leqslant 33.3$. So the highest order that is theoretically visible in the diffraction pattern is the 33rd order.

This might be difficult to see for two reasons. Firstly the angle of diffraction is given by $\sin\theta_{33} = (33 \times 600 \times 10^{-9}\,\text{m})/(20 \times 10^{-6}\,\text{m}) = 0.99$ and so $\theta_{33} = 81.9°$. The distance of the 33rd-order maximum from the optical axis is then given by $s = D\tan\theta_{33} = 2.0\,\text{m} \times \tan 81.9° = 14\,\text{m}$. In other words the 33rd maximum in the diffraction pattern appears 14 m from the optical axis. It is unlikely that a sufficiently large screen would be available! Secondly, the maxima decrease in brightness moving away from the optical axis, so the 33rd maximum is unlikely to be bright enough to see.

Checking (a) The distance comes out in metres, which is as expected. If the wavelength or order of the pattern were larger, the angle would be larger and so the distance from the optical axis would be larger, both of which are as expected. Conversely, if the slits were closer together, the angle and hence the distance of the third maximum from the optical axis would be larger, which is also as expected.

(b) The highest-order maximum is clearly larger than the third order referred to in part (a), so this is encouraging. All units in the calculation explicitly come out as expected, including the dimensionless sine function. If the screen were further away (larger D), then the distance of the 33rd order maximum from the optical axis would increase, as expected.

Example 5.2

A two-lens system comprises a diverging lens (A) of focal length −6.0 cm and a converging lens (B) of focal length +3.0 cm. The two lenses are positioned 3.0 cm apart, with lens A to the left of lens B. An object is placed on the common optical axis of the two lenses with its base at the left focal point of lens A. Determine *where* the final image will be formed, and state whether it will be real or virtual. (The thin-lens approximation may be assumed to hold.)

Solution

Preparation The most important thing to do in setting up problems of this sort is to draw a diagram, as shown in Figure 5.2. Always draw such diagrams *to scale*; in this case a scale of 1 : 1 is appropriate. Clearly, one of lens B's focal points coincides with the centre of lens A, whereas its other focal point coincides with a focal point of lens A.

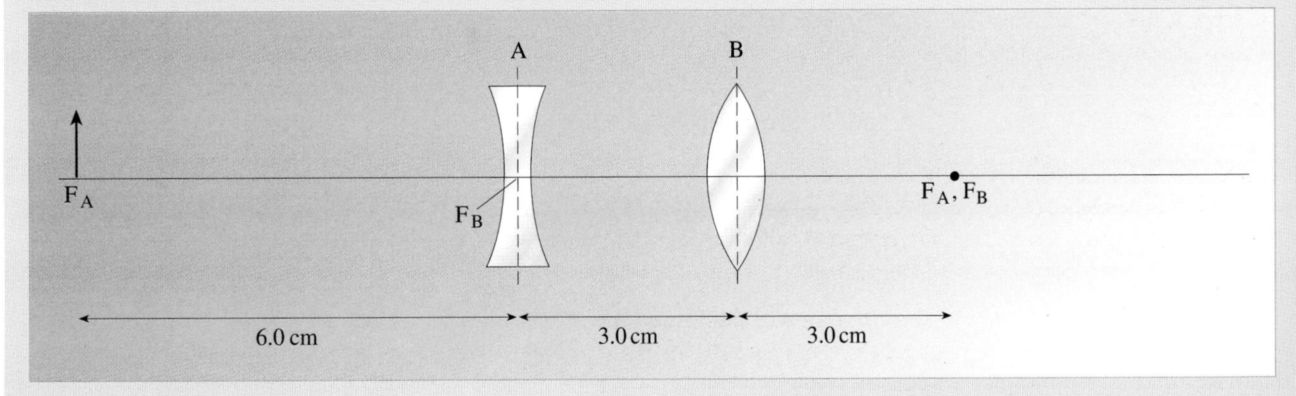

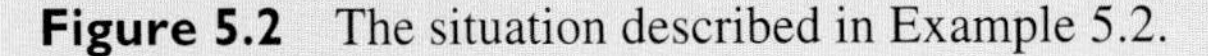

Figure 5.2 The situation described in Example 5.2.

Since the thin-lens approximation is assumed, other things that will be needed are the lens equation:

$$\frac{1}{u} + \frac{1}{v} = \frac{1}{f}$$

and the 'real is positive' convention, which states that:

(i) A converging lens has a positive focal length.

(ii) A diverging lens has a negative focal length.

(iii) Real objects (objects from which rays *diverge*) are assigned positive u-values.

(iv) Real images (formed by *converging* rays) have positive v-values.

(v) Virtual images (images from which rays *appear* to have *diverged*) have negative v-values.

(vi) When a lens converges rays as if to a real image, but the rays are intercepted by a second lens before the image is formed, then what would have been the real image behaves as a *virtual object* for the second lens, and must be assigned a negative u-value.

The first lens will generate an intermediate image, which will in turn act as an object for the second lens. We can calculate the positions of the intermediate and final images using the lens equation, and can later check if the answer is correct by constructing a ray diagram. To do this we shall also need to use information about how rays may be traced through lenses, namely:

(i) All rays entering a lens parallel to its optical axis will, on emerging, pass through the focal point (in the case of a converging lens), or will appear to have originated from the focal point (in the case of a diverging lens).

(ii) Any ray that passes through the focal point before striking a converging lens, or that would have passed through the focal point had it not been intercepted by a diverging lens, will emerge parallel to the optical axis.

(iii) All rays that pass through the optical centre of the lens can be assumed, to a good approximation, to be undeviated by the lens.

For the calculation, let the distance of the object from lens A be u_1, the distance of the intermediate image from lens A be v_1, the separation between the lenses be d, the distance of the intermediate image from lens B (for which it acts as an object) be u_2, and the distance of the final image from lens B be v_2.

Working Knowing u_1 and d, as well as the focal lengths of the two lenses, we can now calculate each of v_1, u_2 and v_2 in turn.

So, using the lens equation for lens A,

$$\frac{1}{u_1} + \frac{1}{v_1} = \frac{1}{f_A}$$

therefore $$\frac{1}{6.0\ \text{cm}} + \frac{1}{v_1} = \frac{1}{-6.0\ \text{cm}}$$

and so $v_1 = -3.0\,\text{cm}$. The intermediate image is 3.0 cm to the *left* of lens A, and is a *virtual* image.

Now, the object distance for lens B is therefore

$$u_2 = d - v_1 = 3.0\text{ cm} - (-3.0\text{ cm}) = 6.0\text{ cm}$$

Hence, the object for lens B is 6.0 cm to the *left* of lens B, and is a *real* object.

Using the lens equation for lens B,

$$\frac{1}{u_2} + \frac{1}{v_2} = \frac{1}{f_B}.$$

Therefore $$\frac{1}{6.0\text{ cm}} + \frac{1}{v_2} = \frac{1}{3.0\text{ cm}}$$

and so $v_2 = 6.0$ cm. Hence, the final image is 6.0 cm to the *right* of lens B, and is a *real* image.

Checking The best check to do in this case is to draw a ray diagram by completing the sketch shown in Figure 5.2. A completed version is shown in Figure 5.3.

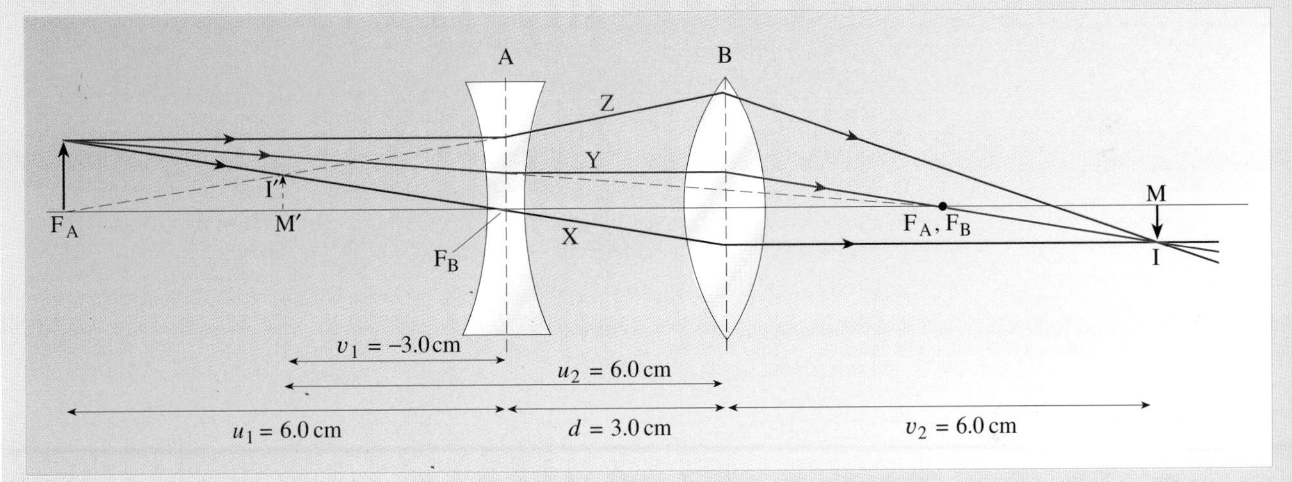

Figure 5.3 The completed ray diagram for Example 5.2.

Three rays are drawn from the top of the object.

The first of these (labelled X) is directed through the centre of lens A, from where it continues undeviated. Since this ray then arrives at lens B from one of its focal points, it emerges parallel to the optical axis.

The second ray (labelled Y) is directed towards the right-hand focal point of lens A. However, before it can reach there it is intercepted by lens A from which it emerges parallel to the optical axis. When this ray is then intercepted by lens B, it is deviated towards the right-hand focal point of lens B (which happens to coincide with the right-hand focal point of lens A).

Now, the third ray from the object (labelled Z) is initially parallel to the optical axis and is diverged on passing through lens A so as to appear to come from its focal point. (By tracing back ray Z to the point where it crosses ray X, the location of the intermediate image (i.e. the object for lens B) can be found: it is indicated by I′M′.) Ray Z then meets lens B, from which it emerges in such a direction as to meet rays X and Y at the point where they cross.

The point where all three rays intersect is the location of the final image IM. Clearly, this image is real (rays converge to it), whereas the intermediate image I′M′ is virtual (rays diverge from it). From the scale diagram, the two image distances are also consistent with the values calculated above.

Example 5.3

Particle A of rest mass m and translational kinetic energy $2mc^2$ hits and sticks to a stationary particle B of rest mass $2m$. Find an expression for the rest mass M of the composite particle C in terms of m only.

Solution

Preparation Since the question talks about 'rest mass' and mentions the speed of light c, it is likely that this question involves relativistic mechanics. We are therefore likely to need the equations for total relativistic energy and relativistic momentum, namely:

$$E_{\text{tot}} = \frac{mc^2}{\sqrt{1-v^2/c^2}} = E_{\text{trans}} + E_{\text{mass}} \qquad (PM \text{ Eqn } 3.22)$$

$$p = \frac{mv}{\sqrt{1-v^2/c^2}}. \qquad (PM \text{ Eqn } 3.20)$$

Note that the second of these equations involves taking the magnitude of each side of *PM* Equation 3.20.

The following relationship between relativistic energy and the magnitude of the linear momentum for a particle of mass m may also be needed:

$$E_{\text{tot}}^2 = p^2c^2 + m^2c^4. \qquad (4.22)$$

Also, this question clearly involves a collision, so a diagram, such as that shown in Figure 5.4 will be useful..

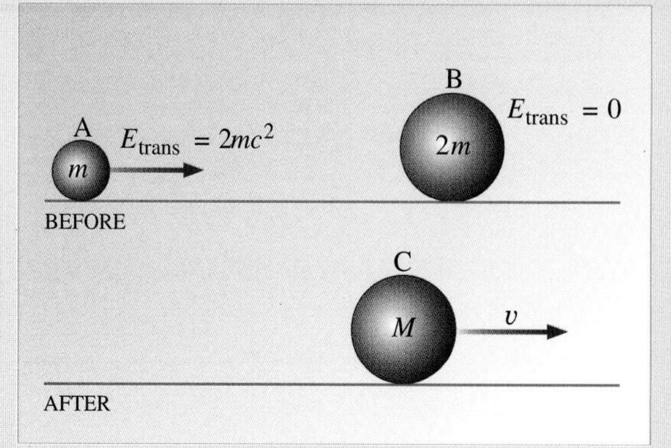

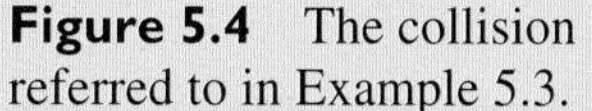
Figure 5.4 The collision referred to in Example 5.3.

The key to solving this problem is to set up the various energies and momenta correctly.

Let the total relativistic energy of particles A and B before the collision be $E_A + E_B$, and the total relativistic energy of particle C after the collision be E_C. Since relativistic energy is conserved, $E_A + E_B = E_C$, but remember that each term can be made up of different amounts of relativistic translational kinetic energy and mass energy.

Similarly, let the magnitudes of the initial and final momenta be $(p_A + p_B)$ and p_C, respectively. Since relativistic momentum is conserved, $p_A + p_B = p_C$.

Let the final speed of the composite particle C be v_C.

Working The way forward will be to calculate the total energy and momentum before and after the collision in terms of the two unknown quantities M and v_C. With two equations (i.e. one for energy and one for momentum), this will be solvable.

Remembering that the total relativistic energy is the sum of the relativistic translational kinetic energy and the mass energy, the initial energy of particle A is $E_A = 2mc^2 + mc^2 = 3mc^2$, whereas the initial energy of particle B is $E_B = 2mc^2$. Therefore, $E_A + E_B = 5mc^2$.

The initial momentum of particle A can be found from $E_A^2 = p_A^2c^2 + m^2c^4$, which in this case gives $(3mc^2)^2 = p_A^2c^2 + m^2c^4$. Dividing through by c^2 and rearranging, $p_A^2 = 8m^2c^2$, so $p_A = \sqrt{8}mc$. Since particle B is stationary, $p_B = 0$.

Total relativistic energy is conserved, so $E_A + E_B = E_C$, and therefore

$$5mc^2 = \frac{Mc^2}{\sqrt{1 - v_C^2/c^2}}. \tag{5.1}$$

Relativistic linear momentum is conserved, so $p_A + p_B = p_C$, and therefore

$$\sqrt{8}mc = \frac{Mv_C}{\sqrt{1 - v_C^2/c^2}}. \tag{5.2}$$

There are several ways of proceeding from here in order to combine these two equations to work out M. One way is as follows.

Dividing Equation 5.1 through by c^2 and squaring the result gives

$$25m^2 = \frac{M^2c^2}{c^2 - v_C^2}.$$

Rearranging further to make speed the subject,

$$v_C^2 = c^2 - \frac{M^2c^2}{25m^2}. \tag{5.3}$$

Similarly, squaring Equation 5.2 gives

$$8m^2c^2 = \frac{M^2v_C^2c^2}{c^2 - v_C^2}.$$

Dividing this through by c^2 and rearranging, we have

$$8m^2c^2 = v_C^2(M^2 + 8m^2) \tag{5.4}$$

Equations 5.3 and 5.4 contain M and v_C in separate terms, and we can therefore combine them to find an expression for M in terms of m alone. So substituting for v_C^2 in Equation 5.4 using Equation 5.3 gives

$$8m^2c^2 = \left(c^2 - \frac{M^2c^2}{25m^2}\right)(M^2 + 8m^2)$$

Multiplying out the brackets on the right-hand side gives:

$$8m^2c^2 = M^2c^2 + 8m^2c^2 - \frac{M^4c^2}{25m^2} - \frac{8M^2c^2}{25}.$$

Cancelling the $8m^2c^2$ terms and rearranging gives:

$$M^2c^2 - \frac{8M^2c^2}{25} = \frac{M^4c^2}{25m^2}.$$

Dividing through by M^2c^2 we have:

$$1 - \frac{8}{25} = \frac{M^2}{25m^2}$$

and simplifying further:

$$\frac{17}{25} \times 25m^2 = M^2.$$

Finally, this gives:

$$M^2 = 17m^2.$$

So the rest mass of the composite particle is $M = \sqrt{17}m$.

Checking This was a long-winded calculation, but came down to a simple answer at the end. This is always an encouraging indication that the answer is correct! Since m is a mass, clearly $\sqrt{17}m$ also has the unit of mass, so the final answer is dimensionally correct. The final answer of a little over $4m$ is also plausible, given that the rest masses of the individual particles were m and $2m$. If the rest masses of the original particles were larger, then so would be the rest mass of the composite particle, which is also to be expected.

As a final check, putting the value for M into Equation 5.3 gives:

$$v_C^2 = c^2 - \frac{17m^2c^2}{25m^2} = c^2\left(1 - \frac{17}{25}\right) = \frac{8c^2}{25}.$$

So $v_C = \sqrt{8}c/5$, which is about 57% of the speed of light. It is encouraging that this is not greater than the speed of light (if it were, then clearly it would indicate an error somewhere), and it seems a plausible speed for a relativistic calculation.

Now that you have seen a few examples, try the following questions for yourself, following the three-stage problem-solving strategy outlined earlier. Remember to use diagrams wherever they will help you to solve a particular problem.

Question 5.1 A square wire loop of side $b = 5.0\,\text{cm}$ is positioned vertically on a solid workbench. A permanent magnet of strength $B = 2.0\,\text{T}$ is arranged so that it can be dropped with its poles on either side of the loop, as shown in Figure 5.5. Assume that the magnetic field of the magnet is uniform between its poles, which are also square with a side length $b = 5.0\,\text{cm}$, and the magnetic field is zero elsewhere. If the magnet is now dropped under the influence of gravity, sketch a graph to show the variation of induced EMF around the wire loop with time, and calculate the maximum EMF induced in the loop. (Assume that $t = 0$ corresponds to the time when the wire loop just enters the magnetic field region — that is, when the lowest edge of the poles of the magnet is a distance $b = 5.0\,\text{cm}$ above the workbench. You may also assume that $g = 10\,\text{m}\,\text{s}^{-2}$ for the purposes of this question.)

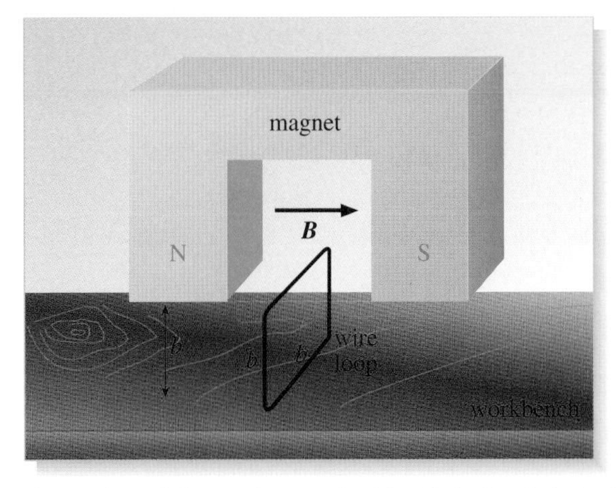

Figure 5.5 The situation in Question 5.1.

Question 5.2 A beam of white light in air is incident on a glass prism as shown in Figure 5.6. The glass prism has angles of 60° at each corner, and refractive indices for red and blue light of $n_{red} = 1.45$ and $n_{blue} = 1.50$, respectively. What is the angle between the red and blue beams that emerge from the prism into the air? (Assume that $n_{air} = 1.00$ for all colours.)

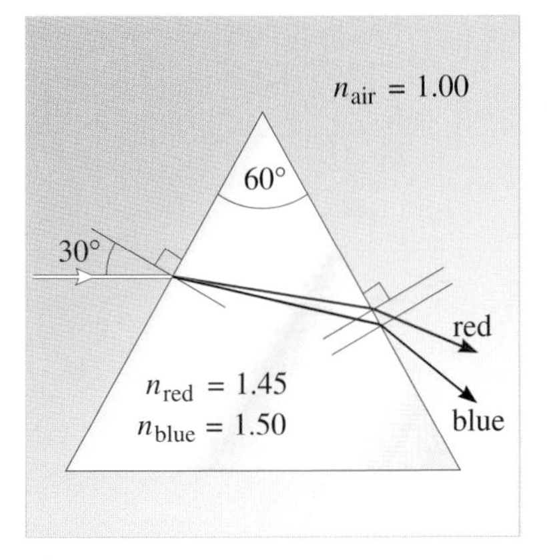

Figure 5.6 The situation in Question 5.2.

Question 5.3 A two-lens system comprises a converging lens (A) of focal length +4.0 cm and a diverging lens (B) of focal length −8.0 cm. The two lenses are positioned 4.0 cm apart, with lens A to the left of lens B. An object is placed on the common optical axis of the two lenses at a distance 8.0 cm to the left of lens A. Determine *where* the final image will be formed, and state whether it will be real or virtual. (The thin-lens approximation may be assumed to hold.)

Question 5.4 In a frame of reference in which they are both at rest, rocket P has a length of 80.0 m and rocket Q has a length of 60.0 m. Imagine that these two rockets fly past observer R in *opposite* directions, and that she measures them both as having the *same* length. If rocket P has a speed of 80% of the speed of light relative to observer R, what is the speed of rocket Q in the frame of reference in which rocket P is at rest? ■

5.4 Basic skills and knowledge test

You should be able to answer these questions without referring to earlier chapters. Leave your answers in terms of π, $\sqrt{2}$, etc., where appropriate.

Question 5.5 A wire loop of area 100 cm^2 sits in a uniform magnetic field of magnitude 0.1 T. The loop is free to turn about an axis that is perpendicular to the direction of the magnetic field. What is the *maximum* magnetic flux that can pass through the loop?

Question 5.6 The loop in Question 5.5 now rotates at a constant angular frequency of 400 s^{-1}. What is the *maximum* EMF induced around the wire loop?

Question 5.7 A transformer is required to supply power to a device, which is designed to run on a 120 V supply, using the 240 V mains supply in the United Kingdom. What is the required ratio of N_2/N_1, where N_1 is the number of windings on the primary coil, and N_2 is the number of windings on the secondary coil?

Question 5.8 What value of capacitance is required to tune an *LC* circuit to a resonant frequency of 100 Hz if the inductor in the circuit has the value 10^{-4} H?

Question 5.9 A wire with mass per unit length $\mu = 1.0 \times 10^{-3}$ kg m^{-1} is oscillating with waves of wavelength 20 cm and frequency 5.0 Hz. What is the tension force being applied to the wire?

Question 5.10 A light ray initially travelling in air encounters the surface of a medium with refractive index $\sqrt{3}$. If the angle of incidence with the surface is 60°, what is the angle of refraction?

Question 5.11 What angles do the incident and refracted *wavefronts* make with the normal to the surface in Question 5.10?

Question 5.12 What would you expect to see if a red laser and a blue laser were simultaneously used as light sources in a Young's two-slit experiment?

Question 5.13 A small refracting telescope in far-point adjustment has an angular magnification of 10. If the objective lens has a focal length of 20 cm, what is the focal length of the eyepiece lens?

Question 5.14 If a myopic patient is prescribed spectacles with lenses of power −2.0 D in the right eye and −1.5 D in the left eye, what focal lengths should be chosen for the two spectacle lenses?

Question 5.15 Which of the two lenses in Question 5.14 would you expect to be heavier, assuming they are both manufactured from the same material?

Question 5.16 Which optical system would you choose to accomplish each of the following tasks: (a) read very small print, (b) examine blood cells, (c) examine a distant ship. What combination of lenses does each optical system consist of?

Question 5.17 Summarize in one or two sentences why the Galilean coordinate transformation equations went unchallenged until Einstein formulated his special theory of relativity.

Question 5.18 State (in one sentence) the principle of relativity.

Question 5.19 An unfortunate UFO pilot crashes vertically downwards to Earth at speed $0.80c$ as measured by the pilot. The wreckage lands near a building 100 m tall as measured by the occupant of the building. What was the height of the building as measured by the pilot during the descent?

Question 5.20 From the point of view of the occupant of the building in Question 5.19, how long does the UFO take to fall from the height of the roof to the ground? ■

5 Interactive questions

Open University students should leave the text at this point and use the interactive question package for *Dynamic fields and waves*. When you have completed the questions, you should return to this text.

The interactive question package includes a random number feature that alters the values used in many of the questions each time those questions are accessed. This means that if you try the questions again, as part of your end-of-course revision for instance, you will find that many of them will have changed, at least in their numerical content.

6 *Physica* problems

Open University students should leave the text at this point and tackle the *Physica* problems that relate to *Dynamic fields and waves*.

Answers and comments

Q1.1 (a) When the loop is entirely within the magnetic field region, only conduction electrons in arms WX and ZY feel a magnetic Lorentz force; but for each electron in the arm WX feeling a force from X to W there is an electron in the arm ZY feeling the same force from Y to Z. The contributions that each electron makes to the current is exactly cancelled by its 'twin' in the opposite arm, and so the ammeter would show that no current flows in the circuit.

(b) When the loop is partially in and partially out of the magnetic field, as shown in Figure 1.11(ii), the problem is the same as the one just shown in Figure 1.8, and so a current flows from W to X.

As the loop continues to move, there will be fewer conduction electrons in the magnetic field region, but the magnetic Lorentz force on each electron still within the magnetic field will be the same, so its push on its neighbours will be the same. Hence the current remains the same until the loop leaves the magnetic field, when it quickly drops to zero.

During the period when the loop is partly within and partly outside the magnetic field:

(c) doubling the velocity $\boldsymbol{v}$ doubles the force $\boldsymbol{F}$; hence it doubles the current;

(d) moving the loop in the opposite direction reverses the velocity, and so reverses the direction of the Lorentz force, causing the current to flow in the opposite direction;

(e) turning the magnet upside down reverses the magnetic field, and so reverses the direction of the Lorentz force, causing the current to flow in the opposite direction.

(f) If the circular loop is moving at a constant velocity within the magnetic field region, no current will flow. As for the rectangular loop, for every conduction electron feeling an electric field push in one direction, there will be another feeling the same push in the opposite direction, so giving a net force of zero. In fact, as long as the loop is in the plane of the magnetic field and is not twisted over itself, no current will flow no matter what its shape.

Q1.2 (a) Once again there will be a changing current in the primary circuit as it quickly falls to zero, so there is (briefly) a changing magnetic field produced by the primary circuit. This produces an electric field in the secondary circuit, so causing a current to flow. When the current in the primary circuit has fallen to zero, it is creating no magnetic field, so the induced current too will fall to zero.

(b) A current will flow in the secondary circuit during the time the voltage is changing from 2 V to 3 V, but the current will then quickly fall to zero and stay there once the voltage change ends.

Q1.3 All the results are obtained using Equation 1.1a: $\phi = AB\cos\theta$.

(a) $B = 0.4\,\text{T}$; $A = 0.1 \times 0.1\,\text{m}^2 = 0.01\,\text{m}^2$; $\theta = 0°$. Remember θ is the angle between the magnetic field and the *normal* to the circuit. In this case $\theta = 0°$, so $\cos\theta = 1$, and therefore,

$$\phi = 0.4 \times 0.01 \times 1\,\text{Wb} = 4 \times 10^{-3}\,\text{Wb}.$$

(b) $B = 10^{-5}\,\text{T}$; $A = 10^{-3}\,\text{m}^2$; $\theta = 0°$.

$$\phi = 10^{-5} \times 10^{-3} \times 1\,\text{Wb} = 10^{-8}\,\text{Wb}.$$

(c) $B = 0.5\,\text{T}$; $A = 4.0\,\text{m}^2$; $\theta = 30°$.

$$\phi = 0.5 \times 4.0 \times 0.866\,\text{Wb} = 1.7\,\text{Wb}.$$

(d) A plane at 60° to the magnetic field has a normal at 30° to the field, so this is the same as (c). Hence $\phi = 1.7\,\text{Wb}$.

Q1.4 The magnitude of the magnetic flux through each of the upper and lower faces of the cube is

$$|\boldsymbol{B}|A\cos\theta = (3.0\,\text{T}) \times (0.1\,\text{m})^2 \times \cos 30° = 2.6 \times 10^{-2}\,\text{Wb}.$$

The magnetic flux through the upper face is *into* the cube, whereas the magnetic flux through the lower face is *out of* the cube. Whatever convention is adopted for the sign of the magnetic flux, these two fluxes will be of opposite sign. So the total magnetic flux through these two faces is zero.

Similarly, the magnitude of the magnetic flux through each of the left-hand and right-hand faces of the cube is

$$|\boldsymbol{B}|A\cos(90° - \theta)$$
$$= (3.0\,\text{T}) \times (0.1\,\text{m})^2 \times \cos 60° = 1.5 \times 10^{-2}\,\text{Wb}.$$

The magnetic flux through the left-hand face is *into* the cube, whereas the magnetic flux through the right-hand face is *out of* the cube, and once again the total magnetic flux through these two faces is zero.

There is no component of magnetic field perpendicular to either the front or back face of the cube, and so they each contribute zero magnetic flux.

Thus, the total magnetic flux through the entire surface of the cube is zero.

Q1.5 (a) When the switch is first closed in the primary circuit, the battery causes a current to build up from zero (the battery is the origin of a primary EMF). As the current increases, it creates an increasing magnetic field in the region around the wire; that is, there is an increasing magnetic flux through the area enclosed by the secondary wire loop. (Remember: it is not the fact that the magnetic flux is *increasing* that is critical, but that it is *changing*.) This changing magnetic flux induces an EMF in the secondary wire, which produces a current surge (since it will be a rapid build up).

The current in the primary circuit soon settles into a steady state, meaning that there is a constant current flowing in it. This current still creates a magnetic field, and hence a magnetic flux through the enclosed area, but now the magnetic field is constant in time, and so is the magnetic flux. Therefore there is no induced EMF in the secondary and so no induced current: the initial current surge dies away.

(b) During the period when the primary EMF is changing, the current is changing too. The magnetic field it creates changes, and the magnetic flux therefore changes, so there is a current surge in the secondary. After the primary EMF stops changing, the magnetic field remains constant, as does the magnetic flux, and the induced current dies away.

(c) Finally, when the switch in the primary is opened, the primary current dies away, and so must the magnetic field it creates. Hence the magnetic flux decreases to zero. But as it is decreasing, it induces a current in the secondary, a surge that then falls to zero.

Q1.6 (a) The magnitude of the induced EMF is given by Faraday's law (Equation 1.2) as: $|V_{\text{ind}}(t)| = |\text{d}\phi(t)/\text{d}t|$. In Example 1.1, the magnitude of the rate of change of magnetic flux during the times the wire loop is entering or leaving the magnetic field region was found to be $|\text{d}\phi(t)/\text{d}t| = avB$. So in this case, the magnitude of the induced EMF is $|V_{\text{ind}}| = (0.1\,\text{m}) \times (0.5\,\text{m}\,\text{s}^{-1}) \times (0.5\,\text{T}) = 0.025\,\text{V}$. From Ohm's law the magnitude of the induced current is therefore

$|i_{\text{ind}}| = |V_{\text{ind}}|/R = (0.025\,\text{V})/(10\,\Omega) = 2.5 \times 10^{-3}\,\text{A}$ or 2.5 mA.

(b) The magnitude of the induced EMF is once again given by Faraday's law (Equation 1.2) as: $|V_{\text{ind}}(t)| = |\text{d}\phi(t)/\text{d}t|$. In Example 1.2, the rate of change of magnetic flux was found to be $\text{d}\phi(t)/\text{d}t = \omega AB \cos \omega t$, so the magnitude of the maximum rate of change of magnetic flux is simply ωAB (as the cosine function varies only between ± 1). Hence, the magnitude of the maximum induced EMF is

$$|V_{\text{ind}}| = (40\,\text{s}^{-1}) \times (2.5 \times 10^{-3}\,\text{m}^2) \times (2.0\,\text{T}) = 0.2\,\text{V}.$$

From Ohm's law the magnitude of the induced current is

$|i_{\text{ind}}| = |V_{\text{ind}}|/R = (0.21\text{V})/(5.01\text{W}) = 0.040\,\text{A}$ or 40 mA.

Q1.7 During the period when the loop is moving into the magnetic field, the magnetic flux through the loop is increasing, so the induced EMF must *oppose* this increase. Therefore the magnetic field due to the induced current must point in the *opposite* direction to the magnetic field of the magnet. By the right-hand grip rule, the current flows *anti-clockwise*. When the loop is fully in the magnetic field, the current is zero until it starts to leave the magnetic field. Now the magnetic flux is decreasing, so the induced EMF must *oppose* this decrease. Therefore the magnetic field due to the induced current must point in the *same* direction as the magnetic field of the magnet. By the right-hand grip rule again, the current flows in the *clockwise* sense.

Notice that the current in the loop flows in the *same* direction (anticlockwise) whether it is entering the magnetic field from the left or from the right. Similarly, the current in the loop flows in the *same* direction (clockwise) whether it is leaving the magnetic field to the right or to the left. This is a consequence of the fact that the magnetic field caused by the induced current due to the induced EMF always opposes the change that causes the induced EMF in the first place.

Q1.8 Before deciding on the direction of the current, you should remind yourself of the direction of the magnetic field generated by a current in the coil (without questioning the cause of this current). In Figure 1.59 the two possibilities are shown, determined by a right-hand grip rule: the current is followed around the coil with the fingers of your right hand, and your thumb points in the direction of the generated field. So for a current coiling into the page (Figure 1.59a), the magnetic field points to the left, whereas for a current coiling out of the page (Figure 1.59b), the magnetic field points to the right.

The cause of the current in the coils is the motion of the bar magnet towards the coil from the left, with its south pole entering the coils first. The magnetic field of the bar magnet is therefore pointing to the left (that is, pointing towards the south pole) The current must flow so that its magnetic field opposes the incoming magnetic field, which will make pushing the magnet more difficult. The coil magnetic field

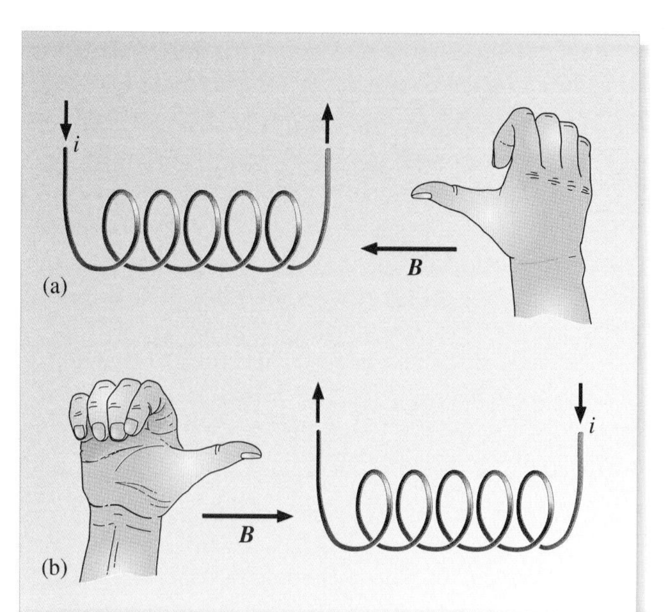

Figure 1.59 The direction of the magnetic field generated by a current in the coil, as determined by a right-hand grip rule, for (a) a current coiling into the page and (b) a current coiling out of the page.

must therefore point to the right, so the current must wind out of the page. The induced current in the circuit in Figure 1.29 must therefore be clockwise.

Q1.9 The inductance of a solenoid is given by Equation 1.10 as $L = \mu AN^2/l$. So in this case,

$$L = \frac{(1.0 \times 10^{-3}\,\mathrm{T\,m\,A^{-1}}) \times (40 \times 10^{-6}\,\mathrm{m^2}) \times (200)^2}{0.05\,\mathrm{m}}$$

$$= 0.032\,\mathrm{H}.$$

The inductance of the solenoid is therefore 32 mH.

Q1.10 (a) If the current in the circuit is to reach 50% of the steady-state value, then Equation 1.14 becomes

$$0.5 \times \left(\frac{V_{\mathrm{bat}}}{R}\right) = \left(\frac{V_{\mathrm{bat}}}{R}\right)(1 - \mathrm{e}^{-Rt/L}).$$

The steady-state current (V_{bat}/R) cancels out to give: $0.5 = 1 - \mathrm{e}^{-Rt/L}$. Rearranging this yields: $0.5 = \mathrm{e}^{-Rt/L}$. Now taking the natural logarithm of each side of this equation gives $\log_e 0.5 = -Rt/L$. So the time needed for the current to reach 50% of the steady-state value is given by

$$t = \frac{-L \times \log_e 0.5}{R}$$

$$= \frac{(-32 \times 10^{-3}\,\mathrm{H}) \times (-0.69)}{22\,\Omega}$$

$$= 1.0 \times 10^{-3}\,\mathrm{s} = 1\,\mathrm{ms}.$$

(b) The steady-state current in the coil is $i_{\mathrm{max}} = V_{\mathrm{bat}}/R = 5.5\,\mathrm{V}/22\,\Omega = 0.25\,\mathrm{A}$. So when the current in the coil is close to the steady-state value, the energy stored in the coil may be calculated from Equation 1.11 as

$$\text{magnetic energy} = \tfrac{1}{2} L i_{\mathrm{max}}^2$$

$$= \tfrac{1}{2} \times (32 \times 10^{-3}\,\mathrm{H}) \times (0.25\,\mathrm{A})^2$$

$$= 1.0 \times 10^{-3}\,\mathrm{J} = 1\,\mathrm{mJ}.$$

(c) Following the same method as in part (a), the time for the current to fall to 50% of its steady-state value can be found from Equation 1.16 as $0.5 = \mathrm{e}^{-Rt/L}$. Taking natural logarithms of both sides gives $\log_e 0.5 = -Rt/L$. So the time needed for the current to fall to 50% of the steady-state value is given by

$$t = \frac{-L \times \log_e 0.5}{R}$$

$$= \frac{(-32 \times 10^{-3}\,\mathrm{H}) \times (-0.69)}{22\,\Omega}$$

$$= 1.0 \times 10^{-3}\,\mathrm{s} = 1\,\mathrm{ms}.$$

Notice that the time for the current to decay from its steady-state value to half of that value is identical to the time taken for the current to grow from zero to half its steady-state value. In other words, the time constant of an inductive circuit is the same for both growth and decay of current.

(d) Using Equation 1.16 again, the time for the current to fall to 25% of its steady-state value can be found from $0.25 = \mathrm{e}^{-Rt/L}$. Taking natural logarithms again gives $\log_e 0.25 = -Rt/L$. So the time needed for the current to fall to 25% of the steady-state value is

$$t = \frac{-L \times \log_e 0.25}{R}$$

$$= \frac{(-32 \times 10^{-3}\,\mathrm{H}) \times (-1.39)}{22\,\Omega}$$

$$= 2.0 \times 10^{-3}\,\mathrm{s} = 2\,\mathrm{ms}.$$

You can probably see that to fall by another factor of two (to 12.5% of the steady-state value) would take about 3 ms, and so on: every halving of the current will take a further 1 ms in this example. The value of this characteristic time clearly depends on the resistance and the inductance of the circuit in question, but the overall behaviour is a feature of *all* exponential decays.

Q1.11 Figure 1.40a–d is repeated in Figure 1.60a–d with the directions of force, velocity and magnetic field shown for positive charges in the arms ZW and XY. In each case

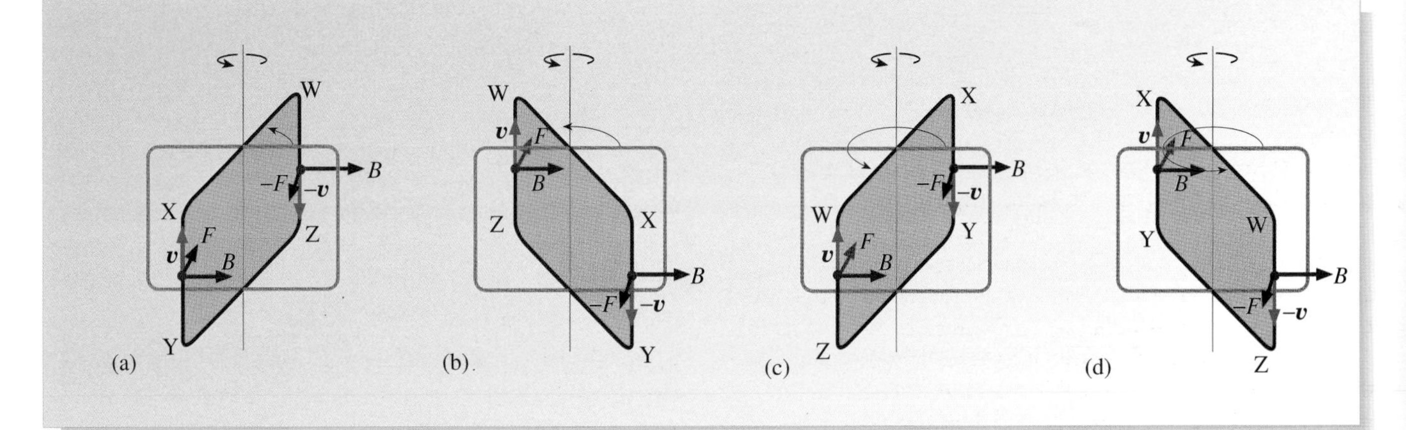

Figure 1.60 (a)–(d) A simple AC generator with the coil in four positions, one-quarter of a cycle apart. The direction of the Lorentz force and velocity of a positive charge in the arms ZW and XY are shown in each case.

the opposing torque must act *clockwise* when viewed along the rotation axis from above. Using the right-hand rule, the direction of motion of the positive charges must be as follows: (a) from W to Z and from Y to X, (b) from Z to W and from X to Y, (c) from Z to W and from X to Y, and (d) from W to Z and from Y to X. Therefore the current flows in the direction ZYXW in (a) and (d), and in the direction WXYZ in (b) and (c): it reverses direction twice per cycle.

Q1.12 (a) The DC generator in Figure 1.43 could be turned into a DC motor simply by feeding in a constant current via the split ring.

(b) When the coil is parallel to the poles of the magnet, as shown in Figure 1.61, the Lorentz forces on the charges in arms WZ and XY of the wire are all in the plane of the coil itself. There is therefore no torque acting on the coil at this point, which occurs twice per cycle.

In a real DC motor, the inertia of the coil is able to keep it turning past these zero torque points. Also, a real motor will have many turns of wire at varying angles to the magnetic field. So there will always be some turns which are contributing a torque, whatever the orientation of the coil to the magnetic field.

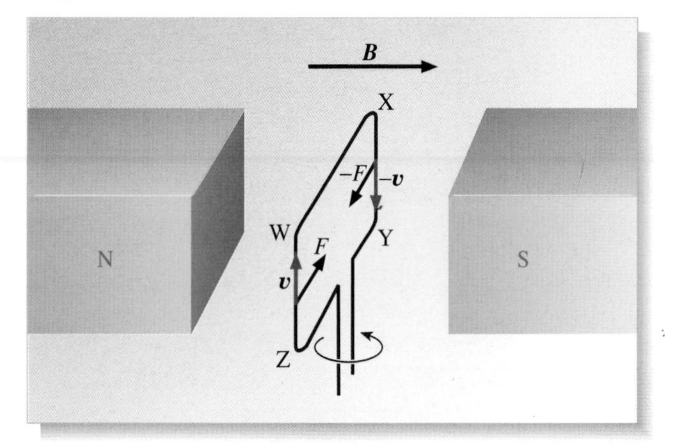

Figure 1.61 A DC motor at the position where zero torque acts on the wire loop.

Q1.13 (a) The two kinds of energy in an *LC* circuit are electrical energy (stored in the capacitor) and magnetic energy (stored in the inductor). As the charge oscillates back and forth in the circuit, the energy will be mostly in the inductor (energy = $\frac{1}{2}Li^2$) when maximum current flows, and the energy will be mostly in the capacitor (energy = $\frac{1}{2}CV^2$) when maximum voltage exists across the circuit. These changes are directly analogous to those occurring in mechanical oscillating systems.

(b) All real electrical circuits have a resistance, even if that is just due to the wire connecting the components. In any electrical circuit, therefore, energy will be lost due to the current flowing through a resistor. Energy is dissipated by a resistor at a rate i^2R and will be transformed into thermal energy.

Q1.14 The natural frequency of an *LC* circuit is given by $f = 1/2\pi\sqrt{LC}$ (Equation 1.30). So to tune the circuit to this particular frequency requires a capacitance of

$$C = \frac{1}{L(2\pi f)^2}$$

$$= \frac{1}{(32\times10^{-3}\,\text{H})\times(2\pi\times200\times10^3\,\text{s}^{-1})^2}$$

$$= 2.0\times10^{-11}\,\text{F}.$$

The capacitor must therefore be adjusted to a value of 20 pF.

Q1.15 (a) Using Equation 1.44, the number of turns on the secondary winding can be expressed as $N_2 = N_1(|V_2(t)|/|V_1(t)|)$. Since the time-varying EMF in each coil is of the form $V(t) = V_{\text{max}}\cos\omega t$, this can be simplified to give $N_2 = N_1(V_{2,\text{max}}/V_{1,\text{max}})$. In this case therefore, the number of turns needed on the secondary coil is

$$N_2 = 5000\times\frac{12\,\text{V}}{240\,\text{V}} = 250.$$

(b) Using Equation 1.39, the mutual inductance of the transformer can be found as

$$M = \frac{|V_2(t)|}{|\text{d}i_1(t)/\text{d}t|}$$

When the current in the primary is undergoing its maximum rate of change, the EMF in the secondary will have its maximum value, so $M = 12\,\text{V}/300\,\text{A s}^{-1}$ or 40 mH.

Q1.16 The magnetic Lorentz force has magnitude quB, where q is the conduction charge, u is the speed and B is the magnitude of the magnetic field. According to the right-hand rule, its direction will be from right to left. The force causing the wire to move is from left to right, so this is in agreement with Lenz's law.

Q1.17 (a) Since the ions move through the region without change in velocity, the resultant force acting on them must be zero; that is, the Lorentz force

$$\boldsymbol{F} = q(\boldsymbol{\mathscr{E}} + \boldsymbol{v}\times\boldsymbol{B}) = \boldsymbol{0}.$$

In other words, the electric force $q\boldsymbol{\mathscr{E}}$ and the magnetic force $q(\boldsymbol{v}\times\boldsymbol{B})$ must act in opposite directions. Figure 1.57 shows that the electric field is acting upwards, so the electric force must also act upwards. Therefore the magnetic force must act downwards. Using the right-hand rule, with the force acting downwards, and the velocity from left to right, the magnetic field must be in a direction out of the plane of the paper towards you, as shown in Figure 1.57.

The two forces will therefore cancel each other provided that their magnitudes are the same. This requires that

$$q|\boldsymbol{\mathscr{E}}| = q|\boldsymbol{v}||\boldsymbol{B}|\sin 90°.$$

Since the charge q cancels from both sides and $\sin 90° = 1$, the magnitude of the magnetic field is:

$$|\boldsymbol{B}| = \frac{|\mathscr{E}|}{|\boldsymbol{v}|} = \frac{2\times10^2\,\mathrm{N\,C^{-1}}}{5\times10^3\,\mathrm{m\,s^{-1}}} = 4\times10^{-2}\,\mathrm{T}$$

or 40 mT.

(b) The magnetic flux is given by

$$\phi = AB\cos 0° = (1\,\mathrm{m}^2)\times(4\times10^{-2}\,\mathrm{T}) = 4\times10^{-2}\,\mathrm{Wb}$$

or 40 mWb.

Note that neither the charge nor the mass of the ions comes into the problem, so it would make no difference what the ion beam were composed of. Note also that if the ions were to have a greater speed, the compensating magnetic field strength, and hence the magnetic flux, would be proportionately smaller. This is a consequence of the fact that the magnetic Lorentz force is proportional to the ion velocity, whereas the electric Lorentz force is not.

Q1.18 From Equation 1.1, the magnetic flux ϕ through a loop of area A whose plane is perpendicular to $\boldsymbol{B}$ (i.e. $\cos\theta = 1$) will be

$$\phi(t) = A(B_0 + kt^2)$$

so

$$\begin{aligned}\frac{\mathrm{d}\phi(t)}{\mathrm{d}t} &= A\frac{\mathrm{d}}{\mathrm{d}t}(B_0 + kt^2)\\ &= A(0 + 2kt)\\ &= 2Akt.\end{aligned}$$

From Faraday's law (Equation 1.2) the magnitude of the induced EMF is then

$$|V_{\mathrm{ind}}(t)| = \left|\frac{\mathrm{d}\phi(t)}{\mathrm{d}t}\right| = 2Akt.$$

The magnitude of the induced current can be found from $|i_{\mathrm{ind}}(t)| = |V_{\mathrm{ind}}(t)|/R$.

So the magnitude of the induced current is given by $|i_{\mathrm{ind}}(t)| = 2Akt/R$.

Q1.19 (a) Although the magnetic field is uniform and static, the magnetic flux through the circuit increases, because the moving rod increases its area. If the rod is moving with constant speed v, the area increases at a rate of $\mathrm{d}A(t)/\mathrm{d}t = lv$. So the magnitude of the rate of change of magnetic flux is

$$\begin{aligned}\left|\frac{\mathrm{d}\phi(t)}{\mathrm{d}t}\right| &= \left|\frac{\mathrm{d}(AB)}{\mathrm{d}t}\right|\\ &= B\left|\frac{\mathrm{d}A(t)}{\mathrm{d}t}\right| = Blv.\end{aligned}$$

From Equation 1.2, the induced EMF will be of magnitude Blv, so from Ohm's law, the induced current will be of magnitude $i = Blv/R$.

(b) In order to induce a current of a particular value, the rod must move with a speed $v = iR/Bl$. So in this case,

$$v = \frac{(1.5\,\mathrm{A})\times(3.0\,\Omega)}{(5.0\,\mathrm{T})\times(0.1\,\mathrm{m})} = 9.0\,\mathrm{m\,s^{-1}}.$$

Q1.20 A coil that supplies current to an external circuit, represents a current-carrying conductor moving in a magnetic field, and the charges within it are therefore subject to the magnetic Lorentz force. This opposes the rotation of the coil and has to be overcome by an external force.

Q1.21 The maximum medium-wave frequency is 1.5×10^6 Hz and the maximum long-wave frequency is 3.0×10^5 Hz. So, in changing from medium-wave to long-wave operation, the upper limit to the frequency range *decreases* by a factor of 5. The lower limit of the frequency range also decreases by a factor of 5. In switching between the two modes of operation, the range of capacitance remains fixed.

Now, Equation 1.30 may be rearranged to give an expression for the inductance of an LC circuit in terms of its natural frequency and its capacitance: $L = 1/C(2\pi f)^2$. In other words, the inductance is proportional to $1/f^2$. So, if the capacitance can only be varied over the same range, but the frequencies required are 5 times smaller, the inductance required is 25 times larger.

The inductance of a solenoid is proportional to the square of the number of turns (Equation 1.10). So to increase the inductance by a factor of 25 means that the number of turns must be increased by a factor of 5. In order to receive long-wave broadcasts, therefore, the solenoid must have $5\times100 = 500$ turns of wire.

Q2.1 (a) If the wave crests are closer together, then, as long as the wave is travelling at the same speed as before, a shorter time interval will elapse between each crest breaking on the beach and the next. The period of the wave will therefore be shorter. Since the frequency of a wave is equal to the reciprocal of its period, the frequency of the wave will be higher.

(b) If the waves are travelling more slowly across the sea, but with the wave crests the same distance apart as before, there will be a longer time interval between each crest breaking on the beach and the next. The period of the wave will therefore be longer and the frequency of the wave will be lower.

Q2.2 If the distance to the shoal of fish is d, then the sound wave must travel a distance $2d$ (i.e. there and back again) in 0.20 s. Since distance = speed × time, $2d = (1500\,\mathrm{m\,s^{-1}}) \times (0.20\,\mathrm{s})$, so $d = 300\,\mathrm{m}/2$. The shoal of fish is therefore 150 m away from the dolphin.

Q2.3 y can be calculated from Equation 2.1a with $A = 1$ Pa and $k = 2\pi/\lambda = 2\pi/1.2$ m. The calculated values are shown in Table 2.6 and the graph is shown in Figure 2.86.

Table 2.6 Calculated values for y (Q2.3).

x/m	kx	$\sin kx$	y/Pa
0.1	$\pi/6$	0.50	0.50
0.2	$\pi/3$	0.87	0.87
0.3	$\pi/2$	1.0	1.0
0.4	$2\pi/3$	0.87	0.87
0.5	$5\pi/6$	0.50	0.50
0.6	π	0	0
0.7	$7\pi/6$	−0.50	−0.50
0.8	$4\pi/3$	−0.87	−0.87
0.9	$3\pi/2$	−1.0	−1.0

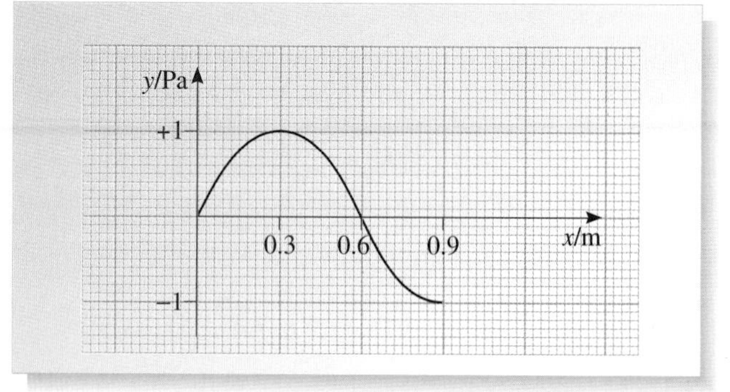

Figure 2.86 Graph to show the variation of y with x (for Q2.3 answer).

Note that, although the graph sketched from the calculated values is only a part of a sine wave, it is consistent with an amplitude of 1 Pa and a wavelength of 1.2 m.

Q2.4 The two representations are shown in Figure 2.87. You may have drawn your curves displaced to the left or right; any position of the sinusoidal curve is acceptable because the initial displacement was not specified. However, the maximum and minimum values of the sinusoidal curves are +2 cm and −2 cm, respectively, the 'repeat' distance in graph (a) is $\lambda = 10$ cm, and the 'repeat' time in graph (b) is $T = 0.05$ s.

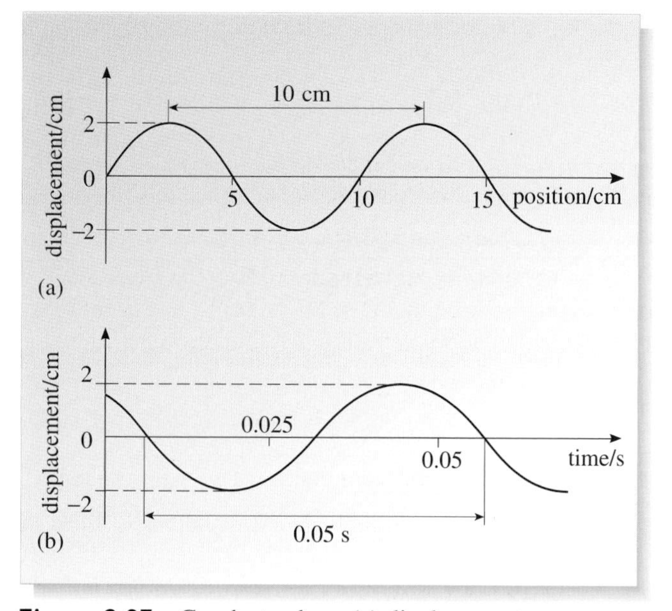

Figure 2.87 Graphs to show (a) displacement versus position at a certain instant, and (b) displacement versus time at a certain position (for Q2.4 answer).

Q2.5 Comparing the description of the wave with the wave equation of a travelling wave (Equation 2.3a), it can be seen that:

(a) the amplitude $A = 2$ m;

(b) $k = 3\,\mathrm{m^{-1}}$, but $k = 2\pi/\lambda$ so the wavelength, $\lambda = 2\pi/k = (2\pi/3)$ m;

(c) $\omega = 4\,\mathrm{s^{-1}}$, but $\omega = 2\pi/T$ so the period, $T = 2\pi/\omega = 2\pi/4\,\mathrm{s} = \pi/2$ s;

(d) the phase constant, $\phi = 0$.

Q2.6 (a) Substituting in Equation 2.6 gives

$$v = \sqrt{\frac{\text{axial modulus}}{\text{density}}}$$

$$= \sqrt{\frac{8.5\times10^{10}\,\mathrm{N\,m^{-2}}}{2.7\times10^{3}\,\mathrm{kg\,m^{-3}}}}$$

$$= 5.6\times10^{3}\,\mathrm{m\,s^{-1}}.$$

So the speed of seismic P-waves through granite is about $5.6\,\mathrm{km\,s^{-1}}$.

(b) If the density does not change but the axial modulus increases by a factor of 2, the speed of P-waves will increase by a factor of $\sqrt{2}$. Similarly, if the axial modulus increases by factors of 3 and 4, the speed will increase by factors of $\sqrt{3}$ and $\sqrt{4}$ (= 2), respectively.

(c) Conversely, if the axial modulus does not change but the density increases by factors of 2, 3 and 4, the speed of P-waves will *decrease* by factors of $\sqrt{2}$, $\sqrt{3}$ and $\sqrt{4}$ (= 2), respectively.

Q2.7 The speed of waves on a string is given by Equation 2.7

$$v = \sqrt{\frac{F_T}{\mu}}.$$

Doubling the tension will increase the speed by a factor of $\sqrt{2}$, so the new speed will be

$$(2.0\,\text{m s}^{-1} \times 1.4) = 2.8\,\text{m s}^{-1}.$$

Since the period remains the same (0.5 s), the frequency will also remain the same ($f = 1/T = 1/0.5\,\text{s} = 2.0\,\text{Hz}$). The wavelength is given by $v = f\lambda$ (Equation 2.5), so

$$\lambda = \frac{v}{f} = \frac{2.8\,\text{m s}^{-1}}{2.0\,\text{s}^{-1}} = 1.4\,\text{m}.$$

Q2.8 (a) The intensity falls off as the inverse square of the distance (Equation 2.8a). Thus, if r is the distance at which the shout could just be heard

$$\frac{8 \times 10^{-5}\,\text{W m}^{-2}}{1 \times 10^{-12}\,\text{W m}^{-2}} = \frac{r^2}{(1\,\text{m})^2}$$

$$r^2 = 8 \times 10^7\,\text{m}^2$$

$$r \approx 9\,\text{km}.$$

(b) Since the intensity is proportional to the square of the amplitude (Equation 2.9), the amplitude is proportional to the square root of the intensity. Hence the ratio of the amplitudes is

$$\sqrt{\frac{1 \times 10^{-12}\,\text{W m}^{-2}}{8 \times 10^{-5}\,\text{W m}^{-2}}} = \sqrt{1.25 \times 10^{-8}} \approx 10^{-4}.$$

Of course, this is a rather idealized calculation. In practice, some sound energy would be absorbed by the air, and the intensity would be affected by wind and convection currents. In addition, background noises would increase the minimum intensity level of the sound one could detect.

Q2.9 Rearranging Equation 2.18 gives

$$V = \frac{v\Delta f}{2 f_{em} \cos\theta}$$

$$= \frac{1500\,\text{m s}^{-1} \times 280\,\text{Hz}}{2 \times 3.0 \times 10^6\,\text{Hz} \times \cos 45°}$$

$$= 0.099\,\text{m s}^{-1}.$$

So the speed of the blood flow is about $0.1\,\text{m s}^{-1}$.

Q2.10 (a) The pulse would be inverted at the bound-ary. As the two pulses met, the resultant displacement would be similar to that shown in Figure 2.23, apart from the fact that in part (d) of the Figure the two displacements would cancel completely.

(b) Figure 2.88 shows sketches of the resultant displacement.

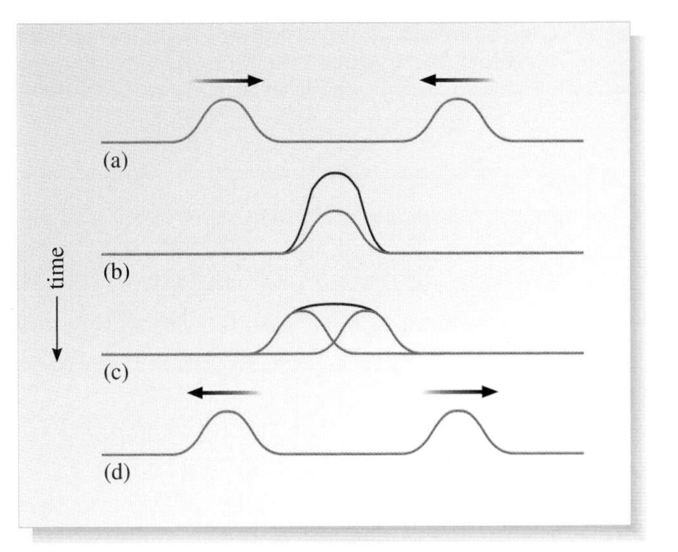

Figure 2.88 Using the principle of superposition to predict the resultant displacement as two wave pulses, travelling in opposite directions, meet and continue on their way (for Q2.10). The sum of the displacements from the individual pulses is shown in red.

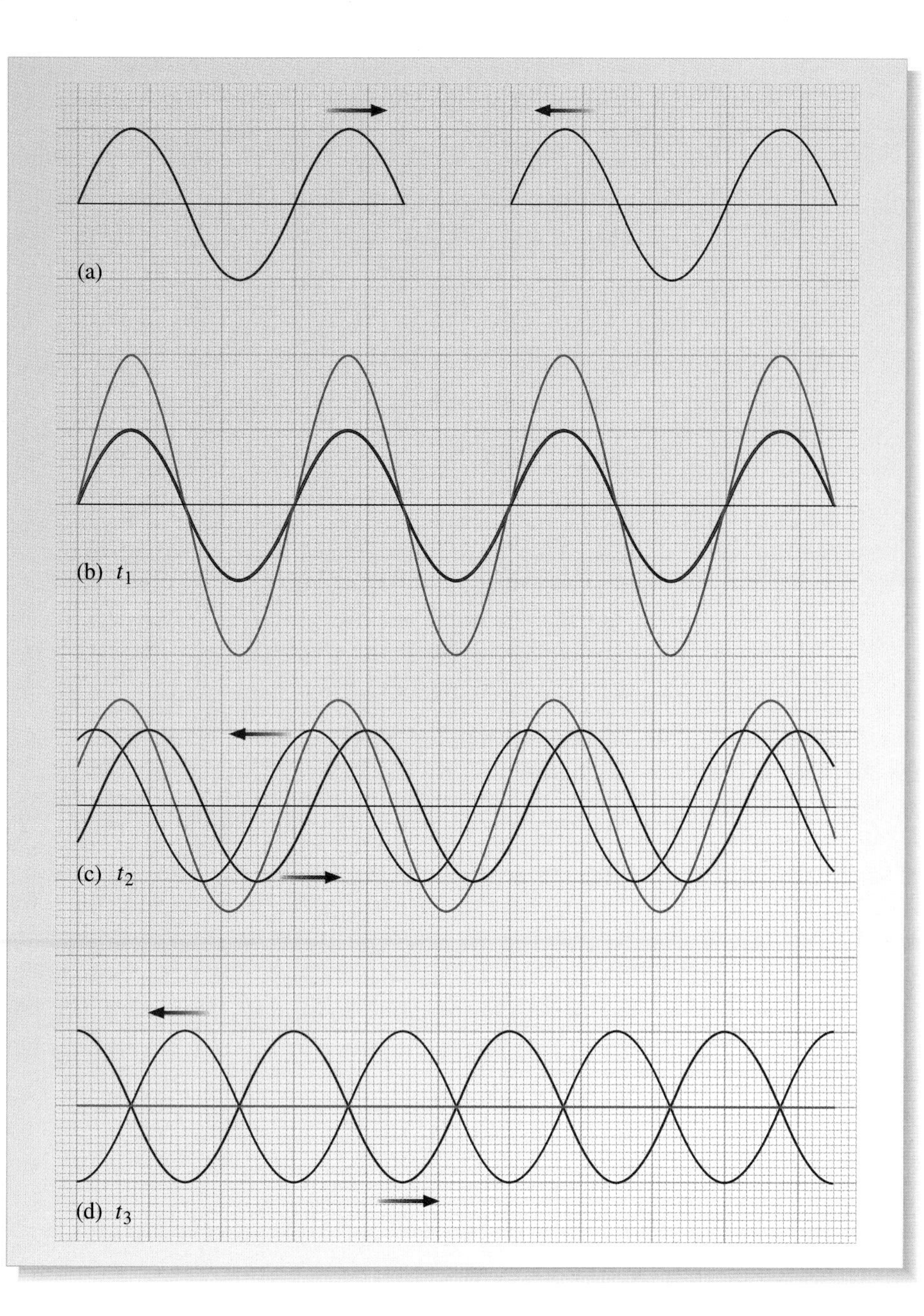

Figure 2.89 Using the principle of superposition to predict the resultant displacement due to identical sinusoidal travelling waves meeting (for Q2.11). The sum of the displacements at various times is shown in purple.

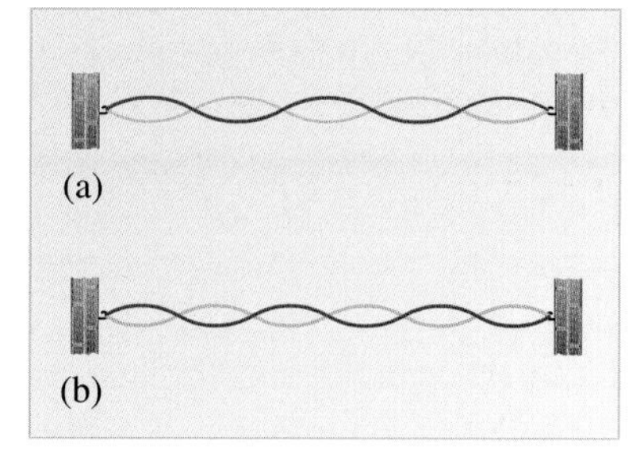

Figure 2.90 The next two standing waves after those shown in Figure 2.27 (for Q2.12). (a) is the fifth harmonic and (b) is the sixth harmonic.

Q2.11 Figure 2.89 shows sketches of the resultant displacement.

Q2.12 (a) Figure 2.90 shows the next two standing waves.

(b) The fifth harmonic has $L = 5\lambda/2$, so

$$\lambda = 2L/5 = 2 \times 3.0\,\text{m}/5 = 1.2\,\text{m}.$$

The sixth harmonic has $L = 3\lambda$, so

$$\lambda = L/3 = 3.0\,\text{m}/3 = 1.0\,\text{m}.$$

Q2.13 The fundamental mode has

$$\lambda = 2L = 2 \times 0.32\text{ m} = 0.64\text{ m}$$

$$f = \frac{n}{2L}\sqrt{\frac{F_\text{T}}{\mu}}$$

$$= \frac{1}{2 \times 0.32\text{ m}}\sqrt{\frac{68\text{ N}}{3.8 \times 10^{-4}\text{kg m}^{-1}}}$$

$= 660$ Hz to two significant figures.

The second harmonic has

$$\lambda = L = 0.32\ \text{m}$$

$$f = \frac{n}{2L}\sqrt{\frac{F_T}{\mu}}$$

$$= \frac{2}{2 \times 0.32\ \text{m}}\sqrt{\frac{68\ \text{N}}{3.8 \times 10^{-4}\,\text{kg m}^{-1}}}$$

$= 1300$ Hz to two significant figures.

Note that the question gives a value for μ in terms of g m^{-1}, so you must first express it in terms of kg m^{-1} to obtain the correct answer.

Q2.14 Initially, as no light is transmitted, the two Polaroid sheets must be *crossed*; that is, one of them only allows through light that is polarized in one direction, whereas the second only allows through light that is polarized in a direction perpendicular to the first. As the second sheet of Polaroid is rotated, the amount of light transmitted increases steadily up to a maximum at a rotation angle of 90°. At this point, the two sheets of Polaroid each allow through light that is polarized in the *same* direction. Then, as the second sheet of Polaroid is rotated further, up to 180° from its original position, the transmitted light decreases to zero once again. At this point the two sheets of Polaroid are once again crossed.

Q2.15 (a) Rearranging Equation 2.22 gives

$$\lambda = \frac{c}{f}.$$

So the wavelength of the radio waves, at the frequency used by Radio 5, is

$$\lambda = \frac{3.00 \times 10^8\ \text{m s}^{-1}}{909 \times 10^3\ \text{Hz}} \doteq 330\,\text{m}.$$

(b) We can also rearrange Equation 2.22 to give

$$f = \frac{c}{\lambda}.$$

So the frequency of the X-rays is

$$f = \frac{3.00 \times 10^8\ \text{m s}^{-1}}{0.021 \times 10^{-9}\ \text{m}} = 1.43 \times 10^{19}\ \text{Hz}.$$

Q2.16 Equation 2.26 can be rearranged to give

$$\sin r = \frac{v_2}{v_1}\sin i = \frac{1.88 \times 10^8\ \text{m s}^{-1}}{3.00 \times 10^8\ \text{m s}^{-1}} \times \sin 45° = 0.443.$$

Thus, $r = 26.3°$. Note that this is *less* than the angle of incidence, as expected.

Q2.17 The completed sketch is shown in Figure 2.91. As the incident ray strikes the glass face of the triangle normally, it continues to travel in the same direction inside the glass. It then strikes the glass–air interface at an (internal) angle of incidence of 45°. This is a situation where the wave tries to cross the boundary *from* glass *into* air, so total internal reflection can occur. It will happen if the angle of incidence is *greater* that the critical angle. From Equation 2.30,

$$\sin i_{\text{crit}} = \frac{n_2}{n_1} = \frac{1.00}{1.60} = 0.625.$$

Hence $i_{\text{crit}} = 38.7°$. As the angle of incidence is 45°, total internal reflection *will* occur. The angle of reflection will be 45°, so the ray will be bent through a right-angle. This means it will strike the second face of the triangle at an angle of incidence of 45°, so once again total internal reflection will occur, and the ray will be deflected through a second right-angle. It will then strike the 'base-face' normally, and so will pass straight out into the air, travelling in a direction exactly opposite to that in which it was originally incident.

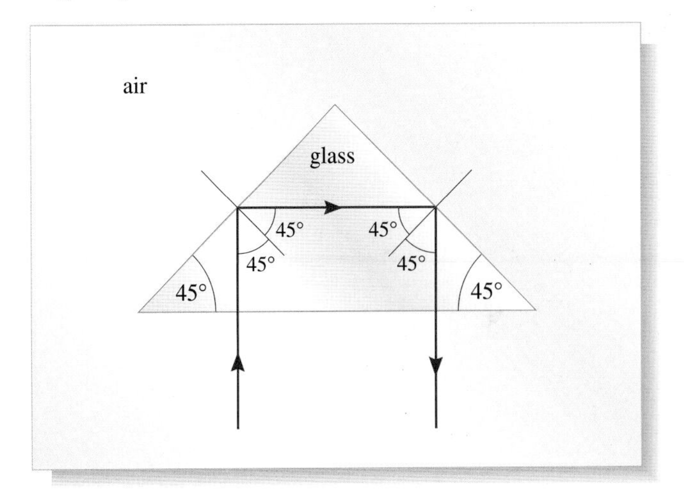

Figure 2.91 The completed version of the sketch for Q2.17.

Q2.18 The diffraction grating has 700 lines per mm, so $d = 1/(700\,\text{mm}^{-1}) = 1.43 \times 10^{-3}\,\text{mm}$ or $1.43 \times 10^{-6}\,\text{m}$. Using Equation 2.31, $n\lambda = d\sin\theta_n$ with $n = 2$, $\theta_n = 62.4°$ and $d = 1.43 \times 10^{-6}\,\text{m}$:

$$2\lambda = 1.43 \times 10^{-6}\,\text{m} \times \sin(62.4°)$$

$$\lambda = \frac{1.43 \times 10^{-6}\,\text{m} \times 0.886}{2} = 6.33 \times 10^{-7}\,\text{m}.$$

So the wavelength of the laser light is 633 nm.

Q2.19 (a) Rearranging Equation 2.32 gives $w = \lambda/\sin\theta$. The wavelength $\lambda = 589.0$ nm, so

for $\theta = 20.0°$,

$$w = \frac{589.0 \times 10^{-9}\,\text{m}}{\sin 20.0°} = 1.72 \times 10^{-6}\,\text{m} = 1720\,\text{nm};$$

for $\theta = 40.0°$,

$$w = \frac{589.0 \times 10^{-9}\,\text{m}}{\sin 40.0°} = 9.16 \times 10^{-7}\,\text{m} = 916\,\text{nm};$$

for $\theta = 60.0°$,

$$w = \frac{589.0 \times 10^{-9}\,\text{m}}{\sin 60.0°} = 6.80 \times 10^{-7}\,\text{m} = 680\,\text{nm}.$$

Note that the diffraction angle *increases* as the aperture width *decreases.*

(b) From Equation 2.32, with $\lambda = 546.1$ nm,

when $w = 1.72 \times 10^{-6}\,\text{m}$,

$$\sin\theta = \frac{546.1 \times 10^{-9}\,\text{m}}{1.72 \times 10^{-6}\,\text{m}} = 0.3175, \text{ so } \theta = 18.5°;$$

when $w = 9.16 \times 10^{-7}\,\text{m}$,

$$\sin\theta = \frac{546.1 \times 10^{-9}\,\text{m}}{9.16 \times 10^{-7}\,\text{m}} = 0.5962, \text{ so } \theta = 36.6°;$$

when $w = 6.80 \times 10^{-7}\,\text{m}$,

$$\sin\theta = \frac{546.1 \times 10^{-9}\,\text{m}}{6.80 \times 10^{-7}\,\text{m}} = 0.8031, \text{ so } \theta = 53.4°.$$

Notice that the diffraction angle *decreases* as the wavelength *decreases* for any particular slit width.

Q2.20 The cells behave as a reflection diffraction grating, with grating spacing $d = 500$ nm. Equation 2.31 holds for light incident at an angle of incidence of 90°. The maximum possible value of $\sin\theta$ is 1, so the *maximum* wavelength that can be observed in the first-order diffraction maximum is

$$\lambda_{\text{max}} = \frac{d(\sin\theta)_{\text{max}}}{1} = \frac{500\,\text{nm}}{1} = 500\,\text{nm}.$$

Only wavelengths below 500 nm can be seen in first-order diffraction maxima from the grating when it is illuminated at 90° incidence. All such wavelengths are in the green-to-violet part of the spectrum (see Figure 2.44).

The first-order diffraction maxima occur at slightly different angles for each value of λ. Thus, the feathers appear to change colour slightly as they are viewed at different angles, but they are always within the range of green to violet.

Q2.21 The soap film appears black as a result of *destructive* interference between the waves reflected from the top and the bottom of the film. Since we can assume that the top and bottom of the film are zero distance apart at the thinnest point of the film, the two reflected waves are π radians out of phase. This is because the light waves are inverted when reflected at the air–film boundary (since the speed of light is less in the film than in air), but not at the film–air boundary.

The oil film appears bright at its thinnest point as a result of *constructive* interference between the waves reflected at the top and bottom of the film; that is, the two reflected waves are in phase. The light waves are inverted at both the air–oil and oil–water interfaces; that is, the speed of light is less in the oil than it is in air, but also less in water than it is in the oil.

Q2.22 The angular wave number $k = 2\pi/\lambda$, where $\lambda = vT$ (from Equation 2.4), so

$$k = \frac{2\pi}{vT} = \frac{2\pi}{0.15\ \text{m s}^{-1} \times 5.0\,\text{s}} = 8.4\,\text{m}^{-1}.$$

The angular frequency $\omega = \dfrac{2\pi}{T} = \dfrac{2\pi}{5.0\,\text{s}} = 1.3\ \text{s}^{-1}$.

So, using Equation 2.3, $y = A\sin(kx - \omega t)$, with $A = 3.0\,\text{cm} = 0.030\,\text{m}$, the required function is

$$y = 0.030\sin(8.4x - 1.3t)$$

where y and x are in metres and t is in seconds.

Q2.23 The frequency is unchanged when sound travels from one medium to another, so a 1.00 kHz sound wave in air will still be a 1.00 kHz sound wave in water. The wavelength in water can be found from Equation 2.5:

$$\lambda = \frac{v}{f} = \frac{1480\ \text{m s}^{-1}}{1000\ \text{Hz}} = 1.48\,\text{m}.$$

Q2.24 From Equation 2.8a, $I \propto 1/r^2$.

Given that $I_1 = 10^{-3}\,\text{W m}^{-2}$, $r_1 = 0.25$ m and $I_2 = 3 \times 10^{-6}\,\text{W m}^{-2}$, the distance r_2 corresponding to an intensity I_2 can be found from

$$\frac{I_1}{I_2} = \frac{r_2^2}{r_1^2}.$$

Therefore

$$r_2^2 = \frac{I_1 \times r_1^2}{I_2} = \frac{10^{-3}\,\text{W m}^{-2} \times (0.25\,\text{m})^2}{3 \times 10^{-6}\,\text{W m}^{-2}} = 21\ \text{m}^2$$

giving $r_2 = 4.6$ m.

Q2.25 When the train is approaching the observer, the shift in frequency is given by Equation 2.13 as

$$\Delta f = f_{\text{em}}\left(\frac{V}{v - V}\right)$$

$$= 1500\,\text{Hz} \times \left(\frac{25\,\text{m}\,\text{s}^{-1}}{330\,\text{m}\,\text{s}^{-1} - 25\,\text{m}\,\text{s}^{-1}}\right) = 123\,\text{Hz}.$$

So the frequency heard by the observer is

$$(1500\,\text{Hz} + 123\,\text{Hz}) = 1623\,\text{Hz}.$$

When the train is receding from the observer, the shift in frequency is given by Equation 2.13, but replacing V with $-V$. So

$$\Delta f = f_{\text{em}}\left(\frac{-V}{v + V}\right)$$

$$= 1500\ \text{Hz} \times \left(\frac{-25\ \text{m s}^{-1}}{330\ \text{m s}^{-1} + 25\ \text{m s}^{-1}}\right)$$

$$= -106\ \text{Hz}.$$

So the frequency heard by the observer is

$$(1500\,\text{Hz} - 106\,\text{Hz}) = 1394\,\text{Hz}.$$

The change in frequency as the train passes the observer is (1623 Hz − 1394 Hz) = 229 Hz, or 230 Hz to two significant figures.

Q2.26 Snell's law states that the angle of refraction r is given by

$$\frac{\sin 55°}{\sin r} = \frac{v_{\text{air}}}{v_{\text{glass}}}$$

Hence

$$\sin r = \frac{v_{\text{glass}}}{v_{\text{air}}} \times \sin 55°$$

For the blue light

$$\sin r_{486} = \frac{1.80 \times 10^8\ \text{m s}^{-1}}{3.00 \times 10^8\ \text{m s}^{-1}} \times \sin 55° = 0.4915$$

so $\quad r_{486} = 29.44°$.

For the red light

$$\sin r_{656} = \frac{1.82 \times 10^8\ \text{m s}^{-1}}{3.00 \times 10^8\ \text{m s}^{-1}} \times \sin 55° = 0.4970$$

so $\quad r_{656} = 29.80°$.

The angular separation is therefore

$$29.80° - 29.44° = 0.36°.$$

Note that this question does not require you to use the wavelength values you were given. Note also that when working with sines (and cosines) you should keep at least four significant figures throughout the calculation and then quote the answer to an appropriate number of significant figures (two in this case, to match the two significant figures for the angle of incidence).

Q2.27 (a) Substituting the specified conditions relating to Figure 2.64 into the refraction equation $\sin i/\sin r = n_2/n_1$, gives

$$\frac{\sin 27.0°}{\sin 45.0°} = \frac{1.00}{n_1}.$$

(Note that in going from glass to air, n_2 represents the refractive index of air.) Hence

$$n_1 = (\sin 45.0°)/(\sin 27.0°) = 1.56.$$

(b) The critical angle is given by

$$\sin i_{\text{crit}} = \frac{n_2}{n_1} = \frac{1.00}{1.56} = 0.641.$$

So the critical angle is $i_{\text{crit}} = 39.9°$.

Q2.28 The string will be flat when the maxima of one wave coincide with the minima of the second. The next time this will occur is when the waves have moved one wavelength with respect to each other; that is, when *each* wave has moved half a wavelength. Thus, each wave moves half a wavelength in 0.5 s, so $f = 1$ Hz.

Rearranging Equation 2.5 gives

$$\lambda = \frac{v}{f} = \frac{0.2\ \text{m s}^{-1}}{1\,\text{Hz}} = 0.2\,\text{m}.$$

After 0.25 s, the waves have travelled one half-wavelength with respect to each other, which means that the maxima (and minima) of the two waves will be coincident. Thus, a sine wave will be instantaneously observed with twice the amplitude of the individual waves.

Q2.29 Equation 2.31 can be rearranged to give

$$\sin\theta_n = \frac{n\lambda}{d}.$$

Now, $\sin\theta_n$ must be less than or equal to 1, so $n\lambda/d \leqslant 1$, and $n \leqslant d/\lambda$.

In this case, $\lambda = 450\,\text{nm} = 4.5 \times 10^{-7}\,\text{m}$, and there are 600 lines per millimetre so

$$d = \frac{1}{600 \times 1000}\,\text{m}.$$

Hence

$$\frac{d}{\lambda} = \frac{1}{6.0 \times 10^5 \times 4.5 \times 10^{-7}} = 3.7.$$

n must be an integer, so the maximum number of orders is three.

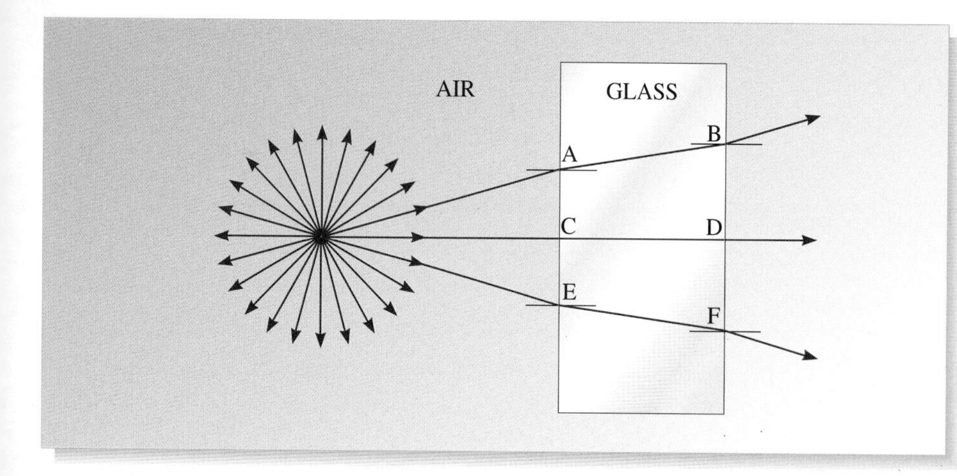

Figure 3.51 Rays are bent towards the normal as they pass from air into glass, and away from the normal as they pass back from the glass into air.

Q3.1 Snell's law says that the rays of light will change direction as they cross from medium 1 into medium 2, according to the equation

$$\frac{\sin i}{\sin r} = n_{21} = \frac{n_2}{n_1}.$$

Glass has a refractive index *greater* than that for air (i.e. the wave speed is slower in glass than air), so that on entering the glass block (when $n_1 = n_{\text{air}}$ and $n_2 = n_{\text{glass}}$), n_2/n_1 is greater than 1, and hence i must be greater than r; therefore, all rays must be bent *towards* the normal. No bending occurs for rays that strike the interface normally (i.e. along the direction of the normal): these rays will continue along the direction of the normal after they have crossed the interface. (This follows because, if $i = 0$ then $\sin i = 0$ so that, by Snell's law, $\sin r = 0$, and therefore $r = 0$.)

When the rays leave the glass block, however, $n_1 = n_{\text{glass}}$ and $n_2 = n_{\text{air}}$, so that n_2/n_1 is now *less* than 1, and r must be greater than i. So, on exit from the glass, the rays are bent *away* from the normal (except for normally incident rays, which pass straight out of the glass undeviated.) Your completed sketch should have looked something like Figure 3.51.

Q3.2 At the first interface:

$$\frac{\sin i_{\text{air}}}{\sin r_{\text{glass}}} = \frac{n_{\text{glass}}}{n_{\text{air}}}. \tag{3.29}$$

At the second interface:

$$\frac{\sin i_{\text{glass}}}{\sin r_{\text{air}}} = \frac{n_{\text{air}}}{n_{\text{glass}}}. \tag{3.30}$$

Since the right-hand side of Equation 3.29 is the reciprocal of the right-hand side of Equation 3.30, it follows that

$$\frac{\sin i_{\text{air}}}{\sin r_{\text{glass}}} = \frac{\sin r_{\text{air}}}{\sin i_{\text{glass}}}. \tag{3.31}$$

But it is clear from Figure 3.4 that $r_{\text{glass}} = i_{\text{glass}}$, so that $\sin r_{\text{glass}} = \sin i_{\text{glass}}$. Hence, from Equation 3.31, $\sin i_{\text{air}} = \sin r_{\text{air}}$, or $i_{\text{air}} = r_{\text{air}}$, as required.

This result has to be true, as you can see by imagining the light travelling in the *reverse* direction along the rays. (Recall the principle of the reversibility of light paths from Chapter 2, Section 6.1.)

Q3.3 The focal length of a converging lens is that distance over which parallel rays are brought to a focus. Hence, if parallel rays are allowed to fall onto the lens, all you then have to do is measure the distance from the lens to the point at which all the rays come to a sharp focus. A very good approximation to a source of parallel light rays is the Sun. So if you focus sunlight onto a piece of card (or similar surface), adjusting the distance between the lens and the card until the sharpest, smallest spot of light is obtained, this lens-to-card distance will then be the focal length of the lens.

As an aside, you should note that any methods that involve you *looking through* the lens, say at a distant object, are doomed to fail! Such systems contain not one, but two lenses — the lens of interest *and your eye-lens*. And as you will see later, your eye-lens is cleverer than most!

You should also note that the method of focusing parallel rays will not work with a diverging (concave, negative) lens, since the focal point in this case is not 'real'. That is, the rays of light are not focused down to a real spot, but only appear to diverge *as if* from an imaginary spot. More complicated methods are needed for such lenses.

Q3.4 When the object O is *inside* the focal distance, the bending power of the lens will not be sufficient to overcome the high divergence of the incident rays to render them even parallel. The rays will still be diverging after they leave the lens, though not as strongly as when they entered it. As can be seen from Figure 3.52, these output rays will now be diverging *as if* from a point to the

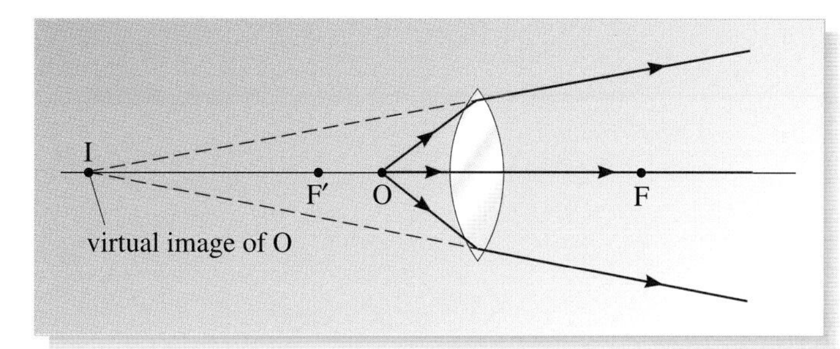

Figure 3.52 When an object is positioned a distance *less* than the focal length from a converging lens, the emergent rays are still diverging; the image I is said to be *virtual.*

left of the lens. This will be the image I of the object O. You will see as you read on in the main text of Chapter 3 that an image like this, in which the rays do not really pass through the point but only *appear* to diverge from the point, is called a *virtual* image.

Q3.5 In all three cases $1/v + 1/u = 1/f$.

(a) Here, the object is a distance f in front of the lens; hence $u = +f$. The lens equation reduces to $1/v + 1/f = 1/f$, or $1/v = 0$. It therefore follows that $v = 1/\text{zero} = \text{infinity}$: the image is at infinity, so the output beam is parallel.

(b) The input beam is parallel: $u = \infty$. The lens equation becomes $1/v + 1/\infty = 1/f$. As $1/\infty$ equals zero, so $1/v = 1/f$, or $v = f$. The image is real (v is positive since f is positive) and a distance f beyond the lens.

(c) Again, $1/u = 1/\infty = \text{zero}$. Hence $1/v + \text{zero} = 1/f$, so $v = f$. But remember that in this case f is negative (it's a diverging lens), so that v will be negative. Hence, the image will be a distance f from the lens, but it will be a virtual image located in front of the lens.

Q3.6 Your sketch should look something like Figure 3.53. The diverging lens, being concave in shape, delays the wavefront *less* at its centre than at its perimeter. The emergent wavefronts, therefore, will acquire a curvature that makes them *appear* to have diverged from the focal point F to the left of the lens.

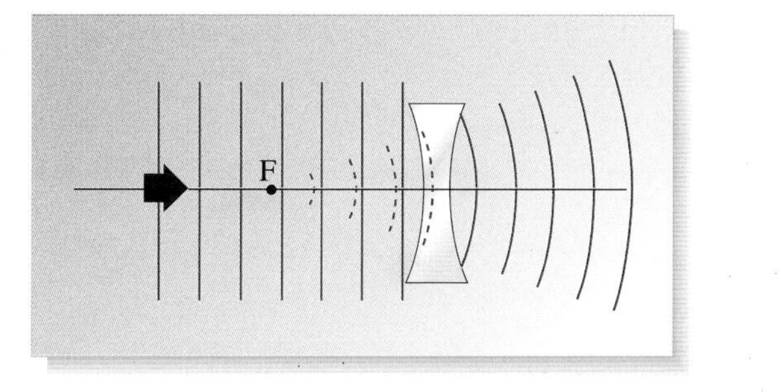

Figure 3.53 A diverging lens converts plane wavefronts into *diverging* spherical wavefronts.

Q3.7 As you already know, the image will be real and in the same plane as the object; however, as you can see from Figure 3.54, it is inverted. It will also be the same size as the object.

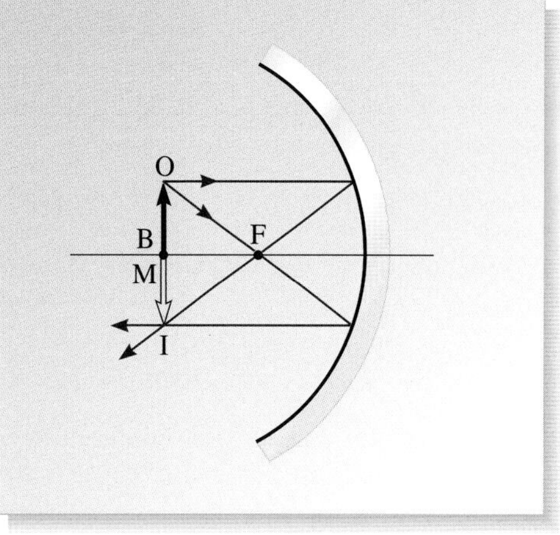

Figure 3.54 An extended object placed at the centre of curvature of a converging mirror produces an inverted, same-size image, also located in the plane passing through the centre of curvature.

Q3.8 The suggested construction is shown in Figure 3.55. The point where the optical axis intersects the mirror is labelled A. But the optical axis must be perpendicular to the mirror surface at A. Hence, the angle of incidence i of ray OAI must equal the angle of reflection R. It therefore follows that triangles OBA and IMA are *similar* ($i = R$, they each have a right-angle, at B and M, respectively, and hence the third angle must be the same in each triangle).

So, $\dfrac{\text{IM}}{\text{MA}} = \dfrac{\text{OB}}{\text{BA}}$ or $\dfrac{\text{IM}}{\text{OB}} = \dfrac{\text{MA}}{\text{BA}}$.

But MA is v, and BA is u, so

$$\text{magnification } m = \frac{\text{IM}}{\text{OB}} = \frac{v}{u}$$

as required.

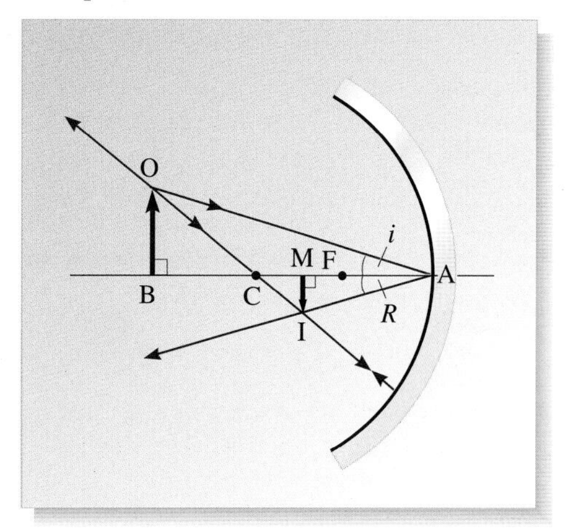

Figure 3.55 Answer to Q3.8.

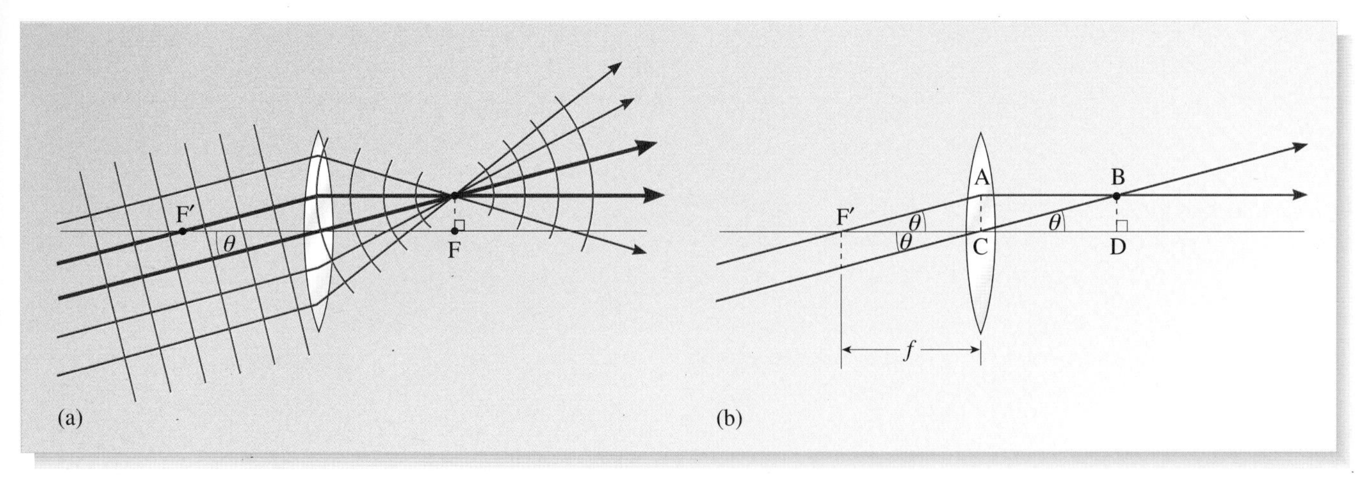

Figure 3.56 (a) Inclined plane wavefronts are focused by a converging lens to an off-axis image point in the focal plane beyond the lens; (b) geometry and trigonometry show that this image point is located a distance $f \tan \theta$ from the optical axis.

Q3.9 You should have sketched something like Figure 3.56a. Note that the rays are depicted by red lines and the wavefronts by yellow lines. The plane wavefronts are converted by the lens into spherical waves that converge down to a point in the focal plane. Two key rays are included among the inclined parallel rays: one passing undeviated through the centre of the lens, and one that passes through the focal point (on the input side) to emerge parallel to the optical axis. Note that all the other emergent rays are then everywhere perpendicular to the (spherical) wavefronts.

To work out expressions for the location of the image point, a simplified diagram is shown in Figure 3.56b, showing only the two key rays and the optical axis. All the angles shown as θ in this diagram are equal. B is the image point of the inclined plane wave; D is the point on the axis perpendicularly below B. From the diagram, triangles ACF′ and BDC are identical. Hence, CD = F′C = f. The image point is in the focal plane beyond the lens, and the point D coincides with F. Using trigonometry in the triangle BDC gives $\tan \theta = \mathrm{BD}/f$, or $\mathrm{BD} = f \tan \theta$. The image point is located a distance $f \tan\theta$ from the optical axis.

Q3.10 An incident parallel beam of light is equivalent to an object distance of infinity for the first lens; the image produced by this lens will therefore be at f_1. But this image will constitute a *virtual object* for the second lens, with the object distance given by $u_2 = -f_1$ (assuming the lens separation is negligible). Hence, for the second lens:

$$\frac{1}{v_2} - \frac{1}{f_1} = \frac{1}{f_2}$$

or

$$\frac{1}{v_2} = \frac{1}{f_1} + \frac{1}{f_2}.$$

But as the input to the two-lens combination is parallel light, the final image distance must be the same as the *equivalent* focal length, f_{total} say, for the two lenses taken together. Therefore, f_{total} can be calculated from

$$\frac{1}{f_{\text{total}}} = \frac{1}{f_1} + \frac{1}{f_2}.$$

Q3.11 Since $f = 25\ \text{mm} = 0.025\ \text{m}$,

$$P = 1/f = 1/(0.025\ \text{m}) = 40\ \text{D}.$$

Q3.12 (a) Light from the *normal* near point (at 250 mm) must *appear* to come from the long-sighted near point. Hence, for the added lens, v must equal $-d_{\text{np}}$ when $u = 250\ \text{mm}$ (note the image is virtual). Thus,

$$\frac{1}{f} = \frac{1}{0.25\ \text{m}} - \frac{1}{d_{\text{np}}}$$

or

$$P = 4\,\text{D} - \frac{1}{d_{\text{np}}}$$

where P is the power of the required lens (in dioptres).

(b) If $d_{\text{np}} = 1.50\ \text{m}$, then the power of the correcting lens must be

$$P = 4\,\text{D} - (1/1.50)\,\text{D}$$
$$= 4\,\text{D} - (2/3)\,\text{D}$$
$$= +3.33\,\text{D}$$

which is equivalent to a focal length of +300 mm. (Make sure you understand why.)

Q3.13 The calculation is straightforward. Linear magnification m is given by v/u. Here $v = 17\ \text{mm}$ and $u = 250\ \text{mm}$, so $m = 0.068$. Hence, the diameter of the image will be $0.068 \times 22\ \text{mm}$ (since 22 mm is the object diameter), which is 1.50 mm, as stated in the text.

Q3.14 For the normal (standard) eye, d_{np} is assumed to be 250 mm. Hence in far-point adjustment, $M = d_{np}/f = 250\,\text{mm}/50\,\text{mm} = 5$. This is usually written as a magnifying power of ×5 (times 5). In near-point adjustment, the magnifying power will be ×6.

Q3.15 The image seen through the microscope will appear *inverted*. As is clear from Figure 3.34, the *retinal* image will be upright, but an upright retinal image is interpreted by the brain as being upside-down. The magnifying-lens part of the microscope (the eyepiece) does *not* invert the image (a magnifying glass doesn't, does it?); the inversion is caused by the objective lens, which produces a real, magnified *and inverted* image I′M′ of the original object.

Q3.16 The magnifying power M_e of the eyepiece is independent of the position of the object. In far-point adjustment, it is given by d_{np}/f_e, which in this case is 250 mm/50 mm or ×5. In near-point adjustment, it will be one greater than this — that is, ×6 (compare the second terms in Equations 3.20 and 3.21). Also in near-point adjustment,

$$1/u_e = 1/f_e + 1/d_{np} = 1/(50\,\text{mm}) + 1/(250\,\text{mm})$$

that is, $u_e = (250/6)\,\text{mm} = 41.7\,\text{mm}$.

(a) When $u_o = 20\,\text{mm}$ (and f_o is 15 mm),

$$\frac{1}{v_o} = \frac{1}{15\,\text{mm}} - \frac{1}{20\,\text{mm}}$$

$$\frac{1}{v_o} = \frac{4-3}{60}\,\text{mm}^{-1}.$$

Therefore $v_o = 60\,\text{mm}$. Hence,

$$m_o = v_o/u_o = 60\,\text{mm}/20\,\text{mm} = \times 3.$$

In far-point adjustment: the overall magnifying power (angular magnification), M(far point), is $(3 \times 5) = \times 15$; the microscope length is $(v_o + f_e) = (60 + 50)\,\text{mm} = 110\,\text{mm}$.

In near-point adjustment: the overall magnifying power, M(near point), is $(3 \times 6) = \times 18$; the microscope length is now $(v_o + u_e) = (60 + 41.7)\,\text{mm} = 101.7\,\text{mm}$ (using the value of u_e calculated above).

(b) When $u_o = 18\,\text{mm}$, applying $1/v_o + 1/u_o = 1/f_o$ gives $v_o = 90\,\text{mm}$. Hence, $m_o = v_o/u_o = 90\,\text{mm}/18\,\text{mm} = \times 5$.

In far-point adjustment: M(far point) $= (5 \times 5) = \times 25$; the microscope length is $(90 + 50)\,\text{mm} = 140\,\text{mm}$.

In near-point adjustment: M(near point) $= (5 \times 6) = \times 30$; the microscope length is $(90 + 41.7)\,\text{mm} = 131.7\,\text{mm}$.

Notice the dramatic increase required in the length of the microscope (30 mm) in order to accommodate a shift of only 2 mm in the object position.

Q3.17 (a) In far-point adjustment, and looking at the stars, the telescope has 'parallel light' input and 'parallel light' output. Hence the length of the telescope is given by the sum of the two focal lengths, that is

$$\text{length} = f_o + f_e = 20\,\text{m} + 8\,\text{mm} \approx 20\,\text{m}.$$

The magnifying power (angular magnification) is given by f_o/f_e; that is

$$M = \frac{20 \times 10^3\,\text{mm}}{8\,\text{mm}} = \times 2500.$$

(b) The beam diameter D_e into which the light passing through the objective is concentrated by the time it reaches the eyepiece can be found from the similar triangles ratio $D_o/D_e = f_o/f_e$. Thus, $D_e = (1\,\text{m} \times 8\,\text{mm})/20\,\text{m} = 0.4\,\text{mm}$. This is much less than D_p, so all the light does enter the eye, and the light-gathering power is given by $(D_o/D_p)^2$. Hence, the light-gathering power $= (1000\,\text{mm}/8\,\text{mm})^2 = 15\,625$ or about 16 000.

Q3.18 If 80 mm is the *minimum* value v can take, then this value must correspond to the *maximum* required value of u, i.e. to u = infinity. Hence this value of v must be equal to the focal length of the lens, so f must be 80 mm. The lens equation can now be used to find the v-value corresponding to $u = 1\,\text{m}$ (1000 mm). Thus

$$\frac{1}{v} = \frac{1}{f} - \frac{1}{u} = \frac{1}{80\,\text{mm}} - \frac{1}{1000\,\text{mm}}$$

giving $v = 87.0\,\text{mm}$.

In other words, the distance between the lens and the film plane must be adjustable from 80 mm to 87 mm, a range of 7 mm.

Q3.19 (a) An aperture-stop specification of $f/8$ (or $F = 8$) means that the *aperture diameter* is $f/8$, where f is the lens's focal length. Here, $f = 48\,\text{mm}$, so the aperture diameter $D = 48\,\text{mm}/8 = 6\,\text{mm}$.

(b) If the time exposure is reduced by a factor of 2, the intensity (and therefore the aperture area) must be increased by 2 to keep the total exposure ($I\,\Delta t$) constant. Increasing the aperture area by 2 means increasing the aperture *diameter* by $\sqrt{2}$, to approximately 8.5 mm; this is achieved by changing the aperture setting one position to the next *smaller* F-number — that is, to $f/5.6$ (or $F = 5.6$).

Do make sure that you are quite clear about the distinctions between exposure, exposure time and intensity. Exposure is the product of the intensity and exposure time (refer to Equation 3.27).

(c) If $D = 3\,\text{mm}$, then $F = f/D = 48/3 = 16$; that is, the aperture setting would be $f/16$ or $F = 16$. Reference to Table 3.1 shows that this is *two* stops down (i.e. an area smaller by a factor of 4) than the original $f/8$ stop. Hence to compensate, the time exposure should be *increased* by two positions (i.e. by a factor of 4) to 1/30 s.

Q3.20 Refer to Figure 3.57. Triangle $BB'F_1$ is similar to triangle $CC'F_1$. Therefore $BB'/f_1 = CC'/(f_1 - d)$. Since $f_1 = 50$ mm, $d = 30$ mm, and BB' is equivalent to the beam diameter D, then $D/50 = CC'/20$ or $CC' = (2/5)D$. But triangles $AA'I$ and $CC'I$ are also similar; hence $CC'/v_2 = AA'/f_{equiv}$. Since v_2 has already been calculated to be 100 mm, and since $AA' = D$ and $CC' = (2/5)D$, then

$$f_{equiv} = \frac{D}{(2/5)D} \times 100 \text{ mm} = 250 \text{ mm}.$$

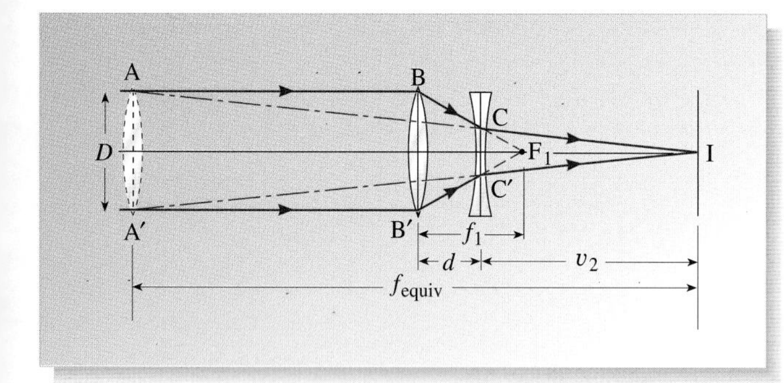

Figure 3.57 Using geometry to calculate the value of f_{equiv} for a telephoto lens.

This analysis indicates how you might go about deducing a generalized equation to calculate the equivalent focal length of a pair of lenses separated by a distance d. You will not be required to derive, or even use, such an equation. However, you might be interested to know that it is:

$$\frac{1}{f_{equiv}} = \frac{1}{f_1} + \frac{1}{f_2} - \frac{d}{f_1 f_2}.$$

Perhaps you'd like to check that this really does give $f_{equiv} = 250$ mm for the example here.

Q3.21 The distance between the optical centre of the mirror and the optical centre of the lens must be 150 mm (that is, 200 mm − 50 mm). Imagine for a moment, the rays reversed in direction. Then the narrow parallel beam falling on the diverging lens will diverge as if from the focal point of this lens — that is, as if from 50 mm on the input side of the lens. But for the rays to leave the mirror as a (broad) parallel beam, they must have originated from a point 200 mm in front of the mirror. In other words, the mirror's focal point and the lens's focal point must coincide. This can happen only if the distance between the lens and the mirror is (200 − 50) mm.

Q3.22 The three sketches you should have drawn are shown in Figure 3.58a–c.

For I to behave as a *real* object for the second lens, this lens must be placed to the *right* of I. If the distance u_2 between I and this second lens is greater than f_2, as in (a), then a *real* final image I_2 is formed; if the distance u_2 is less than f_2, as in (b), then a *virtual* final image I_2 is formed.

For I to behave as a virtual object, the second lens must be placed to the *left* of I, as in (c). This then produces a *real* image I_2 closer to the second lens than the first image I. From (c), you can see that the second lens converges the rays yet more strongly: clearly this can *never* produce a virtual image. The same thing can be seen from the lens equation,

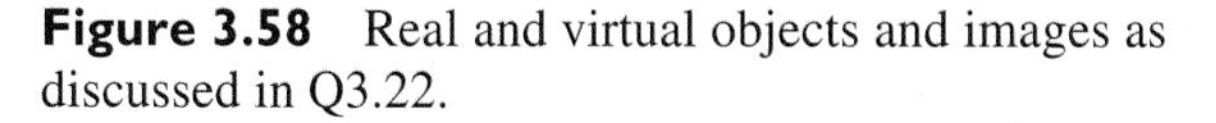

$1/v = 1/f - 1/u$, by noting that, since u here is negative, the right-hand side of this expression must always be positive.

The image distance v could become negative if f were more *negative* than u. In other words, the second lens would have to be a strong diverging lens, as shown in Figure 3.58d.

Figure 3.58 Real and virtual objects and images as discussed in Q3.22.

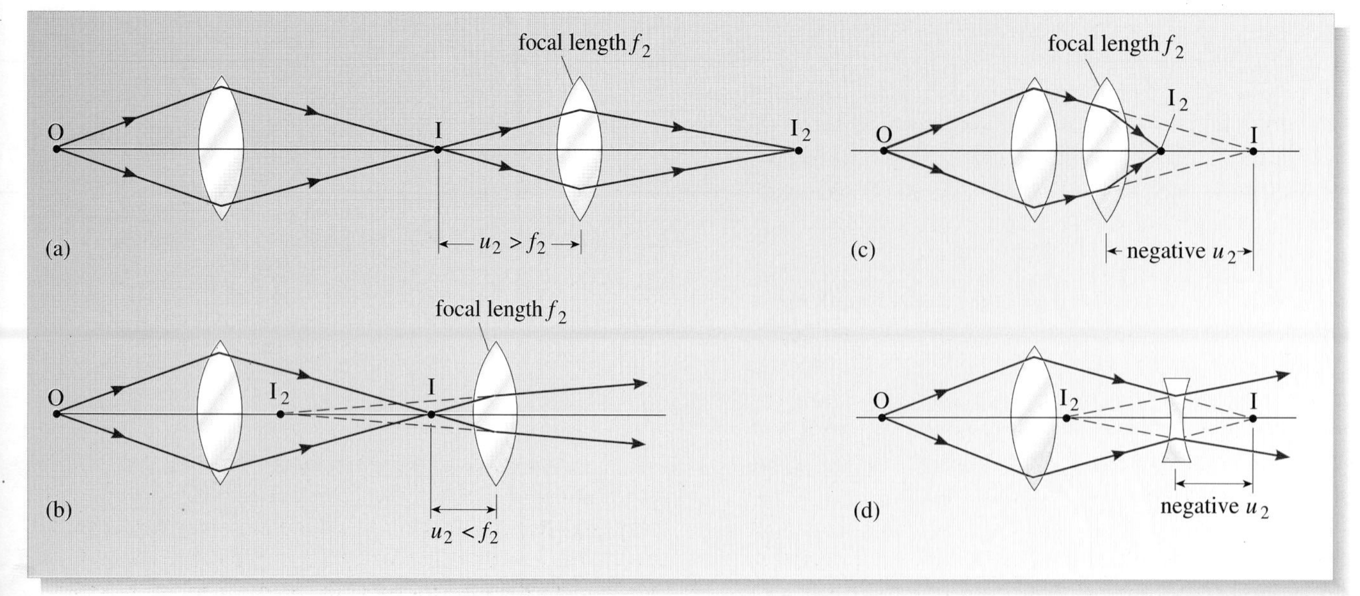

Q3.23 (a) The diverging lens has a *negative* focal length. The object is real and must be assigned a positive value. Substituting in the lens equation $1/v + 1/u = 1/f$,

$$\frac{1}{v} + \frac{1}{100\,\text{mm}} = -\frac{1}{100\,\text{mm}}$$

or

$$\frac{1}{v} = -\frac{1}{100\,\text{mm}} - \frac{1}{100\,\text{mm}} = -\frac{2}{100\,\text{mm}}$$

so $v = -50\,\text{mm}$.

Don't be seduced into thinking that a point source at the focal point always produces parallel rays: a *diverging* lens will diverge the rays *even more*! As the calculation shows, the image is virtual, and positioned midway between the lens and its focal point.

(b) A converging lens has a positive focal length. The image is required to be virtual, and a distance 50 mm in front of the lens; that is, $v = -50\,\text{mm}$. Substituting in the lens equation

$$-\frac{1}{50\,\text{mm}} + \frac{1}{u} = \frac{1}{100\,\text{mm}}$$

or

$$\frac{1}{u} = \frac{1}{100\,\text{mm}} + \frac{1}{50\,\text{mm}} = \frac{3}{100\,\text{mm}}$$

so $u = 33.3\,\text{mm}$.

In order to produce the required virtual image, the object has to be closer to the lens than the image, and well inside the lens's focal length.

Q3.24 Figure 3.59 is an example of the sort of sketch you should have drawn. To determine the position of the image, the key rays chosen here are one that would have passed through the mirror's centre of curvature C, and one travelling initially parallel to the axis, which is reflected in such a way that, when projected back, it passes through the focal point F. Alternatively, you could have chosen the ray that leaves O heading straight towards F, but on intersecting the mirror is reflected back parallel to the axis; the backwards projection of this parallel line would also have passed through I (try it).

In this case the *virtual* image is: (a) located somewhere between F and the mirror, (b) upright, and (c) reduced in size relative to the object. (d) If OB is moved nearer the mirror, the ray initially parallel to the axis will not be altered in any way, but the ray 'passing through' C will now be inclined more steeply, with the result that it will intersect the ray projected back through F *nearer the mirror*. Hence, the image will be nearer the mirror, still upright and, although larger than is the case depicted in the diagram, still reduced in size relative to the object.

Conversely, if OB is moved further away from the mirror, the line OC will intersect the ray projected back through F further away from the mirror, but still somewhere between the mirror and F (the image only moves as far back as F when the object is moved off to infinity); the image will still be upright, and will be smaller than in the case depicted in the diagram.

A convex mirror (which is what this mirror is) makes a good car wing-mirror. The image will always be upright and located behind the mirror (which is what we are used to with ordinary plane mirrors), but it will be reduced in size. This is quite a good safety feature, because it means that the convex mirror offers us a wider 'field of view'—and thus better visibility—than the same-sized plane mirror.

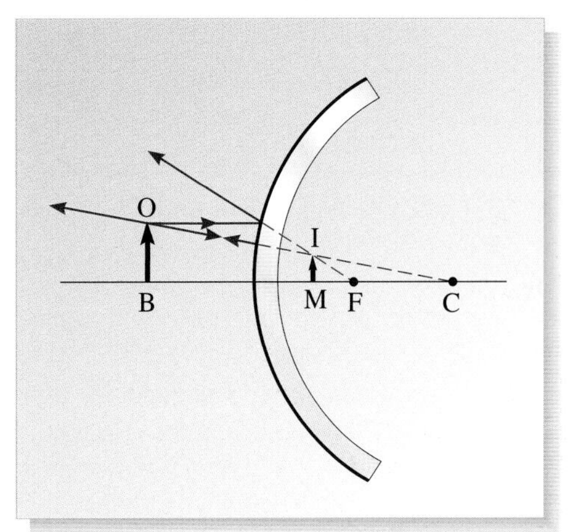

Figure 3.59 A diverging (convex) mirror always produces an upright, reduced-size, virtual image, located somewhere between the mirror and the focal point F.

Q3.25 If the radius of curvature of the mirror is 120 mm, then its focal length f must be half this; that is, $f = 60\,\text{mm}$. Since this is a diverging mirror, f is negative.

(a) Using $1/v + 1/u = 1/f$,

$$\frac{1}{v} = -\frac{1}{60\,\text{mm}} - \frac{1}{90\,\text{mm}} = \frac{-3-2}{180\,\text{mm}} = \frac{-5}{180\,\text{mm}}$$

or $v = -36\,\text{mm}$.

So, the image is located 36 mm *behind* the mirror.

(b) The magnification is given by (the magnitude of) v/u, which here is 36 mm/90 mm, or 0.4 times; that is, the image is *reduced* in height by the factor 2.5 (i.e. 1/0.4). Since the object's height is 10 mm, clearly the image height must be 4 mm.

Q3.26 The ray construction to locate the point image is shown in Figure 3.60 (overleaf). Note that once again the ray that passes through the mirror's centre of curvature has been chosen, but this time, just for a change, it has been paired with the ray that strikes the mirror at the point where the mirror intersects the optical axis; since the optical axis is normal to the mirror at this point, this ray will be reflected at an angle θ above the axis (angle of incidence equals angle of reflection).

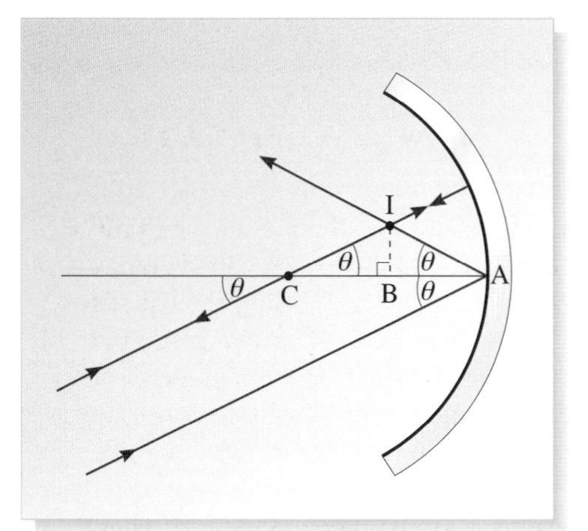

Figure 3.60 The inclined plane wave is imaged at the point I. Triangles IBC and IBA are identical, so AB must equal CB, and thus B must coincide with the mirror's focal point F. Hence, AB $= f$ and IB $= f \tan\theta$.

The point B is the location, on the optical axis, of the perpendicular dropped from the image point I. Therefore, triangles IBC and IBA are identical, and in particular the length CB equals AB. Hence B is midway between A and the mirror's centre of curvature and must therefore coincide with the focal point F. So AB equals f, which here equals 100 mm; the point image is in the focal plane of the mirror. Also, the distance IB equals $f \tan\theta$, which here is 100 mm × 0.087 = 8.7 mm.

Q3.27 (a) Using the lens equation for an object located at the person's near point (of 100 mm) gives:

$$1/f_{np} = 1/(16.7\text{ mm}) + 1/(100\text{ mm}).$$

This gives a power at the near point of $P_{np} = 69.88$ D, which is equivalent to a focal length f_{np} of 14.31 mm. The corresponding calculation for the far point gives:

$$1/f_{fp} = 1/(16.7\text{ mm}) + 1/(167\text{ mm}).$$

This is a far-point power P_{fp} of 65.87 D, equivalent to a focal length f_{fp} of 15.18 mm.

(b) F-number is defined by $F = f/D$, where D is the diameter of the lens. In bright light, $D = 2.0$ mm, and so the corresponding F-number ranges from (14.3/2.0) to (15.2/2.0) — that is, from $F = 7.2$ to $F = 7.6$. In dim light, when the eye is dark-adapted, D is 8.0 mm. Now the corresponding F-numbers range from (14.3/8.0) to (15.2/8.0) — that is, from $F = 1.8$ to $F = 1.9$.

(c) To move the far point out to infinity, the *combined* focal length must be equal to 16.7 mm, the same as the eye-lens–fovea distance (i.e. when the object is at infinity, the image distance equals the focal length). Making the simplifying assumption that the spectacle-lens and the eye-lens can be treated as if they were in contact, then

$$P_{combined} = P_{fp} + P_{spec} \text{ or } 1/f_{combined} = 1/f_{fp} + 1/f_{spec}.$$

These equations indicate that the correcting spectacle-lens should be diverging, with a power of −6.0 D, and therefore a focal length of −167 mm. Note this means that rays from the *uncorrected* far point ($d_{fp} = 167$ mm) appear to originate from infinity — that is, the normal far point.

Q3.28 The text would have to be 3.5 times taller to be just readable. Hence, the lens has to have a magnifying power, $M = \alpha_{IM}/\alpha_{OB}$, of ×3.5.

(a) Maximum magnifying power is achieved in near-point adjustment, when $M = (d_{np}/f) + 1$. In this adjustment, M will be ×3.5, when $d_{np}/f = 2.5$ — that is, when $f = 100$ mm (taking d_{np} as 250 mm). The virtual image will be at the near point, and, using $1/u = 1/f - 1/v$, the object will be at a distance $u = (1/100\text{ mm} + 1/250\text{ mm})^{-1} = 71.4$ mm in front of the lens. If the object is moved further away from the lens, so that the image moves away from the near point towards infinity, the magnifying power will be reduced (it is *always* given by $M = d_{np}/u$), and the text will become unreadable again.

(b) To ensure that the text is readable at *all* virtual image distances, the magnifying power must be equal to ×3.5 in *far-point* adjustment — that is, when $M = d_{np}/f$. Thus, f must be equal to 250 mm/3.5, or less. The lens would have to have a focal length less than or equal to 71.4 mm.

Q3.29 The distance between the two lenses is always $v_o + u_e$ (see Figure 3.34), and the magnifying power of the eyepiece is always d_{np}/u_e.

In far-point adjustment, $u_e = f_e$, and hence $M_e = d_{np}/f_e = 250/100 = \times 2.5$. Also, v_o must equal the separation of the lenses minus f_e — that is, $v_o = (160 - 100)$ mm $= 60$ mm. Substituting this value of v_o into the lens equation for the objective lens gives $u_o = 12.0$ mm. This is how far the object must be in front of the objective lens in far-point adjustment. The linear magnification of the intermediate image in this case is $m_o = v_o/u_o = 60/12.0 = \times 5.0$. The *overall magnifying power* of the microscope is now $m_o \times M_e$; that is, M(far point) $= 5.0 \times 2.5 = \times 12.5$.

In near-point adjustment, the magnifying power of the eyepiece is greater by one than its magnifying power in far-point adjustment — that is, $M_e = \times 3.5$ here. But M_e is still given by d_{np}/u_e, so it follows that $u_e = (250\text{ mm})/3.5 = 71.43$ mm. So in this adjustment, $v_o = (160 - 71.43)$ mm $= 88.57$ mm. Applying the lens equation to the objective lens now gives $u_o = 11.27$ mm. This is the required distance of the object from the objective in near-point adjustment. The linear magnification of the objective (again given by v_o/u_o) is this time m_o = (88.57)/(11.27) = ×7.859. Hence, the *overall* magnifying power of the microscope is 7.859 × 3.5—that is, M(near point) = ×27.5.

Notice that we have used four significant figures throughout this calculation, only rounding the final answer to three significant figures, which was the same accuracy as the quantities given in the question.

Q3.30 (a) Refer to Figure 3.36. If there were no eyepiece lens, then the closest the eye would be able to get to I′M′ and still see this image in focus would be d_{np}, the near-point distance. Hence, the angle subtended at the eye in this case would be given by $\alpha_{IM} = I'M'/d_{np}$ (assuming the angle is small). The value of α_{OB}, however, will still be $\alpha_{OB} = I'M'/f_o$. Therefore, the magnifying power of the objective lens alone will be $M_o = \alpha_{IM}/\alpha_{OB} = f_o/d_{np}$, as required.

(b) When the telescope is in *near-point* adjustment, the intermediate image I′M′ in Figure 3.36 will no longer be in the focal plane of the eyepiece lens, but instead will be a distance u_e from the lens such that a virtual image will be produced at the eye's near-point distance, such that $v_e = -d_{np}$. In other words, the lenses will now be separated by a distance $f_o + u_e$ (i.e. the telescope will have a shorter length), where u_e is given by $1/u_e = 1/f_e + 1/d_{np}$. (Compare this with the equivalent adjustment for the compound microscope shown in Figure 3.34.) The magnifying power is still α_{IM}/α_{OB}, and therefore, assuming both these angles are small,

$$\begin{aligned} M(\text{near point}) &= \alpha_{IM}/\alpha_{OB} \\ &= f_o/u_e \\ &= f_o(1/u_e) \\ &= f_o(1/f_e + 1/d_{np}) \\ &= f_o/f_e + f_o/d_{np}. \end{aligned}$$

From this result, which is one of the forms of the equation quoted in the question, you can see that the near-point magnifying power is greater than the far-point magnifying power by an amount equal to the magnifying power of the objective lens alone.

It is an easy matter to rearrange the above equation into the form

$$M(\text{near point}) = \frac{f_o}{d_{np}}\left(\frac{d_{np}}{f_e} + 1\right)$$

The term outside the brackets is the magnifying power of the objective lens alone (see part (a) of this question), whereas the term in the brackets is the magnifying power of the eyepiece lens alone *in near-point adjustment*. So, in near-point adjustment, the *total* magnifying power is given by the product of the independent magnifying powers of the objective and eyepiece lenses, as stated in the question. If this is also true in *far-point* adjustment, when the magnifying power of the eyepiece is $M_e = d_{np}/f_e$, then

$$\begin{aligned} M(\text{far point}) &= M_o \times M_e \\ &= \left(\frac{f_o}{d_{np}}\right) \times \left(\frac{d_{np}}{f_e}\right) = \frac{f_o}{f_e} \end{aligned}$$

which is the same as you saw earlier. Everything is consistent!

Q3.31 In far-point adjustment, the magnifying power of this telescope is f_o/f_e = 500/50 = ×10. This is rather small for an astronomical telescope. Using the equation $M_{total} = M_o \times M_e$ from Question 3.30, you can see that this magnifying power of ×10 is made up from an M_e of ×5 (since d_{np}/f_e = 250/50), and an M_o of ×2 (since f_o/d_{np} = 500/250). Even in near-point adjustment, the magnifying power of the telescope would increase only to ×12 (M_e would increase by 1 to ×6). The value of M_e is not unreasonable for a single-lens eyepiece; larger values would be likely to introduce noticeable aberrations. But a magnifying power of ×2 for the objective is much too small. Also this lens has an F-number of 5 (i.e. an aperture diameter of f/5) and is likely to have been rather expensive. A longer focal-length lens would have a larger F-number, would therefore probably be cheaper, and would give a bigger magnifying power.

If you use $(D_o/D_p)^2$ to calculate the light-gathering power relative to the dark-adapted eye, you get a value of $(100/8)^2$ or just over 156. But this is not achievable with the design specifications quoted. Using the similar triangles argument illustrated in Figure 3.38, a diameter D_o of 100 mm over a focal length f_o of 500 mm will give a diameter D_e of 10 mm over the focal length f_e of 50 mm. But since the eyepiece lens has a diameter of only 8 mm, some of the light entering the objective will be lost. Increasing the diameter of the eyepiece to 10 mm would be of no benefit: although all the light would now pass through the eyepiece, it would still overspill the pupil of the dark-adapted eye. In fact, the light-gathering power of this telescope is limited to the value $(f_o/f_e)^2$ = 100. Looked at another way, this is equivalent to saying that only the central 80 mm of diameter of the objective lens contributes light that enters the dark-adapted eye. Again the obvious improvement would be to increase the focal length of the objective. Increasing the focal length to 1 m would double the magnifying power to ×20, would ensure a light-gathering power of 156 (the diameter D_e of the light at the eyepiece lens would now be 5 mm), and would still keep the telescope at a manageable length of 1050 mm.

Q3.32 (a) An aperture diameter of f/22 is *smaller* than an aperture diameter of f/2.8; therefore, the intensity at the film plane is *reduced* when the stop is switched from f/2.8 to f/22. Reference to Table 3.1 shows that this change of aperture diameter corresponds to a change of 6 stops, each of which changes the intensity by a factor of 2. Hence, the total change of intensity is by a factor of 2^6, or 64 times. (Alternatively, calculate $(22/2.8)^2 \approx 62$.)

(b) (i) To 'freeze the motion', you should choose the *shortest* possible exposure time — in this case 1/250 s. This is two positions (i.e. approximately a factor of 4) smaller in exposure time than the 'optimized' 1/60 s, so the aperture diameter must be *increased* two positions from the 'optimized' *f*/8 position to *f*/4. So, use 1/250 s and *f*/4.

(ii) For the largest depth of field, you must use the *smallest* aperture diameter, which with this camera would be *f*/22. But *f*/22 is three stops smaller than *f*/8, so you would have to increase the exposure time by three positions (from 1/60 s) to compensate, thereby keeping the exposure the same. Unfortunately, you can increase the exposure time by only *two* positions, to 1/15 s. Consequently, you would have to make do with 1/15 s and *f*/16. If you were to use 1/15 s and *f*/22, your exposure would be half the 'optimum' exposure.

Q4.1 Consider first the situation when the lift is stationary at the top of the skyscraper. The forces acting on the cork when it is being held below the surface of the water by the workman's finger are:

(i) the force acting downwards on the cork due to the workman's finger;

(ii) the force of gravity on the cork — that is, its 'weight' acting downwards;

(iii) the upward force equal to the 'weight' of displaced water, as given by Archimedes' principle.

There is a net (unbalanced) force upwards as a result of (ii) and (iii), because the cork is less dense than the water. This means that, unless it is constrained by the downward force exerted by the workman's finger (i) it bobs up to the surface of the water, and floats at a level such that the volume of water displaced has the same mass as the cork. All this takes place in a frame of reference attached to the surface of the Earth — that is, in an inertial frame.

Now consider the situation when the lift falls freely towards the centre of the Earth. It accelerates with acceleration of magnitude g, and the workman is now in a non-inertial frame. In this non-inertial frame, the workman, bucket, water and cork are in 'free-fall', which has the effect of 'switching off' the effect of the acceleration due to gravity. In such a non-inertial frame therefore, we might expect effects that are contrary to our everyday experience in an inertial frame. In the non-inertial frame in question, there is no unbalanced force on the cork and it stays exactly where it is, submerged in the water, when the workman removes his finger. So the correct answer is (b). Indeed, the workman could remove the cork and carefully place it in front of his eyes, where it would remain, due to its 'weightless' condition! (But so also is the workman and the bucket.)

Q4.2 The observer in the other frame of reference should find that an expression of the same form is valid in his frame of reference,

that is, $$\frac{d^2x'}{dt'^2} = \frac{k'}{(x' - X')^2}$$

Q4.3 No, your friend's claim is not justified. He can certainly distinguish his frame of reference from yours by observing the motion of the bridge in his frame of reference, but that does *not* discredit the principle of relativity. The point is that the principle of relativity does *not* claim that all inertial frames are totally indistinguishable. (Indeed, it is obvious that frames of reference that have their origins in two *different* adjoining rooms could be distinguished by something as simple as the colour of the curtains nearest to the origin of each frame of reference.) The principle of relativity merely claims that the *laws of physics* take the same form in all inertial frames. So, there is no particular frame of reference singled out by the laws of physics. The speed of the bridge in your friend's frame of reference is *not* controlled by a law of physics: it is just a direct consequence of his viewpoint. It can be called an artefact *of* his frame of reference rather than a result discovered *in* his frame of reference.

Q4.4 Both beams will strike the planet at the same time because the speed of light is independent of the speed of its source. The speeds of the spacecraft given in the question are irrelevant: all that matters is that the lasers are fired simultaneously, while they are both at the same distance from the planet.

Q4.5 In a vacuum, the speed of light relative to a detector is always c. So the speed of the light emitted by the laser is c not only relative to the detector orbiting the Earth, but also relative to the detector on board the spacecraft. The fact that the two detectors are moving relative to one another is irrelevant, as is the fact that the detector on board the spaceship can be regarded as measuring the speed of the light relative to its source. All that matters in both cases is the speed relative to the detector.

Q4.6 The viewpoint of your friend is in direct contradiction to the principle of relativity. The point is that we are dealing with two uniformly moving systems and two inertial frames (that of the observer who sees the clock move past her, and that of the observer who moves uniformly with the clock). Suppose you and your friend had identical wristwatches. Since all physical laws must be the same in *all* inertial frames, this means that the mechanisms of your watch and your friend's watch must run in the same way in both frames of reference. The motion does not affect the watches in the slightest. As stressed in the text, the fact that the time interval between the two events is different in the two frames of reference is connected with what we mean by a time interval, and has nothing to do with the mechanics of the clock.

Q4.7 According to an observer on the Earth, light takes 4.2 years to travel from the Earth to Alpha Centauri. So an astronaut travelling at $0.9c$ would require

$$\Delta T = \frac{4.2\text{ years}}{0.9} = 4.7\text{ years}$$

to complete the journey according to an observer on Earth. However, the astronaut is moving relative to the observer on Earth, so the astronaut's clock is running slow compared with clocks on Earth. The interval of time that would elapse in the astronaut's frame of reference is

$$\begin{aligned}\Delta T_0 &= \Delta T\sqrt{1-(V^2/c^2)}\\ &= 4.7\times\sqrt{1-0.9^2}\\ &= 4.7\times\sqrt{0.19}\\ &= 2.0\text{ years.}\end{aligned}$$

Q4.8 (a) The return leg of the round trip would take just as long as the outward leg. So according to the astronaut the round trip would take 2×2.0 years = 4.0 years.

(b) According to the twin who stayed on Earth, the round trip would take 2×4.7 years = 9.4 years.

Q4.9 (a) Using the Lorentz contraction formula, $L = L_0\sqrt{1-V^2/c^2}$, the height of the mountain in the frame of reference in which the muons are at rest is

$$L = 4050\,\text{m}\times\sqrt{1-0.98^2} = 806\,\text{m}.$$

(b) Travelling at $0.98c$, the muons take

$$\begin{aligned}t &= \frac{806\,\text{m}}{0.98\times 3.00\times 10^8\,\text{m}\,\text{s}^{-1}}\\ &= 2.74\times 10^{-6}\,\text{s}\end{aligned}$$

or about 2.7 μs to travel this distance, in the frame of reference in which they are at rest.

(Notice that, in the frame of reference in which the mountain is at rest, the time taken is

$$\begin{aligned}t &= \frac{4050\,\text{m}}{0.98\times 3.00\times 10^8\,\text{m}\,\text{s}^{-1}}\\ &= 13.8\times 10^{-6}\,\text{s}\end{aligned}$$

or about 5 times longer.)

Q4.10 Clearly the y- and z-coordinates of both events in both frames of reference are zero, so only the x-coordinates and t-coordinates need be calculated. The first thing to do is to calculate the Lorentz factor,

$$\begin{aligned}\gamma &= \frac{1}{\sqrt{1-V^2/c^2}}\\ &= \frac{1}{\sqrt{1-4^2/5^2}}\\ &= \frac{1}{\sqrt{9/25}} = \frac{5}{3}.\end{aligned}$$

So, using Equation 4.17b, the coordinates of Event 1 in frame of reference B are

$$\begin{aligned}x_1' &= \gamma(x_1 - Vt_1)\\ &= \frac{5}{3}\times\left(12\times 10^8\,\text{m} - \left(\frac{4}{5}\times 3.00\times 10^8\,\text{m}\,\text{s}^{-1}\times 7\,\text{s}\right)\right)\\ &= -8.0\times 10^8\,\text{m}\end{aligned}$$

$$\begin{aligned}t_1' &= \gamma(t_1 - Vx_1/c^2)\\ &= \frac{5}{3}\times\left(7\,\text{s} - \frac{\frac{4}{5}\times 12\times 10^8\,\text{m}}{3.00\times 10^8\,\text{m}\,\text{s}^{-1}}\right)\\ &= 6.3\,\text{s.}\end{aligned}$$

Similarly, for Event 2

$$\begin{aligned}x_2' &= \gamma(x_2 - Vt_2)\\ &= \frac{5}{3}\times\left[30\times 10^8\,\text{m} - \left(\frac{4}{5}\times 3.00\times 10^8\,\text{m}\,\text{s}^{-1}\times 11\,\text{s}\right)\right]\\ &= 6.0\times 10^8\,\text{m}\end{aligned}$$

$$\begin{aligned}t_2' &= \gamma(t_2 - Vx_2/c^2)\\ &= \frac{5}{3}\times\left(11\,\text{s} - \frac{\frac{4}{5}\times 30\times 10^8\,\text{m}}{3.00\times 10^8\,\text{m}\,\text{s}^{-1}}\right)\\ &= 5.0\,\text{s.}\end{aligned}$$

So, in frame of reference B, Event 1 occurs *after* Event 2: in other words, the order of the two events in frame of reference B is opposite to the order in frame of reference A.

Q4.11 The Lorentz factor in this case is

$$\gamma = 1/\sqrt{1-V^2/c^2} = 1/\sqrt{1-0.5^2} = 1/\sqrt{3/4} = 2/\sqrt{3}.$$

So, using Equation 4.17b, the time coordinate of Event 1 in frame of reference B is

$$t_1' = \gamma[t_1 - Vx_1/c^2]$$

$$= \frac{2}{\sqrt{3}} \times \left(8\,\text{s} - \frac{\frac{1}{2}\times 15\times 10^8\,\text{m}}{3.00\times 10^8\,\text{m}\,\text{s}^{-1}}\right)$$

$$= \frac{11}{\sqrt{3}}\,\text{s} = 6.35\,\text{s}$$

and the time coordinate of Event 2 in frame of reference B is

$$t_2' = \gamma[t_2 - Vx_2/c^2]$$

$$= \frac{2}{\sqrt{3}} \times \left(8\,\text{s} - \frac{\frac{1}{2}\times 21\times 10^8\,\text{m}}{3.00\times 10^8\,\text{m}\,\text{s}^{-1}}\right)$$

$$= \frac{9}{\sqrt{3}}\,\text{s} = 5.20\,\text{s}.$$

So the two events are *not* simultaneous in frame of reference B. In particular, Event 1 will occur at a time $t_1' = 6.35$ s and Event 2 will occur at a time $t_2' = 5.20$ s.

Q4.12 The question must be rephrased so that the situation can be identified with two frames of reference A and B in standard configuration. To do this, imagine that the astronaut in the observation tower is at the origin of frame of reference A, and spaceship X is at the origin of frame of reference B. Then frames of reference A and B are in standard configuration with $V = 3c/4$. The x-component of the velocity of spaceship Y in frame of reference A is $v_x = -3c/4$. The question posed is then: What is the x'-component of the velocity of spaceship Y in frame of reference B? From Equation 4.20,

$$v_x' = \frac{v_x - V}{1 - Vv_x/c^2}$$

$$= \frac{-(3c/4) - (3c/4)}{1 + (9c^2/16c^2)}$$

$$= -\frac{6c/4}{25/16} = -\frac{96c}{100}.$$

The negative sign indicates that spaceship X measures Y to be racing *towards it* at $0.96c$.

Q4.13 Substituting the expressions for the magnitude of $\mathbf{p}$, $p = mv/\sqrt{1-v^2/c^2}$ (*PM*, Equation 3.20) and $E_{\text{tot}} = mc^2/\sqrt{1-v^2/c^2}$ (*PM*, Equation 3.22) into Equation 4.22:

$$\frac{m^2c^4}{1-v^2/c^2} = \frac{m^2v^2c^2}{1-v^2/c^2} + m^2c^4.$$

Adding the two terms on the right-hand side yields

$$\frac{m^2v^2c^2}{1-v^2/c^2} + \frac{m^2c^4 - m^2v^2c^2}{1-v^2/c^2}$$

$$= \frac{m^2c^4}{1-v^2/c^2}.$$

Thus, the left- and right-hand sides of the equation are identical, which verifies the validity of Equation 4.22.

Q4.14 (a) The mass energy of the proton is

$$E_{\text{mass}} = m_\text{p}c^2 = (1.67\times 10^{-27}\,\text{kg})\times(3.00\times 10^8\,\text{m}\,\text{s}^{-1})^2$$

$$= 1.50\times 10^{-10}\,\text{J}.$$

(b) If the relativistic translational kinetic energy of the proton is equal to its mass energy, then from the definition of the total relativistic energy

$$E_{\text{tot}} = m_\text{p}c^2 + m_\text{p}c^2 = 2m_\text{p}c^2.$$

Using *PM*, Equation 3.22 therefore,

$$E_{\text{tot}} = 2m_\text{p}c^2 = \frac{m_\text{p}c^2}{\sqrt{1-v^2/c^2}}.$$

So, $\sqrt{1-v^2/c^2} = 1/2$, or $1 - v^2/c^2 = 1/4$, and therefore $v^2/c^2 = 3/4$. The speed a proton must travel at in order to have its relativistic translational kinetic energy equal to its mass energy is therefore $v = \sqrt{3}c/2 = 2.6\times 10^8\,\text{m}\,\text{s}^{-1}$.

(c) If the relativistic translational kinetic energy is *three* times as great as the mass energy, the total relativistic energy is

$$E_{\text{tot}} = 3m_\text{p}c^2 + m_\text{p}c^2 = 4m_\text{p}c^2$$

$$= 4\times 1.50\times 10^{-10}\,\text{J}$$

$$= 6.0\times 10^{-10}\,\text{J}.$$

Q4.15 The first thing to do is to calculate the Lorentz factor:

$$\gamma = 1/\sqrt{1-V^2/c^2} = 1/\sqrt{1-3^2/5^2}$$

$$= 1/\sqrt{16/25} = 5/4.$$

So, using Equation 4.17b, the coordinates in frame of reference B are

$$x' = \gamma(x - Vt)$$

$$= \frac{5}{4}\times\left[18\times 10^8\,\text{m} - \left(\frac{3}{5}\times 3.00\times 10^8\,\text{m}\,\text{s}^{-1}\times 10\,\text{s}\right)\right]$$

$$= 0$$

$$y' = y = -1\,\text{m}$$

$$z' = z = 0$$

$$t' = \gamma(t - Vx/c^2)$$

$$= \frac{5}{4} \times \left(10\,\text{s} - \frac{\frac{3}{5} \times 18 \times 10^8\,\text{m}}{3.00 \times 10^8\,\text{m}\,\text{s}^{-1}} \right)$$

$$= 8\,\text{s}.$$

So the new coordinates are: $x' = 0$, $y' = -1$ m, $z' = 0$, $t' = 8$ s.

Q4.16 (a) Imagine that the space station is at the origin of frame of reference A, and the spaceship is at the origin of frame of reference B. The frames of reference A and B are in standard configuration with $V = c/2$. The x-component of the velocity of the missile in frame of reference A is $v_x = 3c/4$. The question may be rephrased as 'What is the x'-component of the velocity of the missile in frame of reference B?'. The answer follows from Equation 4.20, and it is

$$v'_x = \frac{v_x - V}{1 - Vv_x/c^2}$$

$$= \frac{(3c/4) - (c/2)}{1 - (3c^2/8c^2)}$$

$$= \frac{c/4}{5/8} = 0.4c.$$

(b) In this case, you must picture the origin of frame of reference B to be travelling towards the origin of frame of reference A. This situation is easy to deal with if the V that appears in Equation 4.20 is put equal to $-c/2$. (Remember, V is really a velocity component.) So, it follows that the speed of approach of the missile, according to the spaceship is

$$v'_x = \frac{v_x - V}{1 - Vv_x/c^2}$$

$$= \frac{(3c/4) + (c/2)}{1 + (3c^2/8c^2)}$$

$$= \frac{5c/4}{11/8} = \frac{40c}{44}$$

$$= 0.91c.$$

Q4.17 This solution to this question will be presented in the format of the problem-solving technique

Preparation The formula for the relativistic velocity transformation will be required in this problem:

$$v'_x = \frac{v_x - V}{1 - Vv_x/c^2}:$$

In this expression, v'_x is the velocity component of an object in a frame of reference moving with x-component velocity V relative to a frame of reference in which the velocity component of the object is v_x.

Working (a) Because the neutrino has zero rest mass, then just like a photon, it must always travel at the speed of light, c. Moreover, it will travel at this speed in *all* inertial frames, so that the speed of the neutrino in the laboratory frame of reference is also c.

(b) Use the Lorentz velocity transformation equation to calculate the speed of Q in the laboratory frame of reference:

$$v'_x = \frac{v_x - V}{1 - Vv_x/c^2}.$$

In this problem:

$v_x = -v_Q$, the x-component of Q's velocity in P's frame of reference;

$V = -0.95c$, the velocity of the laboratory relative to P;

v'_x is the required x-component of Q's velocity in the laboratory frame of reference.

Hence,

$$v'_x = \frac{-v_Q - (-0.95c)}{1 - [-v_Q \times (-0.95c)]/c^2}$$

$$= \frac{0.95c - v_Q}{1 - (0.95v_Q/c)}.$$

Checking (a) You can check that the neutrino moves at speed c in the laboratory frame of reference by applying the relativistic velocity transformation (Equation 4.20). In this case, $v_x = c$ and $V = -0.95c$. Thus, the neutrino's speed relative to the laboratory frame of reference, v'_x, is

$$v'_x = \frac{v_x - V}{1 - Vv_x/c^2}$$

$$= \frac{c - (-0.95c)}{1 - (-0.95c^2/c^2)}$$

$$= \frac{1.95c}{1.95} = c.$$

(b) At ordinary speeds, the answer for (b) should become identical to the Newtonian solution (i.e. the Galilean velocity transformation). Thus, with $v_Q \ll c$, the denominator of the equation obtained as answer to part (b) becomes unity, and we have $v'_x = 0.95c - v_Q$, as expected.

Q4.18 (a) It is necessary to use the formula for Lorentz contraction, Equation 4.13, $L = L_0\sqrt{1 - V^2/c^2}$, where L is the length of the pole in frame of reference A, and L_0 is the length of the pole in the frame of reference in which it is at rest, B. In this case

$$10\,\text{m} = 20\,\text{m} \times \sqrt{1 - V^2/c^2}$$

So, $\sqrt{1 - V^2/c^2} = 1/2$, or $1 - V^2/c^2 = 1/4$, and therefore $V^2/c^2 = 3/4$. The speed of the runner with respect to the tunnel is therefore $V = \sqrt{3}c/2$, or $2.6 \times 10^8\,\text{m}\,\text{s}^{-1}$.

(b) In order to calculate the time interval between the two events in the runner's frame of reference, it is necessary to transform the time coordinates of Event 1 and Event 2 in frame of reference A to those in frame of reference B. Suppose that Event 1 occurs at $x_1 = 0$ and $t_1 = 0$ in frame A, whereas Event 2 occurs at $x_2 = -L\ (= -10\,\text{m})$ and $t_2 = 0$ in frame A, and we know from part (a) that $V = \sqrt{3}c/2$.

So using $t' = \dfrac{t - Vx/c^2}{\sqrt{1 - V^2/c^2}}$ we have

$$t'_1 = \frac{t_1 - Vx_1/c^2}{\sqrt{1 - V^2/c^2}} = \frac{0 - 0}{1/2} = 0$$

$$t'_2 = \frac{t_2 - Vx_2/c^2}{\sqrt{1 - V^2/c^2}} = \frac{0 + \sqrt{3}L/2c}{1/2} = \frac{\sqrt{3}L}{c}$$

$$= \frac{\sqrt{3} \times 10\,\text{m}}{3.00 \times 10^8\,\text{m}\,\text{s}^{-1}} = 5.8 \times 10^{-8}\,\text{s}.$$

The time interval in frame of reference B is therefore 5.8×10^{-8} s, and Event 2 occurs *after* Event 1.

Q4.19 Remember that, according to stay-at-home Bob, it takes 1 year travelling at the speed of light to travel a distance of 1 light year. So Bob will have to wait $(2 \times 6)/0.8 = 15$ years for Bill to come home and $(2 \times 12)/0.8 = 30$ years for Ben to come home — that is, another 15 years after Bill arrives back.

Now, it make things simpler to calculate $\sqrt{1 - V^2/c^2}$ for both Bill and Ben, namely $\sqrt{1 - 0.8^2} = 0.6$. So, according to Bill, his trip takes only (15 years $\times$ 0.6) = 9 years, whereas, according to Ben, his trip takes only (30 years $\times$ 0.6) = 18 years.

The final triplet to return home is Ben, for whom the experiment has lasted 18 years, and by this time, the experiment for Bob has lasted 30 years. According to Bill, the experiment has lasted 9 years while travelling, plus another 15 years back home waiting for Ben to arrive, giving a total of 24 years.

Q4.20 As noted in the question, since the relative motion is parallel to the x- and x'-axes, so the vector $\boldsymbol{v}$ may be written in terms of its components as $(v_x, 0, 0)$. The field vectors parallel to the direction of motion have components: $\boldsymbol{\mathscr{E}}_\parallel = (\mathscr{E}_x, 0, 0)$, $\boldsymbol{\mathscr{E}}'_\parallel = (\mathscr{E}'_x, 0, 0)$, $\boldsymbol{B}_\parallel = (B_x, 0, 0)$ and $\boldsymbol{B}'_\parallel = (B'_x, 0, 0)$. Similarly, the field vectors perpendicular to the motion are a combination of the y and z components only, and therefore:

$$\boldsymbol{\mathscr{E}}_\perp = (0, \mathscr{E}_y, \mathscr{E}_z),\ \boldsymbol{\mathscr{E}}'_\perp = (0, \mathscr{E}'_y, \mathscr{E}'_z),$$

$$\boldsymbol{B}_\perp = (0, B_y, B_z),\ \text{and}\ \boldsymbol{B}'_\perp = (0, B'_y, B'_z).$$

Now, the cross product rule in terms of components may be written

$$\boldsymbol{a} \times \boldsymbol{b} = (a_y b_z - a_z b_y,\ a_z b_x - a_x b_z,\ a_x b_y - a_y b_x).$$

So $\boldsymbol{v} \times \boldsymbol{B}_\perp = (0,\ -v_x B_z,\ v_x B_y)$

and $\boldsymbol{v} \times \boldsymbol{\mathscr{E}}_\perp = (0,\ -v_x\mathscr{E}_z,\ v_x\mathscr{E}_y)$.

Substituting these components into the relationships in Equation 4.24 yields:

$$\left.\begin{aligned}
\mathscr{E}'_x &= \mathscr{E}_x \\
\mathscr{E}'_y &= \gamma(\mathscr{E}_y - v_x B_z) \\
\mathscr{E}'_z &= \gamma(\mathscr{E}_z + v_x B_y) \\
B'_x &= B_x \\
B'_y &= \gamma\left(B_y + \frac{v_x\mathscr{E}_z}{c^2}\right) \\
B'_z &= \gamma\left(B_z - \frac{v_x\mathscr{E}_y}{c^2}\right)
\end{aligned}\right\} \qquad (4.25)$$

These six equations represent the component transformations between electric and magnetic fields for two inertial frames in standard configuration.

Q5.1 **Preparation** We shall need Faraday's law, $|V(t)| = |d\phi(t)/dt|$, as well as the definition of magnetic flux, $\phi(t) = A(t)B$, which is appropriate in this case. The magnet falls under the influence of gravity, so the distance moved as a function of time is given by $s(t) = \frac{1}{2}gt^2$.

Working First we will calculate the area of the loop in the magnetic field as a function of time, then the magnetic flux, and finally the induced EMF.

The area of the wire loop within the magnetic field is given by

$$A = b \times s(t) = \tfrac{1}{2}bgt^2.$$

So the magnetic flux through the loop which is perpendicular to the magnetic field is

$$\phi(t) = AB = \tfrac{1}{2}bgBt^2.$$

Finally, the induced EMF around the loop is given by Faraday's law as

$$|V_{\text{ind}}(t)| = \left|\frac{d\phi(t)}{dt}\right| = \tfrac{1}{2}bgB \times \frac{d(t^2)}{dt} = bgBt.$$

A graph of the induced EMF as a function of time is therefore a straight line of slope bgB as shown in Figure 5.7.

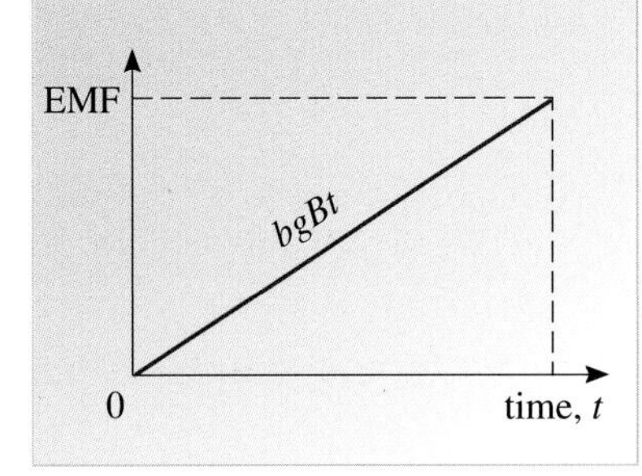

Figure 5.7 The induced EMF around the wire loop as a function of time for the situation described in Q5.1.

The maximum induced EMF will occur when t is a maximum. This will be just at the instant the magnet lands on the workbench — that is, when $s = b = 5.0$ cm.

The time taken to fall a distance b can be calculated from

$$b = \tfrac{1}{2}gt^2$$

so $$t = \sqrt{\frac{2b}{g}} = \sqrt{\frac{2 \times 0.05\ \text{m}}{10\ \text{m s}^{-2}}} = 0.10\,\text{s}.$$

After this length of time, the magnitude of the induced EMF is therefore

$$V_{\text{ind}}(t) = bgBt$$
$$= 0.05\ \text{m} \times 10\ \text{m s}^{-2} \times 2.0\ \text{T} \times 0.10\ \text{s}$$
$$= 0.10\ \text{V}.$$

Checking All units come out correctly as a result of keeping track of them throughout the calculations. If the size of the wire loop and magnet were larger (i.e. if b were greater than 5.0 cm), then clearly the magnitude of the induced EMF would be larger, which is as expected.

Q5.2 Preparation All that is needed here is the correct geometry and Snell's law in the form

$$\frac{\sin i}{\sin r} = \frac{n_2}{n_1}.$$

Diagrams showing the labelled angles for red and blue light are shown in Figure 5.8.

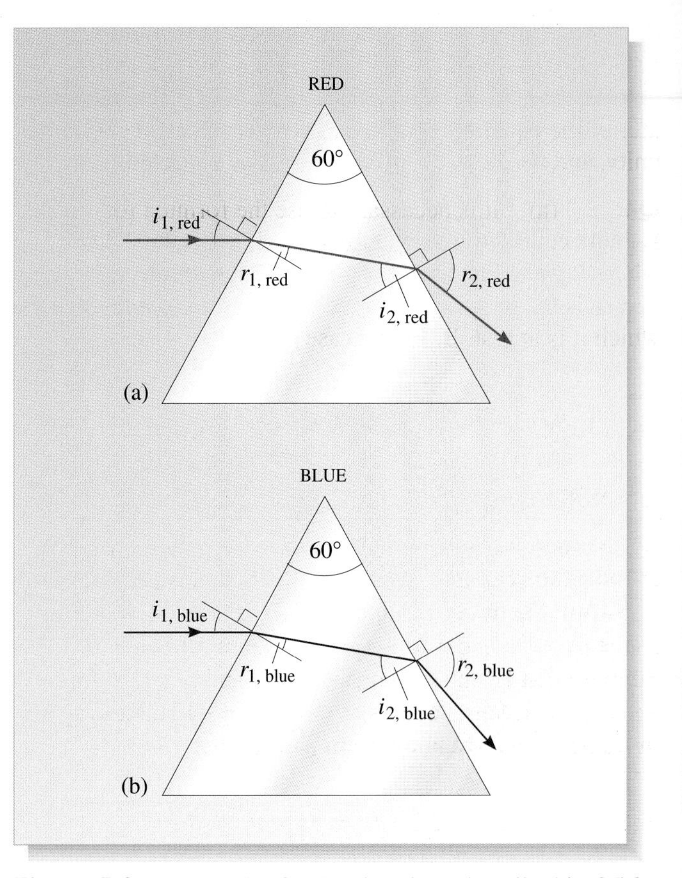

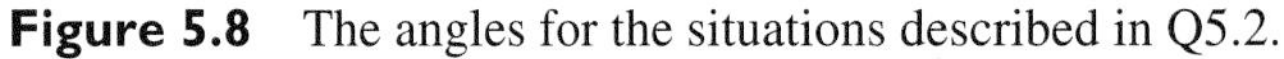

Figure 5.8 The angles for the situations described in Q5.2.

Working We shall work out the angles of refraction at the first interface, then the angles of incidence at the second interface, and finally the angles of refraction at the second interface. The angle between the red and blue beams is then the difference between the two angles of refraction at the second interface.

At the first interface:

$$\frac{\sin i_{1,\text{red}}}{\sin r_{1,\text{red}}} = \frac{n_{\text{red}}}{n_{\text{air}}}$$

$$\sin r_{1,\text{red}} = \sin 30° \times \frac{1.00}{1.45} = 0.3448$$

so $$r_{1,\text{red}} = 20.17°$$

and $$\frac{\sin i_{1,\text{blue}}}{\sin r_{1,\text{blue}}} = \frac{n_{\text{blue}}}{n_{\text{air}}}$$

$$\sin r_{1,\text{blue}} = \sin 30° \times \frac{1.00}{1.50} = 0.3333$$

so $$r_{1,\text{blue}} = 19.47°.$$

So at the second interface

$$i_{2,\mathrm{red}} = 60° - r_{1,\mathrm{red}} = 39.83°$$

$$i_{2,\mathrm{blue}} = 60° - r_{1,\mathrm{blue}} = 40.53°.$$

Therefore

$$\frac{\sin i_{2,\mathrm{red}}}{\sin r_{2,\mathrm{red}}} = \frac{n_{\mathrm{air}}}{n_{\mathrm{red}}}$$

$$\sin r_{2,\mathrm{red}} = \sin 39.83° \times \frac{1.45}{1.00} = 0.9287$$

so $$r_{2,\mathrm{red}} = 68.23°$$

and $$\frac{\sin i_{2,\mathrm{blue}}}{\sin r_{2,\mathrm{blue}}} = \frac{n_{\mathrm{air}}}{n_{\mathrm{blue}}}$$

$$\sin r_{2,\mathrm{blue}} = \sin 40.53° \times \frac{1.50}{1.00} = 0.9748$$

so $$r_{2,\mathrm{blue}} = 77.11°.$$

So the angle between the red and blue beams emerging from the prism into the air is (77.11° – 68.23°) = 8.9°.

Checking The important thing here is that all sines are less than 1.0, and so the angles are all realistic. Both beams are refracted towards the normal at the air–glass interface, and away from the normal at the glass–air interface, as expected.

Q5.3 Preparation As with Example 5.2, the most important thing to do in setting up problems of this sort is to draw a diagram, as shown in Figure 5.9. A scale of 1 : 2 has been used here.

Since the thin-lens approximation is assumed, other things that will be needed are the lens equation:

$$\frac{1}{u} + \frac{1}{v} = \frac{1}{f}$$

and the 'real is positive' convention described in Example 5.2. As in that example, we can calculate the position of the final image using the lens equation, and can later check if the answer is correct by constructing a ray diagram. To do this we shall again need to use information about how rays may be traced through lenses, as outlined earlier.

Let the distance of the object from lens A be u_1, the distance of the intermediate image from lens A be v_1, the separation between the lenses be d, the distance of the intermediate image from lens B (for which it acts as an object) be u_2, and the distance of the final image from lens B be v_2.

Working Knowing u_1 and d, as well as the focal lengths of the two lenses, we can now calculate each of v_1, u_2 and v_2 in turn.

So, for lens A,

$$\frac{1}{u_1} + \frac{1}{v_1} = \frac{1}{f_A}.$$

Therefore $$\frac{1}{8.0\ \mathrm{cm}} + \frac{1}{v_1} = \frac{1}{4.0\ \mathrm{cm}}$$

and so v_1 = 8.0 cm. The intermediate image is 8.0 cm to the *right* of lens A, and is a *real* image.

Now, the object distance for lens B is therefore

$$u_2 = d - v_1 = 4.0\ \mathrm{cm} - 8.0\ \mathrm{cm} = -4.0\ \mathrm{cm}.$$

The object for lens B is 4.0 cm to the *right* of lens B and is a *virtual* object. So for lens B,

$$\frac{1}{u_2} + \frac{1}{v_2} = \frac{1}{f_B}.$$

Therefore $$\frac{1}{-4.0\ \mathrm{cm}} + \frac{1}{v_2} = \frac{1}{-8.0\ \mathrm{cm}}$$

and so v_2 = 8.0 cm. In other words the final image is 8.0 cm to the *right* of lens B, and is a *real* image.

Checking The best check to do in this case is to draw a ray diagram by completing the sketch shown in Figure 5.9. A completed version is shown in Figure 5.10.

Two rays drawn from the top of the object are sufficient to locate the image.

The first ray (labelled X) is directed through the left-hand focal point of lens A. After being intercepted by lens A, it emerges parallel to the optical axis. When this ray is then intercepted by lens B, it is diverged such that it appears to originate from the left-hand focal point of lens B (which happens to coincide with the left-hand focal point of lens A).

The second ray from the object (labelled Y) is initially parallel to the optical axis and, on passing through lens A, is bent towards its right-hand focal point, which coincides with the centre of lens B. It then continues undeviated to intercept ray X at the location of the final image IM.

The intermediate image may be located by continuing ray X parallel to the optical axis. It intercepts ray Y as shown, and this is the location of the intermediate image I′M′. This image too is real (rays converge to it), but it acts as a virtual object for lens B. From the scale diagram, the two image distances are consistent with the values calculated above.

Q5.4 Preparation This a clearly a question involving the concept of length contraction, so we shall need the formula: $L = L_0\sqrt{1 - V^2/c^2}$, where L_0 is the length of the rocket in a frame in which it is at rest, and L is its length as measured from a frame of reference relative to which it is moving at speed V.

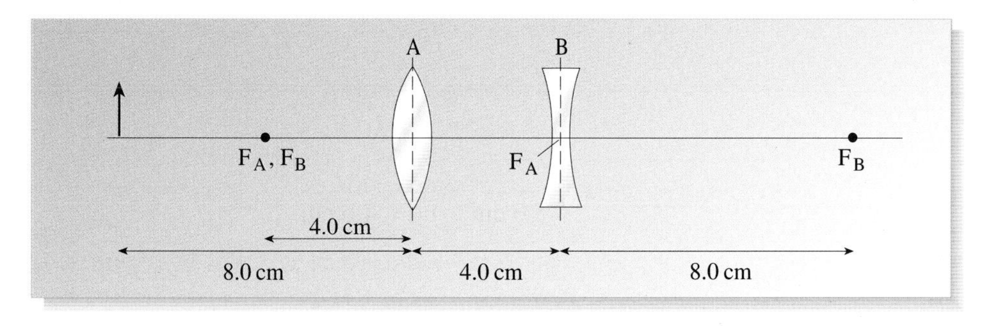

Figure 5.9 The situation described in Q5.3.

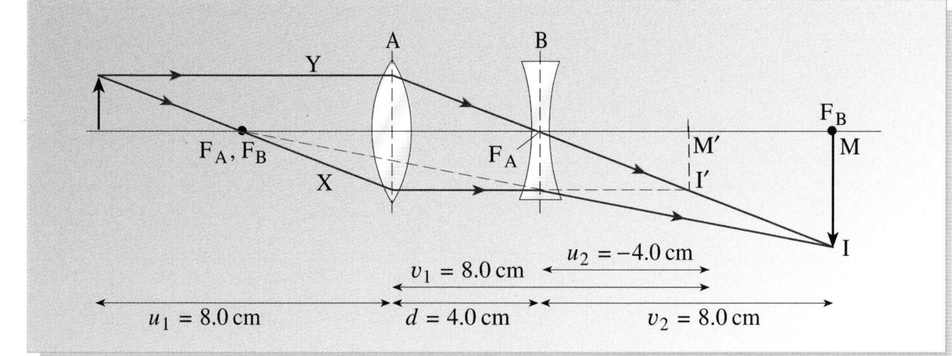

Figure 5.10 The completed ray diagram for Q5.3.

The question will also involve a velocity transformation from one frame of reference to another, so we need the Lorentz velocity transformation formula:

$$v'_x = \frac{v_x - V}{1 - Vv_x / c^2}$$

where v_x is the speed of an object in one frame of reference, v'_x is the speed of the same object in another frame of reference, and V is the relative speed between the two frames of reference.

Let the rest lengths of rockets P and Q be L_P and L_Q, respectively, and the length of each as measured by observer R is L_R.

Let the speed of rocket P relative to observer R be V_P and the speed of rocket Q relative to observer R be V_Q. The speed of rocket Q in the frame in which rocket P is at rest is V_{PQ}.

A diagram illustrating the various quantities that will be referred to is shown in Figure 5.11.

Working There are three stages to consider here, each of which needs to be calculated in turn to arrive at the final answer.

We shall first use the length contraction formula to calculate the length of rocket P as measured by observer R.

Then, using this value as the measured length of rocket Q, we shall use the length contraction formula again to calculate the speed of rocket Q relative to observer R.

Finally, we shall transform the speed of rocket Q as measured by observer R to the frame of reference of rocket P, in order to calculate the speed of rocket Q in the frame in which rocket P is at rest.

The length of rocket P as measured by observer R may be found from:

$$L_R = L_P\sqrt{1 - V_P^2 / c^2}$$

$$= 80.0 \text{ m} \times \sqrt{1 - 0.8^2}$$

$$= 48.0 \text{ m.}$$

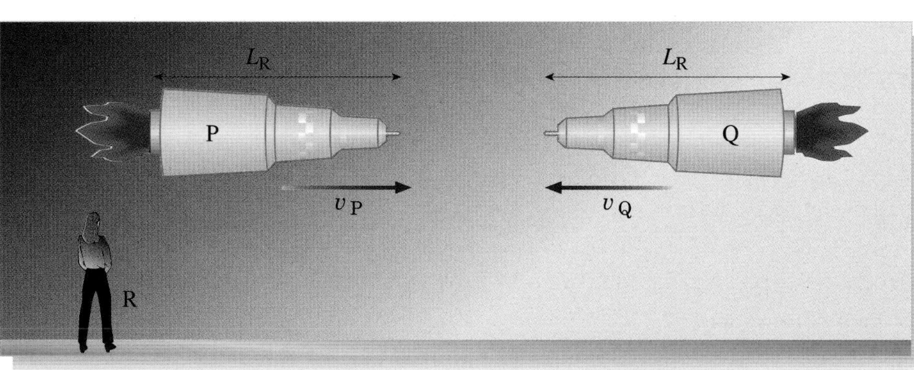

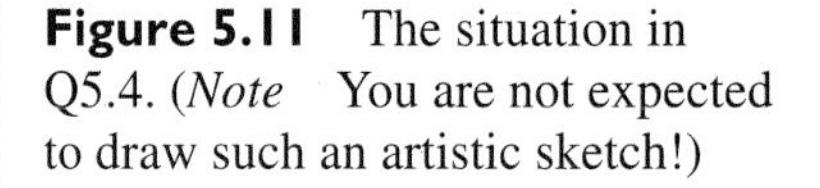

Figure 5.11 The situation in Q5.4. (*Note* You are not expected to draw such an artistic sketch!)

So the length of rocket Q as measured by observer R is also 48.0 m. The speed of rocket Q as measured by observer R can be found from

$$L_R = L_Q\sqrt{1 - V_Q^2/c^2}$$

which can be rearranged as:

$$\frac{V_Q}{c} = \sqrt{1 - \frac{L_R^2}{L_Q^2}}.$$

So $$\frac{V_Q}{c} = \sqrt{1 - \left(\frac{48.0\text{ m}}{60.0\text{ m}}\right)^2} = 0.60$$

which means that the speed of rocket Q as measured by observer R is 60% of the speed of light.

To calculate the speed of rocket Q in the frame in which rocket P is at rest, imagine that observer R is at the origin of frame of reference A, and rocket P is at the origin of frame of reference B. Assume that rocket P is travelling in the positive x-direction. Then frames A and B are in standard configuration with a relative speed between them of $V_P = 0.8c$. The x-component of the velocity of rocket Q in frame of reference A is $V_Q = -0.6c$ (remember the two rockets are travelling in *opposite* directions, so this speed is negative).

The speed of rocket Q in the frame in which rocket P is at rest may now be found from:

$$V_{PQ} = \frac{V_Q - V_P}{1 - V_P V_Q/c^2} = \frac{-0.6c - 0.8c}{1 - (0.8c \times -0.6c)/c^2} = \frac{-1.40c}{1.48} = -0.946c.$$

The negative sign indicates that rocket Q is moving in the *opposite* direction to that in which rocket P is travelling; that is, rocket Q is travelling away from rocket P at a relative speed of about 95% of the speed of light.

Checking The lengths measured by observer R are less than the rest lengths of the two rockets, which is as expected. The speed of rocket P relative to observer R is greater than the speed of rocket Q relative to observer R, which is also as expected, since the length of rocket P is contracted by a greater amount. The speed of rocket Q relative to rocket P is greater than its speed relative to observer R (but less than the speed of light!). This too is as expected; since rockets P and Q are moving away from each other, so their relative speed will be greater than their individual speeds relative to observer R.

Q5.5 The magnetic flux through the loop is given by Equation 1.1a, $\phi = AB\cos\theta$, where A is the area of the loop, B is the magnitude of the magnetic field through the loop and θ is the angle between the magnetic field and the normal to the loop. Maximum magnetic flux occurs when $\theta = 0$, so

$$\phi = (0.01\text{ m}^2) \times (0.1\text{ T}) = 1.0 \times 10^{-3}\text{ Wb}.$$

Q5.6 The magnitude of the EMF around the wire loop is equal to the magnitude of the rate of change of magnetic flux through the loop. Since the loop rotates at a constant angular frequency, we can assume that the magnetic flux takes the form $\phi(t) = AB\sin\omega t$, where ω is the angular frequency at which the loop turns. The rate of change of magnetic flux is $\mathrm{d}\phi(t)/\mathrm{d}t = \omega AB\cos\omega t$, which has a maximum value of ωAB. So the maximum EMF induced around the loop is

$$V_{ind} = (400\text{ s}^{-1}) \times (0.01\text{ m}^2) \times (0.1\text{ T}) = 0.4\text{ V}.$$

Q5.7 The ratio of voltages and the ratio of the number of windings are related according to

$$|V_2(t)| = (N_2/N_1) \times |V_1(t)|$$

Now, the time-varying voltage in each coil of the transformer will each have the form $V(t) = V_{max}\sin\omega t$,

so $$\frac{|V_1(t)|}{|V_2(t)|} = \frac{V_{1,\,max}}{V_{2,\,max}}.$$

Since in this case $V_{1,max}/V_{2,max}$ is required to be $240/120 = 2.0$, we must have $N_2/N_1 = 1/2$.

Q5.8 A resonant LC circuit has a frequency given by the formula $f = 1/2\pi\sqrt{LC}$. In this case we require $f = 100$ Hz, so making C the subject of the equation, we have:

$$C = \frac{1}{4\pi^2 f^2 L}$$

Substituting in the appropriate values and SI units for f and L, we obtain the requirement

$$C = \frac{1}{4\pi^2 \times 10^4 \times 10^{-4}}\text{F}$$

The powers of 10 in the denominator clearly cancel each other, and the final answer is

$$C = \frac{1}{4\pi^2}\text{F}$$

Q5.9 Using the equation $v = f\lambda$, we can work out the speed of the wave by substituting in the values and SI units for the wave frequency and wavelength. Hence: $v = 5.0\,\text{s}^{-1} \times 0.2\,\text{m} = 1.0\,\text{m s}^{-1}$. Knowing the speed of the wave, we can now apply the formula relating this to the tension force and the mass per unit length: $v = \sqrt{F_T/\mu}$. Making the tension force F_T the subject of the equation, this becomes:

$$F_T = v^2\mu.$$

Substituting in the appropriate values and units we obtain

$$F_T = (1.0\,\text{m s}^{-1})^2 \times (10^{-3}\,\text{kg m}^{-1})$$
$$= 1.0 \times 10^{-3}\,\text{kg m s}^{-2}.$$

Q5.10 We are told the refractive index of the medium is $\sqrt{3}$. Therefore the angle of incidence i and the angle of refraction r, are related by $\sin i/\sin r = \sqrt{3}$ We are further told that $i = 60°$, so we know that $\sin i = \sqrt{3}/2$. (Note that you can derive this by applying Pythagoras's theorem to a bisected equilateral triangle as shown in Figure 5.12.) This means

$$\sin r = \frac{\sin i}{\sqrt{3}} = \frac{\sqrt{3}}{2} \times \frac{1}{\sqrt{3}} = \frac{1}{2}$$

Hence $r = \sin^{-1}(1/2) = 30°$ (again this can be derived from Figure 5.12).

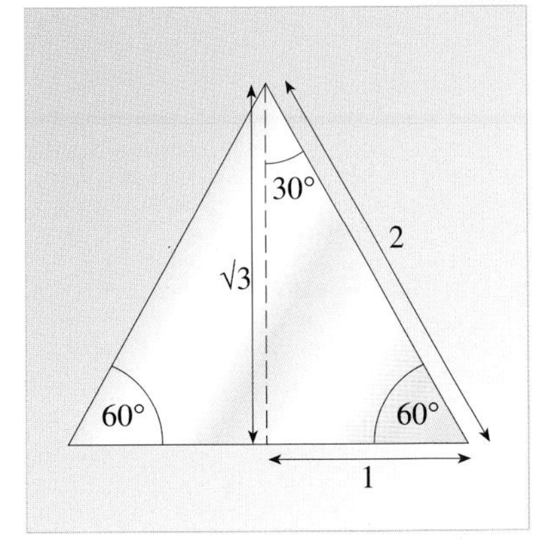

Figure 5.12 The geometry of a bisected equilateral triangle makes it easy to write exact expressions for the trigonometrical functions of 30° and 60°.

Q5.11 The wavefront is always perpendicular to the direction of propagation—that is, perpendicular to the ray—so before encountering the medium, the wavefront makes an angle of 30° with the normal to the surface; after encountering the medium, the wavefront makes an angle of 60° with the normal to the surface.

Q5.12 If we consider only the red laser, we would expect to see a double-slit diffraction pattern of red and dark bands on the screen. Similarly, if we considered only the blue laser, we would expect to see a double-slit diffraction pattern of blue and dark bands on the screen. Since blue light has a shorter wavelength than red light, the spacing between the blue fringes would be smaller than the spacing between the red fringes. If both lasers are simultaneously in position, both light beams will simultaneously undergo diffraction at the slits. Because the wavelengths of the two lasers are different, there will not be a constant phase difference between the red light and the blue light at any position on the screen. This means that there will be no observable coherent interference between the two beams, and we would simply see the red and blue fringes superimposed upon each other. In some places both patterns will be bright, and the screen will reflect a mixture of red and blue light; in other places only a red fringe will be observed; in others only a blue fringe will be observed. Finally in some places both patterns will be dark due to destructive interference, so the screen will be dark.

Q5.13 The angular magnification of a refracting telescope is equal to the ratio f_o/f_e. Hence if the angular magnification is 10, the focal length of the eyepiece lens must be one-tenth that of the objective. Hence

$$f_e = \frac{f_o}{10} = \frac{20\,\text{cm}}{10} = 2.0\,\text{cm}$$

Q5.14 The focal length of a lens in metres is related to its power in dioptres by the formula $f = 1/P$. Hence for the right lens the focal length will be

$$f = \frac{1}{-2.0\,\text{m}^{-1}} = -0.5\,\text{m}.$$

For the left lens the focal length will be

$$f = \frac{1}{-1.5\,\text{m}^{-1}} = -0.67\,\text{m}.$$

Note that these are both *diverging* lenses.

Q5.15 The bending power of a lens depends on its refractive index and on its shape. If the two lenses have the same refractive index, the shape of the right lens must produce more bending than that of the left. This means that the surfaces of the right lens must have more curvature. Since they are *diverging lenses*, the right lens will be thicker at its edge, so it will have a larger volume and hence be heavier than the left lens.

Q5.16 (a) A magnifying glass is the most appropriate optical system for directly reading small print. It consists of a single converging (convex) lens with a focal length less than about 250 mm. (A more modern alternative approach would be to use a photocopying machine to enlarge the document.)

(b) A compound microscope is the best optical system for examining blood cells. In its basic form it consists of just two converging (convex) lenses—a relatively short focal-length objective lens and a longer focal-length eyepiece lens.

(c) A telescope would be the most appropriate optical instrument for visually examining a distant ship. A Keplerian refracting telescope consists of two converging (convex) lenses: a relatively long focal-length objective lens and a shorter focal-length eyepiece lens. (Note, however, that the image formed by such a telescope is *inverted*, and in order to view a distant ship, an erect image would be more useful. Such an image is provided by a Galilean refracting telescope, which has a diverging (concave) eyepiece lens.)

Q5.17 The Galilean coordinate transformation seems perfectly reasonable in everyday situations because we rarely encounter speeds approaching the speed of light, or distances where the light travel time is perceptible. To overturn the Galilean coordinate transformation meant revising Newtonian dynamics, and it was not until the late nineteenth century that empirical evidence demanded this.

Q5.18 The principle of relativity states that the laws of physics can be written in the same form in all inertial frames.

Q5.19 Using the Lorentz contraction formula, $L = L_0\sqrt{1 - V^2/c^2}$, the height of the building in the frame of reference in which the UFO pilot is at rest is

$$L = 100\text{ m} \times \sqrt{1 - 0.80^2}$$
$$= 100\text{ m} \times \sqrt{0.36} = 60\text{ m}.$$

Q5.20 The relative speed of the observer in the building and the UFO is $0.80c$. The time for the UFO to fall the height of the building, as measured by the occupant of the building, is simply given by

$$\text{time} = \text{distance / speed}$$
$$= \frac{100\text{ m}}{0.80 \times 3.0 \times 10^8\text{m s}^{-1}}$$
$$= 4.17 \times 10^{-7}\text{s} = 420\text{ ns},$$

to two significant figures.

Acknowledgements

Grateful acknowledgement is made to the following sources for permission to reproduce material in this book:

Fig. 1.3a pictures furnished by Induction Technology Corporation, Adelanto, California 1–760/246–7333, www.inductiontech.com; *Figs 1.5, 1.9, 1.10, 1.13, 1.36, 1.37* Science Museum/Science and Society Picture Library; *Fig. 1.17a* courtesy of Calvin J. Hamilton, National Solar Observatory/Sacramento Peak; *Figs 1.17b,c, 1.39* Science Photo Library; *Fig. 1.17d* Matra Marconi Space; *Fig. 1.27* Novosti Press agency; *Fig. 1.45a* John Phillips/Photofusion; *Fig. 1.45b* Peter Menzel/Science Photo Library; *Fig. 1.54a* Alan Williams NHPA; *Fig. 1.55* by courtesy of the National Portrait Gallery, London;

Figs 2.1a, 2.83a Dr Peter Vukusic, University of Exeter; *Fig. 2.1b* K. G. Vock/ Okapia OSF; *Fig. 2.2* Images Colour Library; *Fig. 2.3a:* U. S. Army Corps of Engineers; *Fig. 2.3b* Daniel Valla/Survival Anglia OSF; *Fig. 2.6a* Hank Morgan/ Science Photo Library; *Fig. 2.6b* S. I. U./Science Photo Library; *Fig. 2.17* Archiv der Österreichischen Akademie der Wissenschaften; *Fig. 2.28* Jonathan Fisher/ Performing Arts Library; *Fig. 2.29a* Clive Barda/Performing Arts Library; *Figs 2.29a, 2.31b, 2.31c* Bettmann/Corbis; *Fig. 2.31d* Ted Streshinsky/Corbis; *Figs 2.33, 2.50, 2.70* Science Museum/Science & Society Picture Library; *Fig. 2.39* photo supplied by Dr Andrew Tindle/The Open University; *Fig. 2.41* Alfred Pasieka/ Science Photo Library; *Fig. 2.45* NRAO/AUI/Science Photo Library; *Fig. 2.47* Dr. Gene Felfman/NASA GSFC/Science Photo Library; *Fig. 2.48a,b* DERA; *Fig. 2.49a* Pascal Goetgheluck/Science Photo Library; *Fig. 2.49b* Wellcome Trust Medical Photographic Library; *Fig. 2.51a* Deep Light Productions/Science Photo Library; *Figure 2.51b* Mehau Kulyk/Science Photo Library; *Fig. 2.52* Mary Evans Picture Library; *Fig. 2.58a* NRAO/AUI*; Fig. 2.58b and c* photos supplied by Sally Jordan/The Open University; *Fig. 2.62b* Damien Lovegrove/Science Photo Library; *Fig. 2.79* Courtesy of Naomi Williams/Goodfellow; *Fig. 2.80* Courtesy of Dr J. Hinton, Warwick University; *Fig. 2.81* Steve Robinson/NHPA;

Fig. 3.1 Carl Zeiss; *Fig. 3.2a* National Optical Astronomy Observations; *Fig. 3.2b* AURA/STScl; *Fig. 3.31* Tom Morgan/Mary Evans Picture Library; *Fig. 3.33* Ed Young/Science Photo Library; *Fig. 3.35a* from the slide 51 Jupiter's Galilean Satellites by the Astronomical Society of the Pacific, 390, Ashton Avenue, San Francisco, CA 94112, USA; *Fig. 3.35b* Galileo Project JPL/NASA; *Fig. 3.39* UCO/Lick Observatory image; *Figs 3.43a, b* Caltech/Palomar Observatory; *Fig. 3.44* Gemini Observatory;

Fig. 4.1 Simon Fraser/Science and Society Picture Library/Science Museum; *Fig. 4.2* Caltech; *Fig. 4.3* The British Library; *Fig. 4.16* photograph taken by Sven Kohle; *Figs 4.20a, 4.31a, b* Science and Society Picture Library/Science Museum; *Fig. 4.20b* American Institute of Physics; *Fig. 4.39* (Gary Larson © Chronicle Features/Far Works Inc.).

Index